Natural Gas Applications for Air Pollution Control

Natural Gas Applications for Air Pollution Control

Nelson E. Hay, Editor

Published by
THE FAIRMONT PRESS, INC.
700 Indian Trail
Lilburn, GA 30247

Library of Congress Cataloging-in-Publication Data: LCC: 85-45879
Natural gas applications for air pollution control.

Bibliography: p. 313
Includes index.
1. Air-Pollution—Environmental aspects—United States. 2. Gas, Natural—Environmental aspects—United States. I. Hay, Nelson E.
TD883.2.N36 1986 628.5'3 85-45879
ISBN 0-88173-013-0

Natural Gas Applications for Air Pollution Control

Published by The Fairmont Press, Inc.
700 Indian Trail
Lilburn, GA 30247

ISBN 0-88173-013-0 FP
ISBN 0-13-609918-1 PH

Printed in the United States of America.

Distributed by Prentice-Hall, Inc.
A division of Simon & Schuster
Englewood Cliffs, NJ 07632

Prentice-Hall International (UK) Limited, *London*
Prentice-Hall of Australia Pty. Limited, *Sydney*
Editora Prentice-Hall do Brasil, Ltda., *Rio de Janeiro*
Prentice-Hall Canada, Inc., *Toronto*
Prentice-Hall Hispanoamericana, S.A., *Mexico*
Prentice-Hall of India Private Limited, *New Delhi*
Prentice-Hall of Japan, Inc., *Tokyo*
Prentice-Hall of Southeast Asia Pte. Ltd., *Singapore*

Contents

APPENDICES

Acknowledgements

The editor wishes to express his thanks to the authors and copyright holders for permission to publish the works which are collected in this volume. Particular recognition should also be given to Mary Wosoogh, Paul Wilkinson and Mike German for their contribution and support.

While every effort is made to provide dependable information, the publisher, editor and authors cannot be held responsible for any errors or omissions. The discussions of environmental regulations reflect policies in place as of July 1985.

Introduction

WILLIAM D. RUCKELSHAUS
Perkins & Coie
Administrator, U.S. Environmental Protection Agency, 1970–73, 1983–85

The concept of select use—mixing gas with other fuels for improved environmental results—has advantages which are well known. Those advantages are site-specific, and are gaining in acceptance and understanding around the country. There is no question that, under certain circumstances, select use of natural gas with other fuels—whether oil or coal, whether combined in one system or boiler, or averaged under an EPA approved bubble—makes environmental and economic sense.

The product of natural gas combustion is much lower than other fossil fuels in SO_2, the major precursor of acid rain. It is much lower in particulate matter than competing fossil fuels. It is much lower in carbon monoxide than competing fuels. In many cases it is lower in nitrogen oxides than oil, and in most cases than coal. In the case of hydrocarbons, much of the product of combustion of natural gas is a non-reactive hydrocarbon which does not combine with NO_X to form smog.

Natural gas is unquestionably a clean fuel. It is also, in many cases, a cheaper environmental solution than using other fossil fuels and attempting to clean them. To achieve ambient air quality standards, it can be implemented very quickly. Oil-fired facilities often already have access to natural gas, and certainly this access can provide results more quickly than the technological fixes which often take years to put on line. And it is cheaper than the operation and maintenance of these technological controls once they are installed.

For some time, we have resisted using gas because we thought it was in short supply. The assumptions that led to that view now appear to be wrong. If, in the past, it would have been possible for me to say that natural gas was almost unlimited in its abundance there would have been very little doubt in anybody's mind that it was the environmentally preferred fuel versus certain of the other fossil fuels for all kinds of uses. Consequently, we need to look seriously at natural gas as an alternative to other forms of fuel, or to its mixture with other fuels, because of the significant economic and environmental advantages associated with its use.

Significant progress has been made in reducing pollution from stationary sources in this country. Take, for example, the problem of sulfur oxide emissions. The Policy Analysis Group at EPA was asked to make an estimate of where we would be today in terms of sulfur oxide emissions if we had continued to control pollution as we had back in 1970. Where would we be in terms of emissions in the 31 states east of the Mississippi and bordering it on the west—the states normally identified as the major sources of SO_2 and acid rain in the east?

We currently produce about 23 million tons of sulfur oxide a year in that 31-state region. The best estimate EPA could come up with was that about 41 million tons would be emitted annually in the same 31-state area, had we done nothing. So, we have actually reduced the emissions of sulfur oxides from their peak in 1973 by about 23 percent over that 10 year period—a major achievement.

These earlier efforts in the air pollution struggle dealt with pollutants that were close-in to the source and that were causing identified environmental and health problems. The country logically addressed close-in sources of air pollution first so as to reduce the ambient concentrations of these so-called "criteria pollutants" to avoid the impact on public health and the environment. We have made enough progress against our early-identified environmental problems; we should really be moving on to the next generation of problems, of which the long-range transport of air pollution is certainly a major one.

What we have been discovering over the last several years is that at least one of the solutions used to address the close-in problems—namely, dispersion of these pollutants over a wider geographic area by means of tall stacks and other techniques—didn't totally alleviate the environmental impacts. We are now uncovering impacts of the longer-range transport of these pollutants, and this fact is crucial to the acid rain debate.

It is also becoming apparent that the long-range air pollution transport problem does not stop at acid rain, but extends to all air pollutants. Whether it be the precursors to acid rain (SO_2, NO_X), whether it is NO_X that combines with hydrocarbons to form the photochemical oxidants of smog, whether it is particulates (and sulfates make up as much as 50 percent of particulates in some parts of the east)—all are part of the long-range pollutant transport problem.

As research allows us a better understanding of the total impact of this problem, and as we begin to develop strategies for reducing those pollutants, it becomes clear that the use of natural gas is one of the options we should be exploring. And, I wouldn't restrict this solution to the long-range transport problem. Natural gas can be used to control these very same pollutants close-in. While natural gas can address the close-in problems of sulfur oxides, ozones, carbon monoxide, nitrogen oxides and particulates, probably the biggest potential use for natural gas is in addressing the long-range transport problems. Natural gas in a select use mode can be a very useful and cost effective way to achieve ambient level requirements. It needs only be implemented with intelligence around the country.

Why don't we use it more? I think there are a number of reasons. The Fuel Use Act, which was passed back in the mid-70s, was aimed at limiting the use of gas in very large installations, whether industrial boilers or utility boilers, on the grounds that it was in short supply and as a preferred fuel should be reserved for other uses. That may or may not have made sense at the time, but again it was the result of gas price regulations and our perceptions regarding gas supply.

There are many potential uses for natural gas which are still effectively inhibited by the Fuel Use Act. There should be a rigorous review of the logic behind the Fuel Use Act, and if such an examination shows that the thinking of the mid-70s no longer applies, those provisions which prohibit the use of natural gas should be repealed and the marketplace be allowed to judge which type of fuel is preferable.

The Fuel Use Act in its application to natural gas—which is illogical economically, and certainly illogical in its limitation of a tool to control air pollution—is a prime example of what happens when the government intervenes in a given area. There are, of course, things that the market system simply doesn't regulate properly. But where there is an area in which the market will work and which will, if left alone, provide both economically and environmentally sound answers as it would in this case when coupled with environmental regulations—government should step aside.

The Fuel Use Act points up the fact that it is difficult to get the government out of an area once it is in. Advocates are created for the regulatory process, and they have a vested interest in the continuance of that process. There is reluctance on the part of legislative bodies to review their decisions once they have been made. Environmental laws are replete with examples of this kind.

We have found, over the years, that if we develop a government program, particularly a massive program with a target in mind, we may or may not hit the target we're aiming at, but we are certainly going to hit six we haven't aimed at. In any event, it is clearly time to review the logic behind the Fuel Use Act. If it is no longer valid, then repeal it.

There is also a built-in bias in our environmental laws against fuel switching, as opposed to technological fixes as a solution to the problem of air pollution. Environmental laws are full of examples in which Congress has prescribed the solution to a problem as opposed to describing the problem and then permitting the administrating agency to set performance standards and, to the extent possible, maximize choice as to how those standards can be effected.

The percentage reduction, rigid Clean Air Act amendments of 1977 are typical. These, in effect, mandate the use of scrubbers for new coal-fired installations nationwide regardless of their impact on the environment and regardless of the advisability of using lower sulfur content fuels. The fear that lies behind this is that when economic conditions change it is simply too easy to switch back from the preferred fuel, which would then result in a lot of environmental degradation. The whole issue of backsliding often surrounds the legislative debate over environmental laws.

It must be acknowledged that switching back from the preferred fuel to one which has a far greater impact on the environment is always a possibility. On the other hand it is also possible to switch out of a technological fix if economic conditions become severe enough. It is, after all, expensive to operate a scrubber. It is expensive not only in the operation and maintenance of the scrubber itself but in handling the sludge.

There are a lot of reasons why even after technology is installed, a given user of that technology could, in effect, backslide. So, I don't find that argument particularly persuasive as long as we assume that the country is going to continue to regulate the use of various kinds of fuels in protecting public health and the environment.

My own view is that since it is now so apparent that gas supply is not nearly the problem it once was, the argument for getting the government out of the way and simply letting the market dictate fuel use is very persuasive.

It is undeniable that if a natural gas option were more widely available it would provide more cost-effective programs to control SO_2 in the environment. Analysis is needed to determine the most cost-effective approaches in using natural gas. At the same time, legislative changes must be made so that this capability can be expanded.

Since the beginning of our legislative efforts to address the problems of the environment, the intent of the laws becomes obvious if you understand them simply as an expression of a lack of trust between our executive and legislative branches. The Congress gets ever more prescriptive in telling the Executive Branch what to do both in terms of suggesting specific fixes for environmental problems and then providing deadlines which almost inevitably are missed by the environmental agencies and which then create an invitation to sue. The vast majority of environmental decisions are ultimately made in the courts.

Eighty percent of the decisions made by EPA ultimately end up in the courts. If the usual course of action takes effect, it takes three to five years for the parties—both the industry-affected and the environmental groups, with the administrative branch acting as mediator—to work out a negotiated settlement. This is a very inefficient way of making decisions. It detracts from our confidence that we are capable of coping with our own complexities.

Continuing research is important to the select use concept. There are apparent uses for natural gas which are now coming to light—reburning, second generation coal gasification, the use of gas with coal-water. All have promise. The American Gas Association recently testified that there is still a role for government to play in funding this type of research. These kinds of money are really hard to come by today given the pressures on the budget, but it is really short-sighted not to make these social investments in research where the payoff can be so great. Prior to commercialization it is extremely difficult to get the private sector to put very much money into these kinds of things on a consistent basis, and a little bit of stimulation from the government goes a long way.

Well, you might say, "What difference does all this make?" Here is a fuel that is being underutilized; a fuel which could benefit us both economically and environmentally. So what else is new?

I think there are a number of reasons why the subject is important. One of them is this: as the economies of the world become more interdependent, the pressure on our own economic system to adjust to others in the world increases. To the extent that we can make those adjustments with a clear recognition that we are no longer an island isolated from the rest of the world, the better off we are going to be and the more able we will be to compete internationally.

Understanding the economic interdependence of the world, and making sure we are adjusting our own industrial policies not only to make ourselves more competitive but at the same time more sensitive to other values in our society like environment or the protection of public health—doing all of this and at the same time maximizing choice, maximizing freedom, is a big order. The reason it is important is that what's at stake is whether free societies work or not.

If we are not successful in showing through example that we can cope, that we can deal with these problems, then I don't think that ultimately our society will succeed. That's why concepts like select use, and a thousand others dealing with fairly narrow issues, are terribly important if we are to move the country forward toward a more rational industrial policy—a policy that combines a sensible economic and environmental regulatory system, and which succeeds in keeping its eye on major values of maximizing choice, providing as much freedom as possible and at the same time demonstrating to the rest of the world that we can cope, and that our free society works. I think that if the select use concept is successful and, as I said, literally a thousand like it, all of us will be alright and so will our country.

William D. Ruckelshaus

Abbreviations

A.G.A.	American Gas Association
APCA	Air Pollution Control Association
APCD	Air Pollution Control District (i.e., within a state)
AQCR	Air Quality Control Region
AQS	See NAAQS
BACT	Best Available Control Technology
Bbl	Barrels
Bcf	Billion cubic feet
Btu	British thermal unit
CAA	Clean Air Act
CDS	Compliance Data System
CEQ	Council on Environmental Quality, Office of the President
CNG	Compressed natural gas
CO	Carbon monoxide
CO_2	Carbon dioxide
DFO	Distillate fuel oil
DOE	U.S. Department of Energy
EIA	Energy Information Administration, part of DOE
EPA	U.S. Environmental Protection Agency
ERA	Economic Regulatory Administration, part of DOE
FERC	Federal Energy Regulatory Commission
FOIA	Freedom of Information Act, by which EPA furnishes data
FUA	Fuel Use Act (also PIFUA)
GRI	Gas Research Institute
HC	Hydrocarbons, except methane, as per NAAQS
ICS	Intermittent Control Systems
IEMP	Integrated Environmental Management Program
KW	Kilowatts
L48	Lower 48 states
LAER	Lowest Achievable Emission Rate
Mcf	Thousand cubic feet
MMBtu	Million British Thermal Units (million Btu)

MMcf	Million cubic feet
MW	Megawatts
NAAQS	National Ambient Air Quality Standards (also AAQS, AQS)
NEB	National Energy Board, i.e., of Canada
NGPA	Natural Gas Policy Act of 1978
NO_x	Oxides of nitrogen
NSPS	New Source Performance Standards
PIFUA	Powerplant and Industrial Fuel Use Act of 1978 (also FUA)
PSD	Prevention of Significant Deterioration
RACT	Reasonably Available Control Technology
RFO	Residual fuel oil
SIC	Standard Industrial Classification, i.e., of all enterprises
SIP	State Implementation Plan
SO_x	Sulfur oxides (e.g., SO_2)
TSP	Total suspended particulates
Tcf	Trillion cubic feet
VOCs	Volatile organic compounds

Section I
Overview

Selected Authors

Chapter 1

The Role of Natural Gas in Worldwide Environmental Protection

JOHN KEAN
President, International Gas Union
1985–88 Triennium
President and C.E.O.
NUI Corporation

Until recently, concern regarding environmental pollution has been regarded as a special problem—almost a luxury—of the most developed nations. Today that view is changing.

Air quality problems are obvious in most of the world's cities. If anything these problems are worse in developing nations, many of which have a heavy reliance upon diesel-fueled vehicles. Solid wastes, water pollution and water consumption problems are becoming recognized as equally widespread and serious, if not always as visible.

The truly transnational character of environmental degradation has also become recognized. The impacts of many forms of pollution know no international boundaries. For example, in the past decade considerable attention has been given to such topics as ocean pollution and deforestation as threats to the world food and oxygen chains, and problems of all humanity. The phenomenon of "acid rain" is the most recently recognized example. Recently, Canada and eighteen European nations signed an agreement to cut sulfur emissions by at least 30 percent.

In spite of such examples of progress, the transnational pollution issues are, in many respects, the most politically contentious, since the impacts of pollution may not be felt in the nation of origin. It is a challenge to the international community to find the mechanisms to control such transnational pollution, when the nation that must pay for environmental control may not be the direct and immediate beneficiary of such control.

One key factor contributing to the difficulty of pollution control is the concept that pollution control is costly. Clearly, the high capital cost of many forms of control is an impediment for developing nations which are already under serious burdens of international debt. Current developments in the natural gas industry and in gas technology offer an almost serendipitous solution. In many nations, reducing pollution through increased use of natural gas—the cleanest fuel—may actually reduce the cost of providing and using energy.

Much worldwide energy policy has been driven in recent years by the *incorrect* belief in a few basic tenets: that oil and natural gas would be in increasingly short supply, especially in the United States; that oil and natural gas prices would inflate rapidly; that, in spite of these developments, world economic growth would be steady and robust, that central station coal and nuclear fission electric plants would be a highly cost-effective source of energy in this century; and that renewable electric energy would become cost-effective before the turn of the century, with fusion energy to follow. A corollary tenet has been that natural gas is a premium fuel and feedstock, whose highest benefit to producer nations could be achieved through export, either as liquefied natural gas (LNG) or through conversion into such products as ammonia fertilizers and methanol.[1] All of these basic tenets have now been shown to be, in varying degrees, incorrect.

Oil and natural gas are in excess supply both worldwide and in the United States—with prices declining. Natural gas supplies are likely to be adequate for the rest of this century and most of the next century. A variety of new natural gas supply and utilization technologies have emerged. These developments should guarantee the availability and competitiveness of natural gas. The rapidly escalating capital costs of nuclear and coal-fired units as well as associated environmental and regulatory barriers have diminished the attractiveness of these options. In addition, the demand for energy has become less predictable due to conservation, increasing uncertainty in the business cycle, and other factors. Finally, the high capital costs and market fluctuations associated with the LNG, ammonia and methanol trades have meant that netbacks to producer nations from these downstream investments have been less than satisfactory.

Large-scale energy projects with high capital costs and long lead-times have shown themselves to be increasingly unworkable in a late twentieth century environment of uncertainty and rapid change, particularly when the projects involve international trade. Future planning will place a premium on energy alternatives which offer short lead times, low capital costs per unit of energy, and the ability to add capacity in small step-wise increments. Environmental and public health impacts will also be an increasingly important consideration, as population and energy consumption grow.

These developments necessitate a reconsideration of the role of natural gas in energy policy and development. While international trade will certainly play an important role, greatly increased domestic utilization of gas resources will be desirable in many countries.

Environmental Advantages of Natural Gas

Environmental impacts are of concern both at the point of end-use energy consumption and at all points in the chain ("full cycle") of production, processing and distribution. The environmental advantages of natural gas relative to other energy forces are equally impressive from either perspective.

For example, a recent study of pollution associated with residential sector energy consumption in the United States found that when emissions from the full energy cycle are considered, use of natural gas space heating and other appliances results in only 15 to 20 percent of the total air emissions, and less than one percent of both the total water pollutants and noncombustible solid wastes compared to electric appliances.[2] (See Exhibit 1-1.)

Natural Gas Use with Other Fuels for Environmental Purposes

"Select use" is the use of natural gas with other fuels for emissions control purposes in stationary applications.

Japan and the United States have thus far been the leading countries in select use. In Japan the select use concept has been applied primarily as episodic or seasonal switching, although it is well known that Japan uses gaseous fuels heavily in both powerplants and vehicles due to their environmental benefits.

Exhibit 1-1

Full Trajectory Emissions per Household Attributable to Residential Space Heating and Base-Load Energy Usage[1]

(lbs of pollutants/year)[2]

	Existing Powerplants[3] and Conventional Appliances		New Powerplants[6] and High Efficiency Appliances	
	Electric[4]	*Natural Gas*[5]	*Electric*[7]	*Natural Gas*[8]
Particulates	7.67	2.79	13.13	2.20
NO_X	107.64	44.94	122.53	36.60
SO_X	244.63	Neg.	66.94	Neg.
HC	3.86	4.60	2.04	3.42
CO	5.37	2.28	8.53	1.69
Water Pollutants[9]	14.63	Neg.	19.01	Neg.
Solid Waste	2,740.22	Neg.	6,895.94	Neg.

1 Includes emissions from extraction, transportation, and processing of fossil fuels, conversion to electricity (where applicable), distribution, and end-use. Based on energy required to provide space and water heating, cooking and clothes drying.

2 Based on equal appliance output, regardless of the energy source.

3 Electricity is generated from a fuel mix patterned after the 1982 energy mix of electric utilities with pollution control equipment on 15% of the coal-fired powerplants.

4 Utilized electric resistance heating.

5 Uses conventional gas appliances; e.g., furnace efficiency of 65%.

6 Electricity is generated from a new coal central powerplant with pollution control equipment.

7 Utilizes an electric heat pump with a C.O.P. of 2.75.

8 Uses high efficiency (88% efficient) gas furnace and ranges and clothes dryers with electronic ignitions.

9 Includes dissolved and suspended solids and organics.

Neg.—Negligible, less than 0.0005 pounds per year.

In the United States the select use concept originated seven or eight years ago. Today select use is being applied in dozens of applications, assisting the combustion of oil, coal, wood chips and wastes. The two primary motivations of the plant operators have been (1) to reduce costs in plants already in compliance with clean air regulators, and (2) to achieve compliance with air regulations in a least-cost manner.

The simplest application of the first concept has simply been to use gas to permit switching to cheaper, lower quality residual oil or coal, while remaining in compliance. In order to comply, total emissions of the various regulated substances must not increase. Where there is a substantial price differential among fuel grades, and where a small quantity of gas use can offset the emissions from a sizable amount of oil or coal use, the economics can be attractive.

In the second concept, the goal is actual emissions reduction. The least cost means of achieving reduction is a very site-specific matter, but some of the factors which have motivated plant operators to use natural gas include:

- *Economic Reasons.* Natural gas methods tend to have low capital requirements, and in many instances lowest total costs. For example, the cost of SO_2 control using select use of gas has been calculated to vary from \$50/ton of SO_2 removed to over \$1000/ton. In many cases the cost is in the \$200–\$400/ton range. By comparison, ICF, Inc., as quoted in a May 16, 1983 Congressional Research Service study ("Mitigating Acid Rain: Implications for High Sulfur Regions," by Larry B. Parker) found a range of \$250–\$2000/ton for seven alternative control strategies.

- *Immediate Reductions.* Natural gas methods are often fast to implement. Use of gas has been used as an interim measure so that immediate reductions could be achieved, while awaiting scrubber installation.

- *Space Limitations.* Natural gas methods require little space. There are no fuel or waste storage needs, and gas equipment tends to be small. Gas has been used where space limitations prohibited scrubbers and bag houses.

- *Technical Reasons.* The use of gas may reduce or avoid the unit derating associated with other control measures. For instance, switching to lower sulfur coals may cause substantial unit derating, while also increasing particulate emissions and solid waste disposal problems. Gas has been burned simultaneously with coal to reduce problems associated with poor coal quality, such as low volatility.

- *Maintain the Use of Local Fuels.* Especially when combined with other low capital costs methods (e.g., coal washing), the use of natural gas may permit maximum continued use of local coals. It has been estimated, for example, that SO_2 reductions of 50 percent could be achieved with certain high sulfur coals, while maintaining the local coal as 70 percent of the total fuel mix.

- *Supply Security.* Direct access to more than one fuel offers supply security against the threat of strikes, transportation difficulties and other supply disruptions. It also offers fuel price negotiation leverage.

- *Ability to Reduce Many Pollutants Simultaneously.* Only natural gas offers the ability for one control method to simultaneously reduce many pollutants (e.g., sulfur oxides, nitrogen oxides, reactive hydrocarbons, particulates, carbon monoxide, carbon dioxide). Proposals to reduce acid deposition could greatly increase the importance of this capability.

In addition to these "tried-and-true" reasons for using gas, research is now resulting in a number of new technologies which use gas as a compliance method. For instance, reburn technology being developed with funding by the U.S. Environmental Protection Agency and the Gas Research Institute, appears to offer large NO_X reductions with the use of very little natural gas. Ultimately, this method may qualify as Best Available Control Technology (BACT) in the U.S.

In another instance a gas industry-funded project at the University of Florida is investigating the role which gas can play in reducing the derating and emissions associated with converting large oil-fired boilers to burning coal-water mixtures as well as dry pulverized coal.

In a final example, catalytic technology now being commercialized can greatly reduce the already low NO_X emissions associated with natural gas combustion.

All of these uses of natural gas for emissions control have worldwide potential, which is virtually untapped today.

Cost and Environmental Studies

Natural Gas-Fueled Combined Cycle Electric Power Generation

Electric utility planners have become more reluctant to commit to the construction of new large coal-fired and nuclear powerplants due to the high capital costs, long lead times, and numerous environmental and regulatory barriers associated with such projects. These factors, combined with uncertainty over projected levels of electricity demand in the 1990's, have enhanced the outlook for smaller-scale power production alternatives which can be brought on-line quickly and at a lower capital cost per kilowatt of capacity. Construction of these units reduces the risk of over-building, as the demands for electricity and production capability are more closely matched.

One such option available today is combined cycle natural gas technology. A combined cycle gas turbine powerplant combines the operation of a gas turbine, a steam turbine and a heat recovery steam generator. Natural gas (or fuel oil) is fed into a gas turbine which drives a generator. Waste heat from the turbine is captured by the heat recovery steam generator which provides steam for the steam turbine. The steam turbine, in turn, also drives a generator.

Combined cycle power generation does not represent a new technology, although innovations and improvements continue to enhance this method of electricity production. There are currently over one hundred combined cycle gas systems in operation worldwide with a combined capacity of approximately 17,000 MW, (7,000 MW of which is located in the U.S.). The systems range in size from less than 10 MW to several hundred MW. A 2,000 MW plant currently under construction in Japan will be the world's largest combined cycle operation, with 14 systems at one site.

A recent study found that in the United States a new 240 megawatt (MW) combined cycle unit produces electricity at a cost which is only 54 to 92 percent of the cost of electricity produced by a new, comparably sized coal unit.[3] Combined cycle gas units enjoy economic advantages because capital charges (return on investment, capital recovery, taxes, etc.) and non-fuel operation and maintenance expenses are about one-third those of coal units, more than offsetting coal's fuel advantage. (See Exhibit 1-2.) The economic advantage

Exhibit 1-2

Characteristics of a 240 Megawatt Electric Generating Station—Coal-Fired Combined Cycle

	Coal-Fired Steam Plant	Gas-Fired Combined Cycle
Operating Characteristics		
Rated Capacity (MW)	240	240
Capacity Utilization (%)	65	65
Unit Efficiency (%)	34	43
Heat Rate (Btu/kWh)	10,000[2]	8,000[1]
Annual Energy Input (TBtu)	13.744	10.819
Annual Energy Output (TBtu)	4.669	4.669
Economic Parameters		
Installed Capital Cost—USA ($1985/kW)	$1,700[3]	$500[2]
Operation and Maintenance—USA (mills/kWh)	8.5	2.7
—Fixed ($kW/yr)	$19.00[2]	$6.50[2]
—Variable (mills/kWh)	5.2[2]	2.6[2]
Other Factors		
Lead Time—USA (years)		
—160 MW	n.a.[4]	1[5]
—240 MW (Full Load)	9[2]	2[5]
Expected System Life—USA (years)		
—Pollution Control	15	n.a.
—All Other	30[2]	30[2]

[1] General Electric, *STAG Combined Cycle Product Line and Performance Characteristics,* GER 3401, (Schenectady, NY: 1984) pp. 6-7. Based on STAG 207E with two pressure heat recovery steam generator.

[2] Electric Power Research Institute, *Technical Assessment Guide,* (Palo Alto, CA: May 1982) Appendix b-55 and b-79.

[3] Electric Power Research Institute, *Ibid.,* Appendix B-55. Total capital cost scaled up by 35% to account for the higher cost per Kw of constructing a single unit 240 MW facility versus a dual unit 1000 MW facility, as per discussion with EPRI.

[4] n.a.—not applicable.

[5] Gas turbine portion of system operational in 12 months, with remainder on line in 24 months according to: General Electric, *General Electric 60 Hz STAG Combined Cycle Power Plants,* GEA-11387, (Schenectady, NY: 1984) p. 60831.

of combined cycle would be even greater in many developing nations, where the cost of natural gas might be lower than in the U.S. while the capital costs would be greater.

The environmental impacts of the combined cycle unit were calculated in a recent A.G.A. study. (See Exhibit 1-3.) The annual emissions of air pollutants of sulfur dioxide (SO_2), nitrogen oxides (NO_X) and total suspended particulates (TSP) were calculated for the 240 MW coal-fired (2 percent sulfur, 10 percent ash coal) and combined cycle units assuming each would at least meet applicable federal emission standards. Each unit's production of sludge and ash, and its annual water consumption were also calculated. The principal findings were:

- Even when equipped with a flue gas desulfurization system for SO_2 control and an electrostatic precipitator for TSP control, the coal unit emits over 4 thousand tons of SO_2 and 200 tons of TSP per year. Gas-fired combined cycle units emit negligible amounts of SO_2 and TSP—less than 0.3 percent of the coal units combined tonnage.

- Emissions of NO_X from the combined cycle plant are 57 percent of the coal units emissions. These emissions from the combined cycle unit could be further reduced, but unit efficiency would decline.

- The coal plant would produce 60 thousand tons of ash per year and 168 thousand tons of sludge (a semi-solid scrubber by-product). Combined cycle units produce neither of these non-combustible solid wastes.

- Treated water consumption by the combined cycle plant would be 15 million gallons per year, one-half the consumption of the coal plant.

Clearly, from the perspectives of environmental impacts, in addition to resource efficiency, capital efficiency, flexibility and consumer cost, gas-fired combined cycle powerplants have much to offer many nations.

Natural Gas-Fired Cogeneration

"Cogeneration" refers to the sequential production within one system of both electrical (or mechanical) energy and thermal energy.

Exhibit 1-3
Environmental Comparison of Natural Gas-Fired Combined Cycle Generating Station and Coal-Fired Station
(Annual Impacts, 240 Megawatts)

	Coal-Fired Steam Plant	Gas-Fired Combined Cycle
Air Pollutant Emissions (tons/year)		
Sulfur Dioxide	4,123[1]	3[2]
Particulate Matter	206[1]	9[3]
Nitrogen Oxides	2,385[1]	1,352[4]
Other Pollutant Emissions (thousands of tons/year)[5]		
Sludge	168	none
Ash	60	none
Treated Water Consumption (millions of gallons/year)[6]	30	15

1 Based on new source performance standards for coal-fired electric power-plants (2% sulfur coal assumed): SO_2–0.6 pounds per MMBtu, TSP–0.03 pounds per MMBtu, and NO_X–65% reduction. See: 40 *CFR* Subpart Da, Part 60.40a through 60.43a.

2 Data on sulfur dioxide emissions from gas turbines not available, assumed to be equal to SO_2 emissions from gas boiler–.0006 pounds per MMBtu. See: U.S. EPA, *Compilation of Air Pollutant Emission Factors,* (Research Triangle Park, NC: August 1982) p. 1.4-3.

3 Gas turbines operating on distillate oil have particulate emissions of 0.004 pounds per MMBtu according to: Radian Corp. for the U.S. EPA, *Emissions from Stationary Gas Turbines, New Source Performance Standards,* (Research Triangle Park, NC: November 1984) p. 4-78. Emissions when operating on gas would be approximately 39% of this amount based on: U.S. EPA, *Compilation of Air Pollutant Emission Factors,* (Research Triangle Park, NC: August 1977) p. 3.3.1-2.

4 Based on new source performance standards for electric utility stationary gas turbines: NO_X–75 parts per million. See 40 *CFR* Subpart GG, Part 60.330 through 60.333.

5 U.S., Environmental Protection Agency, Research and Development–Energy Minerals and Industry, *Energy/Environment Fact Book,* (Washington, DC: March 1978) p. 24.

6 General Electric, GEA-11387, *60 Hz STAG Combined Cycle Power Plants,* (Schenectady, NY) p. 568316A.

Two-thirds of the energy consumed by conventional electric powerplants is normally lost to the environment.

Cogeneration systems, in contrast, recapture much of the otherwise wasted thermal energy and use this energy for a variety of purposes—e.g., space or water heating and industrial process needs.

Cogeneration systems offer inherent efficiency advantages relative to conventional means of power production, and in addition, they may offer environmental benefits as well. A primary advantage of cogeneration systems is that they are inherently energy efficient. That is, they minimize the waste of primary energy resources—oil, gas and coal.

However, the net impact of cogeneration systems on the environment relative to conventional power production is not clear-cut. In spite of the fact that they consume less fuel, small, dispersed internal combustion or coal-based cogeneration systems may be less environmentally attractive than large central powerplants with tall stacks and pollution control equipment.

For example, the production of NO_X via a diesel fuel-fired internal combustion cogeneration system would be over twice the NO_X level produced by the combination of a scrubber-equipped coal-fired powerplant and a coal-fired boiler (with electrical and steam production capacities equivalent to the cogeneration system).

Gas-fired cogeneration systems are an attractive option from both an environmental and an energy efficiency standpoint. Gas-fired cogeneration systems, such as gas turbines or gas boiler/steam turbine combinations, offer a clean and efficient means of electrical power and thermal energy production.

According to the U.S. Office of Technology Assessment, a gas turbine cogeneration system would require approximately 25 percent less input energy than the combination of a new coal-fired powerplant and an oil-fired boiler producing steam (assuming optimal system sizing.)

In addition, the gas-fired cogenerator would emit less than 1 percent of the SO_X, 27 percent of the particulates, and 50 percent of NO_X produced by the comparably sized conventional coal- and oil-based system (with pollution control equipment).[4]

Although cogeneration technology is not new, the growth of the number of cogeneration systems has accelerated over the past few years. In the United States today, approximately 1,090 commercial and industrial cogeneration facilities either are operating or are in the design/construction phase. The combined electrical output capacity of these units totals approximately 28,000 MW. Over 10,500 MW (about 38 percent) of the total is believed to be fueled by natural gas, with the remainder being powered by other fuels, such as coal, oil, biomass, wood, and wood waste. About 34 percent of the total capacity, and over 70 percent of the natural gas-based capacity, is located in the West South Central region of the United States.[5]

Although the number of these systems is growing, the potential in the U.S. and worldwide has barely been tapped. The fuel input to cogeneration facilities in the U.S. currently accounts for less than one percent of the total fuel input to the approximately 1.8 million commercial and industrial boilers. Of this total, less than one percent (about 10,700) are large industrial boilers (fuel input greater than 50 MMBtu/hr). However, these large industrial boilers account for over 37 percent of the total U.S. commercial and industrial boiler capacity.

A recent A.G.A. analysis found that large-scale natural gas-fueled cogeneration systems are environmentally preferable to coal-based systems with pollution control equipment regardless of the system size. (See Exhibit 1-4.) For example, when compared to a coal-fueled system with a scrubber and an electrostatic precipitator, a gas-fueled steam turbine emits only 6 percent of the sulfur oxides, 10 percent of the particulates, and 28 percent of the nitrogen oxides.

Natural Gas Use in Industrial Boilers

While not a "new" use of gas, it is important to point out that natural gas is economically and environmentally preferable to coal use in most industrial boiler applications. This economic advantage exists, in spite of the generally lower fuel cost for coal relative to natural gas, because of the lower capital (including ancillary pollution control equipment needed for coal use) and non-fuel operating costs of gas systems.

Total air pollutant emissions from natural gas-fired boilers are far lower than from comparably sized coal-fired boilers. For example, when compared to a coal-fueled system with pollution control equipment, a natural gas-fired boiler emits less than one percent of the sulfur dioxide, only 19 percent of nitrogen oxides, 15 percent of the carbon monoxide, 87 percent of the non-methane hydrocarbons, and 89 percent of the particulate matter. (See Exhibit 1-5.)

Exhibit 1-4

Annual Emissions from Cogeneration System Operation
(tons/year)

Based on a steam requirement of 200,000 lbs/hr, assumes a 70 percent capacity utilization.

	Steam Turbine		Gas Turbine with Back-Up	
Pollutant	*Coal-Based*[2]	*Natural Gas-Based*	*Gas Turbine*	*Back-Up Gas Boiler*[1]
Sulfur Oxides	1188	70	3	Neg.[3]
Particulates	100	10	10	5
Carbon Monoxide	40	Neg.[3]	50	9
Nitrogen Oxide	694	197	267	118

1 Since emission factors are based on an equivalent electric output, a back-up boiler is required for the gas turbine in order to reach an equivalent process steam output.

2 Assumes the use of flue gas desulfurization and an electrostatic precipitator in order to meet new source performance standards.

3 Negligible.

Source: Office of Technology Assessment, Congress of the United States, *Industrial and Commercial Cogeneration* (Washington, DC: February, 1983).

Exhibit 1-5

Annual Emissions from Operating a 100 MMBtu/hr Industrial Boiler
(tons/year)

	Particulates	*Sulfur Dioxide*	*Carbon Monoxide*	*Nitrogen Oxides*	*Non-Methane Hydrocarbons*
Natural Gas	0.68	0.17	9.68	33.15	0.77
Coal[1]					
Low sulfur, no scrub	1.53[2]	297.4	63.6	177.9	0.89
Low sulfur, scrub 50%	1.53	148.7	63.6	177.9	0.89
Med. Sulfur, scrub 50%	1.53	446.1	63.6	177.9	0.89
High sulfur, scrub 90%	0.76[3]	168.5	63.6	177.9	0.89

1 The four regions' coal type (Btu content and ash content) were averaged for this calculation.
2 Fabric filter removes 99.8% of TSP.
3 Electrostatic precipitator removes 99.9% of TSP.

Source: Based on emission factors from U.S. Environmental Protection Agency *Compilation of Air Pollutant Factors, Third Edition: Supplement No. 13* (Reserach Triangle Park, NC: August 1982).

Natural Gas-Fueled Vehicles

According to a study by the Georgetown University Center for Strategic and International Studies,[6] worldwide, nearly 400,000 vehicles are operating on natural gas today. Besides its national security advantages in many applications, the use of natural gas as a motor vehicle fuel offers, substantial fuel cost savings to natural gas vehicle (NGV) owners, as well as greater efficiency of energy resource utilization and lower air emissions, than does conventional gasoline, or propane, methanol, or electric vehicle alternatives. In fact, the use of natural gas as a vehicular fuel offers a least-cost solution to the growing air quality problems of the world's major cities.

A particularly promising market segment for natural gas appears to lie in those fleets of motor vehicles which are characterized by relatively high annual fuel consumption due to a high number of miles driven and/or low fuel use efficiency. Even in the United States, where gasoline is relatively inexpensive by world standards, the national average cost to the fleet owner of fueling a standard-sized automobile with natural gas, including the cost of fuel, compression, and filling station capital recovery (but not including vehicle conversion costs) is estimated to be only 72% of the cost of fueling the same car with gasoline—4.3 cents/mile versus 6.0 cents/mile. This average compressed natural gas cost is equivalent to operating on gasoline priced at approximately $0.90/gallon.[7] (See Exhibit 1-6.)

Perhaps even more significant than the economic advantages of natural gas vehicles are the environmental advantages. The NGV produces only 1% to 64% of the air emissions of gasoline-powered vehicles. Furthermore, when the full energy cycle from extraction to end-use is considered, using natural gas as fuel results in substantially lower air emissions than even an electric vehicle, emitting less than 1% of the sulfur oxides and total suspended particulates, 54% of the nitrogen oxides, and 30% of the carbon monoxide of the electric alternative.

Residential Gas Systems

Natural gas has long been recognized to be the most economical fuel for residential space heating, hot water heating, clothes drying and cooking. Recent commercialization of a new generation of highly efficient gas appliances has only enhanced this advantage. For instance, a recent 1983-1993 study found that the national average annual expenditure in the United States for home heating by gas customers is expected to be only about 60% as high as oil or electricity (assuming the use of a heat pump) during most of the forecast period, and less than 50% as great as electric resistance heat.[8]

For baseload, uses, such as water heating, cooking and clothes drying, for which gas and electricity are the primary fuels, natural gas will cost only 30%-40% as much as electricity on an annual basis.

In addition to the economic superiority of natural gas this system is also best from the perspective that it utilizes less primary energy and results in less environmental pollution, than do most alternatives.

Over the total fuel cycle, NO_X, SO_X and suspended particulate emissions associated with residential space conditioning are 300% to 450% with an electric heat pump, and 120% to 220% with an oil

Exhibit 1-6

Fuel Costs, Energy Usage and Air Emissions of Alternatively Fueled Vehicles

Fuel	Total Fuel Cost (cents/mile) ($ 1983)[1]	Total Energy Usage (Btu/mile)[2]	Criteria Emissions (grams/mile)[3]				
			Nonmethane HC	CO	NO_X	SO_X	Particulates
Natural Gas—CNG[4]	4.3	5,940-6,540	0.26	0.03	1.23	Neg.	Neg.
Gasoline	6.0	6,990	0.54	8.35	1.92	0.71	0.08
Methanol	6.2	9,400	0.25	2.90	0.55	Neg.	Neg.
Electricity[5]	4.7	10,920	0.03	0.10	2.28	1.58	0.04
Propane	5.3	5,940	N/A	N/A	N/A	N/A	N/A

1 Fuel costs include fueling station cost recovery. Actual costs will vary significantly depending upon state and local regulations, local labor and construction costs, utility rate structures, tax and other financial considerations. The national average state and Federal fuel tax of $1.48/MMBtu is assumed for gasoline, the national average state tax is assumed for CNG, methanol, and propane, and no road tax is assumed for electricity.

2 Includes all energy used or lost from the point of production to the point of consumption.

3 Assumes 80% of SO_X removed in fossil fuel-fired electric generation.
Assumes 95% of particulates removed in fossil fuel-fired electric generation.
Emissions attributable to battery recharging of electric vehicles are not included.

4 CNG is compressed natural gas.

5 Electric vehicle capacity is limited to 4 passengers; all other costs are for 6 passenger vehicles. This cost for electric vehicles does not include the substantial cost of battery replacement associated with these vehicles.

N/A = Emissions data not available on comparable basis.

furnace/electric air conditioner combination of the emissions with a gas furnace/conventional electric air conditioner combination. Solid wastes are 450% for the electric heat pump compared to gas or oil furnace/conventional electric air conditioner combinations. (See Exhibit 1-7.)

Other Natural Gas Applications

A large number of other new technologies for natural gas use are emerging in residential, commercial and industrial applications. These include natural gas air conditioning (cooling) systems, natural gas heat pumps, natural gas fuel cells, and gas/solar energy systems. A.G.A. analyses have documented the natural gas advantages of cleanliness, efficiency and cost competitiveness for these new technologies.[9,10] Perhaps even more important from an energy planning and environmental perspective, many of these new gas technologies may help avoid the need for new electric peaking capacity to meet summer loads, while also balancing seasonal gas demand.

Conclusion

Natural gas resources are abundant, widely distributed, and highly underutilized. By all measures—air emissions, solid wastes, water pollution, and water consumption—natural gas is by far the most environmentally desirable fossil fuel. Compared to other energy alternatives natural gas systems are also highly capital and resource efficient, requiring significantly lower capital investment to produce, deliver and utilize a unit of energy. Natural gas can and should play a major role in environmental protection in many nations.

Exhibit 1-7

Total System Primary Resource Efficiencies for Residential Space Conditioning

(Percent)

Delivery Efficiency[1]	*Natural Gas*	*Oil*	*Fossil Fuel Generated Electricity* *Gas-Fired*	*Oil-Fired*	*Coal-Fired*
Extraction	99.3%	99.6%	99.3%	99.6%	98.8%
Processing	99.7	87.8	96.7	87.8	92.0
Transportation	96.6	98.0	96.6	98.0	98.1
Conversion	n/a	n/a	31.1	31.9	32.6
Distribution	97.1	98.0	91.6	91.6	91.6
TOTAL	90.1%	84.0%	26.4%	25.0%	26.6%
End-Use Efficiency[1]					
Heating[3,4,5]	87.0	84.0		170.0	
Cooling[5,6]	200.0	200.0		180.0	
TOTAL[2]	116.0%	113.8%		172.6%	
Total System Efficiency	71.8%	65.9%		45.4%	

1 Efficiency refers to the energy used or lost along any step of the system trajectory (processing, transportation, etc.); thus, efficiency may be calculated as the energy output of a particular step divided by the energy input—thereby indicating the amount of energy used or lost in that step.

2 Weighted by heating and cooling loads (75% and 26% respectively) in Philadelphia, representative of a "national average" load. See Appendix B.

3 Recuperative gas condensing furnace; Federal Trade Commission AFUE.

4 Blueray oil furnace; Federal Trade Commission AFUE.

5 The heating season performance factor of an electric heat pump with a C.O.P. of 2.75 ranges from 1.52 (New England) to 2.09 (Pacific Region) with a regional average of 1.70 as weighted by number of households per region according to: Oak Ridge National Laboratory for the U.S. Department of Energy, *Performance and Economics of the ACES and Alternative Residential Heating and Air Conditioning Systems in 115 U.S. Cities,* (Oak Ridge, Tennesee: March 1981). Cooling efficiency, from the same source is 180% invariant with respect to location.

6 Conventional air conditioning assumed to be 10% more efficient than cooling with a heat pump. See: Gordian Associates for the U.S. Department of Energy, *Heat Pump Technology,* (Washington, DC: June 1978).

References

1. Portions of this section are quoted with permission from Nelson E. Hay, *Energy Policy Strategies for Developing Nations: The Role of Natural Gas,* to be published.
2. American Gas Association, *A Comparison of the Full Cycle Emissions of Natural Gas and Electric Residences*, (Arlington, VA, March 30, 1984).
3. American Gas Association, *Natural Gas-Fueled Combined Cycle Electric Power Generation—An Attractive Option*, (Arlington, VA, May 10, 1985).
4. American Gas Association, *Guide to New Natural Gas Utilization Technologies*, (Atlanta, GA, The Fairmont Press, 1985) p. 82.
5. American Gas Association, *Fact Sheet: Estimated Current Natural Gas-Fueled Cogeneration Capacity*, (Arlington, VA, May 25, 1984).
6. Center for Strategic and International Studies Georgetown University, *Natural Gas Vehicles A National Security Perspective*, (Washington, DC, Volume VI, No. 16, 1984), p. ix.
7. American Gas Association, *An Economic, Efficiency, and Environmental Comparison of Alternative Vehicular Fuels: 1983 Update*, (Arlington, VA, September 9, 1983).
8. American Gas Association, *Consumer Costs of Natural Gas, Fuel Oil and Electricity for Residential Use: 1985 Update*, (Arlington, VA, January 23, 1985).
9. American Gas Association, *A Life-Cycle Cost Analysis of the Market Potential of Residential Gas Air Conditioning*, (Arlington, VA, November 30, 1984).
10. American Gas Association, *A Life-Cycle Cost Analysis of Natural Gas Cooling in the Commercial Sector*, (Arlington, VA, June 21, 1985).

Appendix 1-1

Full Trajectory Emissions per Household Attributable to Residential Space Heating and Base Load Energy Usage by Process Step—Existing Powerplants and Conventional Appliances[1]

(lbs of pollutants/average annual uses[2])

	Extraction	*Hauling/ Gathering*	*Processing*	*Transport*	*Conversion to Electricity[3]*	*End Use*	*Total*
Electric Appliances							
Particulates	— —	— —	— —	1.67	6.00	— —	7.67
NO_x	— —	— —	0.34	3.19	104.11	— —	107.64
SO_x	— —	— —	0.25	0.34	280.87	— —	281.46
HC	— —	0.46	0.28	0.26	2.86	— —	3.86
CO	0.09	0.03	0.02	0.38	4.85	— —	5.37
Water Pollutants	7.61	— —	2.48	— —	4.54	— —	14.63
Solid Waste	271.27	— —	1,132.22	— —	1,336.73	— —	2,740.22
Natural Gas Appliances							
Particulates	— —	— —	— —	— —	N/A	2.79	2.79
NO_x	— —	0.65	0.60	34.80	N/A	5.84	41.89
SO_x	— —	— —	— —	— —	N/A	— —	— —
HC	— —	4.60	— —	— —	N/A	— —	4.60
CO	— —	— —	— —	— —	N/A	2.28	2.28
Water Pollutants	— —	— —	— —	— —	N/A	— —	— —
Solid Waste	— —	— —	— —	— —	N/A	— —	— —

[1] Based on electric resistance heating, conventional gas furnace, and gas range and dryer with pilot lights.

[2] Assumes combined average annual thermal heat production of 73 MMBtu by a residential homeowner's space heater, water heater, dryer, and range, regardless of energy source.

[3] Electricity is assumed to be generated from the actual 1982 energy mix: coal (52%), natural gas (14%), hydro (15%), nuclear (13%), and oil (6%).

— — Negligible, less than 0.005 lbs per year.

N/A—Not applicable.

Sources: Hittman Associates, Inc. for the Council on Environmental Quality, *Environmental Impacts, Efficiency, and Cost of Energy Supply and End Use* (Washington, DC: November 1974); Environmental Protection Agency, *Compilation of Air Pollutant Factors, Third Edition* (Research Triangle Park, NC: August 1982); Radian Corporation, *A Western Regional Energy Development Study* (Austin, TX: August 1975); American Gas Association, *Gas Facts 1982* (Arlington, VA: December 1983).

Appendix 1-2

Full Trajectory Emissions per Household Attributable to Residential Space Heating and Base Load Energy Usage by Process Step—New Powerplants and High Efficiency Appliances[1]

(lbs of pollutants/average annual uses[2])

	Extraction	*Hauling/ Gathering*	*Processing*	*Transport*	*Conversion to Electricity[3]*	*End Use*	*Total*
Electric Appliances							
Particulates	--	--	--	3.19	9.94	--	13.13
NO_x	0.19	0.18	--	0.48	121.68	--	122.53
SO_x	--	--	--	0.49	66.45	--	66.94
HC	--	--	--	0.30	1.74	--	2.04
CO	0.15	--	--	0.56	7.82	--	8.53
Water Pollutants	10.77	--	1.94	--	6.30	--	19.01
Solid Waste	355.08	--	1,482.08	--	5,058.78	--	6,895.94
Natural Gas Appliances							
Particulates	--	--	--	--	N/A	2.20	2.20
NO_x	--	0.51	0.48	27.56	N/A	4.63	33.18
SO_x	--	--	--	--	N/A	--	--
HC	--	3.42	--	--	N/A	--	3.42
CO	--	--	--	--	N/A	1.69	1.69
Water Pollutants	--	--	--	--	N/A	--	--
Solid Waste	--	--	--	--	N/A	--	--

[1] Based on electric heat pump, recuperative gas furnace, and gas dryer and range with electronic ignition.

[2] Assumes combined average annual thermal heat production of 73 MMBtu by a residential homeowner's range, water heater, furnace, and clothes dryer, regardless of energy source.

[3] Based on new coal-fired powerplants with pollution control devices.

— — Negligible, less than 0.005 lbs per year.

N/A—Not applicable.

Sources: Hittman Associates Inc., for the Council on Environmental Quality, *Environmental Impacts, Efficiency, and Cost of Energy Supply and End Use* (Washington, DC: November 1974); Environmental Protection Agency, *Compilation of Air Pollutant Factors, Third Edition* (Research Triangle Park, NC: August 1982).

Chapter 2

Select Use for Environmental Compliance

JOSEPH A. CANNON
Pillsbury, Madison & Sutro
Formerly, Assistant Administrator for Air and Radiation, U.S. Environmental Protection Agency

The environmental problems that face us today include acid rain, a plethora of regulations and potential regulations of sulfur dioxide emissions, tall stacks, the Fuel Use Act, nitrous oxides—all of these make the fuel use manager's lot a difficult one.

Threading the regulatory maze while keeping the bottom line black for risk investors is an enormous challenge. In fact one of the most challenging jobs today is balancing the competing goals of environmental compliance and conservation of resources in a cost effective manner.

The select use of natural gas can be an optimum solution to many of the problems of air pollution.

If we define "select" gas use as the combined use of gas with other less environmentally attractive fuels such as coal or oil—either at different times in dual-fired boilers, or in separate boilers fired in parallel—select gas use can produce the following benefits:

- major reductions in emissions of sulfur dioxide
- significantly reduced emissions of nitrogen oxides
- reduced emissions of smog-causing unburned hydrocarbons

- reduced particulate emissions
- reduced volume of solid waste (fly ash, bottom ash, scrubber sludge), with the attendant decrease in land requirements for disposal, and decreased consumption of limestone and other sulfur-capturing materials
- increased flexibility in the siting of new facilities in or near non-attainment areas.

In addition to these environmental benefits, there can be substantial economic benefits, including:

- decreased dependence on imported energy, with the accompany-reduction in sensitivity to energy/fuel supply disruptions
- a potential reduction in actual fuel costs per Btu
- the avoidance of substantial capital expense for air pollution controls.

The need for providing future energy alternatives, combined with the need to achieve a cleaner environment, ultimately will compel the use of natural gas. Without gas the energy options for future power generation are limited to nuclear, coal, foreign imports, or some kind of postponement.

Natural gas is the cleanest fossil fuel. Natural gas combustion produces virtually no sulfur dioxide, particulate matter or solid waste, significantly less nitrogen oxides and water pollution. The simultaneous combustion of gas with less environmentally attractive fuels such as high sulfur coal or oil—the select use of gas—offers great advantages. It can help us in maintenance of air quality standards, and it reduces susceptibility to fuel supply disruptions.

Why Have Fuel Use Planners Not Vigorously Pursued This Option?

1. *Concern about gas supply.* Will gas be available in future years? Will gas prices remain competitive? These are the questions that planners must answer.

I believe there is enough gas, and that there is going to be enough gas. Today we are adding twice as much gas to proven reserves each year as we did a decade ago, and we have a 15-20% deliverability surplus. Developments since the oil crisis of 1973-1974 have changed the state of natural gas supply.

Throughout the 1960s and early 1970s, regulatory bodies, especially pollution control officials, encouraged the use of natural gas. This happened because government regulation held down natural gas prices. The OPEC oil embargo further encouraged use of gas as an alternate energy source. Artificially low prices, however, discouraged new gas production, and demand soon exceeded supply.

Change in national policy was legislated in 1978 with passage of the Natural Gas Policy Act, and similar statutes, which allowed gradually rising gas price ceilings and, at the beginning of 1985, partial deregulation of gas supplies. In the view of many, NGPA did not go far enough, but something is better than nothing.

Subsequent energy conservation and slower economic growth have reduced demand, while a freer market has encouraged development of new supplies. The United States is in the sixth year of an average 18-month deliverability surplus.

Today the outlook for gas supply is greatly enhanced. Even during the bleakest periods the United States has remained 95% self-sufficient in natural gas. For the past four years gas reserve additions have averaged more than 100% of production.

Given the present supply bubble—even with a 2%/year increase in demand, and assuming *no* new discoveries—natural gas supplies will remain constant through the end of this century. Many are bullish on the long-range prospects for increased availability of gas supply.

Major sources of potential supply still exist in the United States. Large resources of methane are known to exist in Devonian shales, in coal seams, tight gas formations in various parts of the country (particularly in the West), in geopressured brines along the Gulf of Mexico Coast, and perhaps in other places. Although, if we exclude the impact of new technologies, we could run short of supply at market-clearing prices 20 years hence, advanced production technology is expected to bring forth sufficient additional volumes of cost-competitive gas to satisfy demand indefinitely. And this does not even consider gas supplies from Mexico or Canada, which are not included in these projections.

The following table demonstrates estimated gas reserves from all 50 states (excluding Alaska's north slope), onshore and offshore. The figures are the most recent available, they are given in Bcf (billion cubic feet). They show a broad consensus among the three groups whose statistics are most widely regarded as objective: the Potential Gas Committee of the Colorado School of Mines, the United States Geological Survey, and the Energy Information Administration.

Table 2-1. Estimated U.S. Domestic Natural Gas Reserves: Conventional and Unconventional Sources

	*present**	*proven*	*probable*	*possible*	*speculative*
PGC[1]	17,000	200,000	184,758	292,939	306,228
USGS[2]	**	191,500	177,500	593,800	
EIA[3]	17,512	200,247	799,400	(mean, high/low)	

*present annual addition to existing supply; 102% of annual consumption

**current figures unavailable. 1979–80 data = 19,000

1 Potential Gas Committee Announced Estimates of U.S. Natural Gas Potential, released 3/25/85

2 *Estimates of Undiscovered Recoverable Conventional Resources of Oil and Gas in the United States,* Geological Survey Circular 860, 1981

3 *U.S. Crude Oil, Natural Gas and Natural Gas Liquids Reserves,* 1983 Annual Report, US/DOE, Energy Information Administration, 1984, and *U.S. Natural Gas Availability: Supply Through the Year 2000,* Office of Technology Assessment report, 2/85

The United States is well enough endowed with indigenous natural gas reserves to enable it to move, without panic, into new patterns of fuel development and use. Further, if much of the staggeringly abundant methane content of the crust of the earth is really of abiogenic origin (not fossil fuel), as some recent research suggests, the opportunity for natural gas development and production could be even greater than we've imagined up to now.

The problem is not that demand will overtake supply; the problem is relative cost. Present gas prices have stabilized. The widely speculated fly-up of gas prices with the advent of deregulation has not occurred. What will be required is to keep demand high enough that exploration and production are not curtailed because of soft markets.

Market demand rather than legislative restriction must be allowed to control supply. If a "market" decision determines how much gas is used, and by whom, demand will be high enough that exploration and production will not be curtailed. The enormous environmental and economic advantages of a free market will have constructive effects on patterns of energy production, choices among energy sources, and choices among long-term investments—in both energy production *and* consumption.

2. *Legislative impediments to natural gas use.* The "Powerplant and Industrial Fuel Use Act" (FUA) and the "Natural Gas Policy Act" (NGPA) were enacted in 1978 in a period when gas was in short supply, *and* under the assumption that natural gas supplies would be inherently scarce at any price. The combined implementation of the FUA and the incremental pricing requirements of the NGPA perpetuate the outdated legislative decision characterizing certain applications as "inferior uses" of gas, rather than permitting fuel managers the right to use gas, as desired, to meet environmental compliance and other requirements.

Provisions of the Fuel Use Act restrict significant boiler fuel uses of gas, thus inhibiting the ability of plant managers to make cost effective decisions. However, there are lots of ways around the Fuel Use Act, and it doesn't bar the use of natural gas in many, many applications. More important than its existence is the psychological barrier it presents in using gas.

The Department of Energy has found a number of ways to permit interesting uses of natural gas, and including some which—four or five years ago—might not have been contemplated as exemptions under the Fuel Use Act. While this legislation does propose a number of restrictions on the use of natural gas, it is not something that completely cuts off the use of natural gas as in the so-called inferior uses.

Bills have been introduced in Congress to eliminate the onerous provisions of both of these statutes—the Fuel Use Act and the Natural Gas Policy Act—which could do away with this negative burden on the use of natural gas.

If we are to continue to have access to a clean combustion fuel source such as natural gas, we must give operators the full range of options, so that demand can remain high, and to encourage new production technologies with which to replace the gradually diminishing conventional supply, at a competitive price.

3. *Burdens or impediments in environmental regulations.* One of the other areas where people can perceive some burden or impediment to the use of natural gas is in the area of environmental regulation.

It is not the policy of the Environmental Protection Agency to discourage the use of natural gas. In earlier years EPA, as well as the Department of Energy, some other governmental agencies, and also non-governmental entities, did have a sense that natural gas shouldn't be used as a boiler fuel: that natural gas, and its molecules, was somehow special, and that it should be treated differently from molecules of other kinds of fuels.

However, there are a number of developments within the EPA that demonstrate not only that there is no kind of prohibition or negative view towards natural gas, but a very positive view towards natural gas.

The very first application of the "bubble policy" involved the use of natural gas.

More than a year ago, the agency proposed Subpart Db regulations which involve the new "first performance" standards for industrial boilers. The proposal adopts an explicit policy to promote the use of natural gas.

Something like that is not inadvertent—these regulations look like natural gas might be a terrific choice. In the view of the agency, these regulations could buy an incentive through the use of natural gas in industrial boilers. This is a very modest step, but it was a step consciously taken by the agency at the time, to signal the interest in the use of natural gas as a means of environmental compliance.

A gradual change is evolving—not just at EPA, but in the consciousness of people's thinking about how to use natural gas as a means of environmental compliance.

There are basically three ways to reduce pollution:

1. Change the pollutant content of the fuel being burned.

2. Collect the pollutant during the burning process.

3. Clean the pollutant emissions in flue gas.

The least expensive way in many applications could be to reduce the pollutant content of the fuel burned. Below is a table which shows the relative emissions of particular hydrocarbons and trace metals of coal, oil, solid waste, wood, and natural gas. It may be surprising to observe that natural gas comes out way less than all the others, by a long, long way. The difference is quite dramatic when comparing gas with basically high sulfur coal or high sulfur oil.

To summarize:

1. The marketplace should determine how much gas is used and where.

Table 2-2. Typical Uncontrolled Emission Factors for Natural Gas and Other Steam Generator Fuels
(pounds per MMBtu heat input)

	SO_X	*Particulate Matter*	NO_X	*Hydrocarbons*	*Trace Metals*
Coal*	5.70	2.54	0.90	0.005	0.009
Oil (residual)**	3.22	0.23	0.39	0.01	0.0002
Solid Waste	0.49	3.22	0.31	– –	– –
Wood	0.02	4.88	0.25	– –	– –
Natural Gas	0.001	0.01	0.23	0.003	0

*Based on high-sulfur (3.5% by weight), high-ash (10.6% by weight) coal burned in a spreader stoker coal-fired steam generating unit.
**Based on high-sulfur oil (3.0% by weight).
(Source: 6/19/85 *Federal Register,* p. 25106, EPA entry)

2. Natural gas use can reduce high capital costs, long lead times, and eliminate the barriers associated with other fuel use options.

3. There *is* plenty of gas out there for our needs for a long period of time. The price is stabilized, and is forecast to remain competitive.

4. It is in the mutual interest of the gas industry and prospective gas users—and the nation—to employ new uses of natural gas—not only to conserve energies, not only for emissions reductions, but to develop technology that offers economic benefits to all of the stake holders.

Section II
The U.S. Regulatory Framework

Selected Authors

Chapter 3

Major Air Pollutants and Their Sources

PAUL L. WILKINSON
NELSON E. HAY
American Gas Association

Sulfur Dioxide

High concentrations of sulfur dioxide in the atmosphere can adversely affect health by irritating the eyes and upper respiratory system, and by possibly reducing air flow through the lungs. In addition, SO_2 can weaken cotton, leather and a variety of other natural fabrics and building materials, such as iron, steel and particularly limestone.[1]

The subject of "acid deposition" is also closely tied to sulfur dioxide emissions. Acid deposition is a controversial issue, but there are points of general agreement regarding this problem. Sulfur oxides (primarily SO_2) are subject to chemical changes in the atmosphere and return to the earth as acid compounds either in rain or snow, or as a dry deposition. There is a general consensus in the scientific community that acid deposition can cause serious ecological damage (e.g., see the Electric Power Research Institute[2] and the National Academy of Sciences[3]). However, uncertainty remains regarding the precise cause and effect relationships of this problem.

The overwhelming majority (80 percent) of SO_2 emissions is attributable to the combustion of fossil fuels in stationary facilities. Coal combustion is responsible for approximately 68 percent of total

SO_2 emissions, compared with 11 percent for oil, and less than 1 percent for natural gas.

National sulfur oxide emissions, primarily in the form of SO_2, total approximately 25 million tons annually, with stationary source fuel combustion accounting for nearly 20 million tons of the SO_2 total. Large coal-burning facilities were responsible for over 16 million tons —80 percent of the stationary source fuel combustion total.[4] The primary "non-combustion" sources of SO_2 emissions included a variety of industrial processes which accounted for 4 million tons in 1981.

Sulfur dioxide emissions from natural gas combustion are virtually zero.

A boiler operating on natural gas emits, on average, only 0.0006 pounds of SO_2 per million Btu (MMBtu) of fuel burned. In contrast, "typical" grades of fuel oil and coal consumed in large boilers (e.g., 1.5 percent sulfur oil; 2 percent sulfur Eastern coal, uncontrolled) emit 1.6 pounds of SO_2 per MMBtu and 3.6 pounds per MMBtu, respectively.[5]

The control of sulfur dioxide emissions from coal-fired boilers via flue gas desulfurization requires very large expenditures for both equipment purchase and operation.

One means of reducing SO_2 emissions is through the flue gas desulfurization, i.e., "scrubbing." The capital cost of SO_2 scrubbers is high—as much as 25 percent of the total cost required to construct a new powerplant. In addition, the non-fuel cost of operating a scrubber-equipped unit may be two to three times the cost of operating a coal-fired unit without a scrubber.

A second method for reducing a facility's total SO_2 emissions—select gas use—requires only minimal capital expenditures and may also offer lower operating costs than scrubbing. Other alternative means of SO_2 reduction include fluidized bed combustion, coal cleaning, coal gasification and switching to low sulfur coal. There are positive and negative aspects to each method of SO_2 control, and the choice for the best solution for a particular facility depends on its specific problems, restrictions and needs.

National Estimate of Sulfur Oxide[1] Emissions By Source
(Millions of Tons per Year)

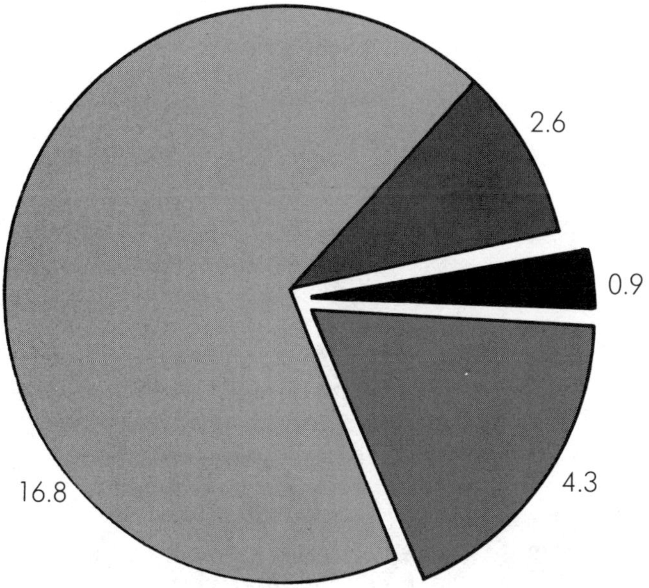

Stationary Source Fuel Combustion	
Natural Gas	Negligible
Fuel Oil	
Coal	
Mobile Sources	
Industrial Processes	
Total	24.8 Million Tons

[1] Primarily sulfur dioxide.

Source: U.S. Environmental Protection Agency, Office of Air Quality Planning and Standards, *National Air Pollution Emission Estimates*, (Research Triangle Park, N.C.: September, 1982).

Sulfur Dioxide Emissions per Million Btu of Fuel Consumed in Large Industrial or Utility Boilers

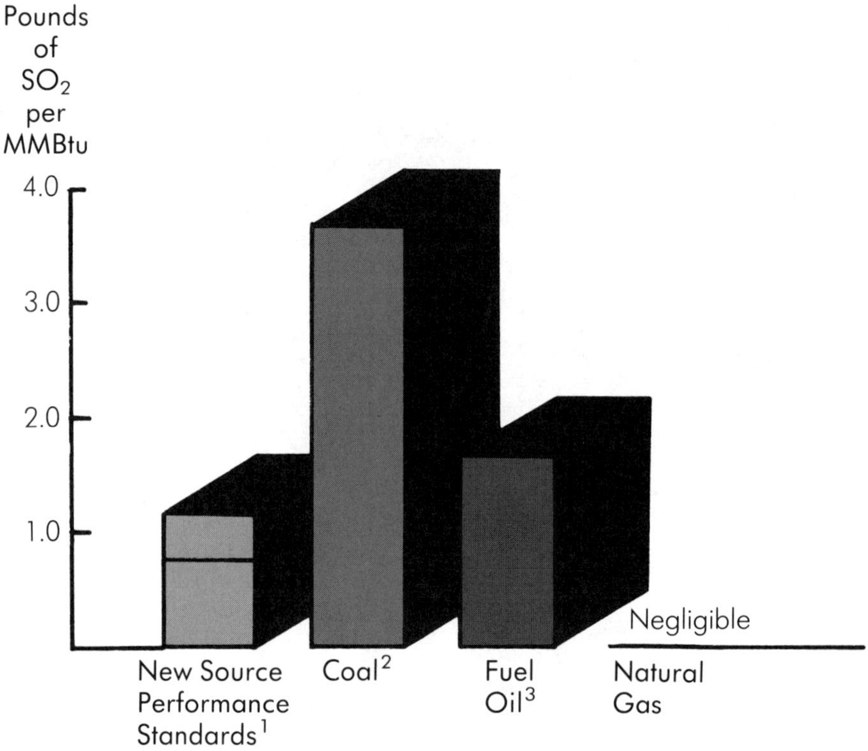

[1]Apply to *new* electric utility powerplants; range reflects difference in standards for oil- and coal-fired units.

[2]Bituminous coal, 2 percent sulfur content, uncontrolled.

[3]Residual fuel oil, 1.5 percent sulfur content, uncontrolled.

Source: U.S. Environmental Protection Agency, Office of Air and Waste Management, *Compilation of Air Pollutant Emission Factors*, (Research Triangle Park, N.C.: August, 1977).

Nitrogen Oxides

Emissions of nitrogen oxides (NO_X) are a critical element in the formation of smog, and have been cited as a possible contributor to "acid rain" formation.

About 90 percent of all NO_X emissions is in the form of NO which, in the presence of sunlight, ozone and hydrocarbons, forms NO_2. The conversion from NO to NO_2 is the first step in a series of photochemical reactions, which, under certain meteorological conditions, produces smog. As mentioned, photochemical smog may have a number of adverse effects on human health and on various materials.

Over 65 percent of all nitrogen oxide emissions stem from highway vehicles and large coal-fired facilities.

Emissions of nitrogen oxides from all U.S. sources total 20.5 million tons, with a fairly even split in the lion's share of this total between transportation sources (9.4 million tons) and stationary source fuel combustion (9.3 million tons).

NO_X emissions from non-combustion industrial processes, solid waste disposal, and miscellaneous sources are fairly modest (1.8 million tons annually. Highway vehicles (gasoline- and diesel-fueled) account for 78 percent of the transportation sector total (7.3 million tons), and large coal-fired facilities account for over two-thirds of the stationary source combustion total (6.4 million tons).

Natural gas-fired pipeline compressor stations emit less than 1 million tons of NO_X annually, approximately 4 percent of total NO_X emissions. Compressor station emissions will continue to decline as advanced natural gas engines and turbines with reduced NO_X emissions and increased fuel efficiencies begin to replace older natural gas engines used for compression purposes. NO_X emissions from all residential sources contribute less than 0.5 million tons per year—about 2 percent of total annual NO_X emissions.[6]

Vehicles fueled with natural gas emit as little as 35 percent as much NO_X as do gasoline-powered vehicles.

The previously cited DOE/EPA testing of vehicles converted from gasoline to natural gas indicated a 40 percent decline in NO_X emissions for one vehicle when converted to natural gas, while no change relative to gasoline operation was found in the second vehicle. A recent Department of Energy analysis of the operating histories of 14 vehicle fleets operating on natural gas (which included thousands of vehicles and millions of miles of travel), found that NO_X emissions were reduced by 50 to 65 percent when operating on natural gas.[7]

National Estimate of Nitrogen Oxide Emissions by Source
(Millions of Tons per Year)

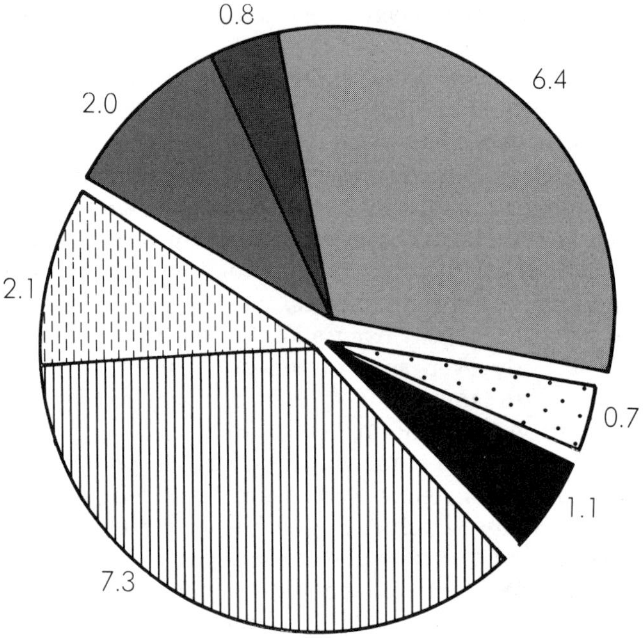

Stationary Source Fuel Combustion
- Natural Gas
- Fuel Oil
- Coal

Mobile Sources
- Highway Vehicles
- Aircraft, Trains and Other

Industrial Processes

Other[1]

Total 20.5 Million Tons

[1] Includes solid waste disposal, open burning, pipeline compressor stations and other miscellaneous sources.

Source: U.S. Environmental Protection Agency, Office of Air Quality Planning and Standards, *National Air Pollution Emission Estimates*, (Research Triangle Park, N.C.: September, 1982).

Nitrogen Oxide Emissions per Mile of Automotive Travel

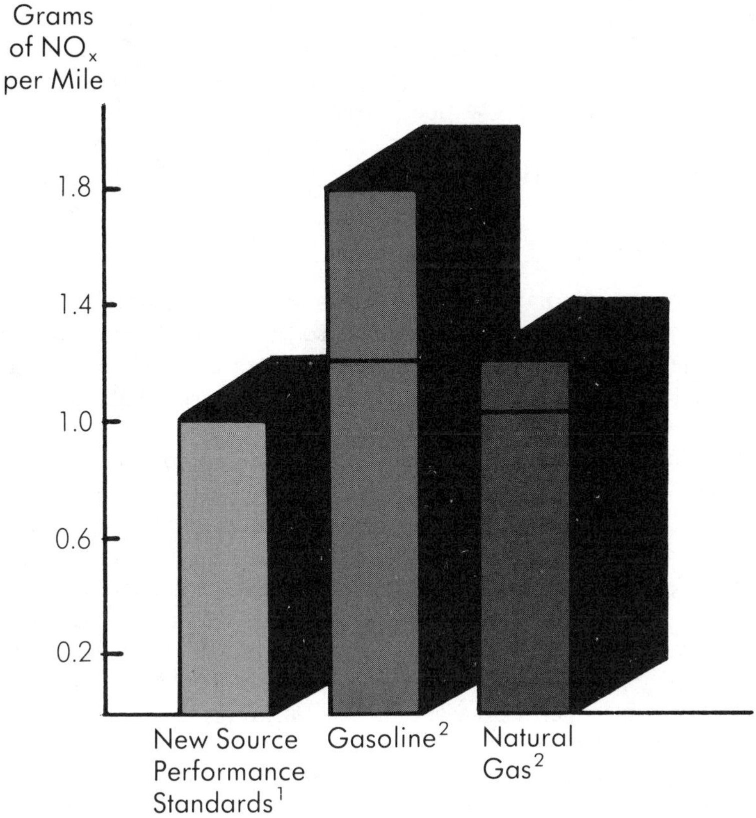

[1]Apply to post-1980 automobiles.

[2]Range for gasoline and natural gas based on testing of two vehicles by the Environmental Protection Agency, a 1979 Chevrolet Impala and a 1980 Dodge Diplomat. NO_x emissions from vehicles which could meet NSPS when operating on gasoline would be further reduced if converted to natural gas.

Source: U.S. Department of Energy, Assistant Secretary for Conservation and Renewable Energy, *Assessment of Methane-Related Fuels for Automotive Fleet Vehicles*, (Washington, D.C.: February, 1982).

NO_X emissions from large boilers operating on coal may be over three times as great as emissions from gas-fired boilers.

Depending on the coal and boiler type, NO_X emissions from large coal-fired boilers range from 0.9 to 2.6 pounds per MMBtu of fuel combusted. In contrast, comparably sized gas-fired units emit only about 0.7 pounds of NO_X per MMBtu. Large oil-fired boiler NO_X emissions are approximately equal to, or somewhat greater than gas-fired boiler NO_X emissions.[8]

Particulate Matter

The term "particulate matter" refers to small, man-made or natural substances which are light enough to become airborne. Examples of man-made particulates include smoke from burning coal or automobile exhausts, and dust from industrial processes. Natural particulates include dust, pollen and sea salt.

Particulate matter at high concentrations can impair respiratory functions, reduce visibility and discolor the air. Fine particulates (less than 1.0 micron in diameter) are the most damaging to human health. Particulate matter also may affect our climate, as it is believed that high concentrations of particulates reflect solar radiation back into space and change precipitation patterns. Particulate matter can cause damage to many materials used by man. It can discolor paint, weaken fabric, fade textiles, corrode metals, etc. Particulates blown by the wind also can cause erosion of buildings, bridges, and other man-made structures.[9]

Approximately 3.9 million tons of particulate matter are released into the atmosphere annually as a result of fuel combustion in the United States, with 80 percent of this amount attributable to coal-fired boilers, wood-burning stoves and fireplaces, and highway vehicles.

A total of 9.4 million tons of particulates are emitted each year, with 3.9 million tons attributable to fuel combustion (transportation and stationary sources) and 4.1 million tons attributable to industrial processes. The primary sources of combustion-related particulate emissions are highway vehicles (1.2 million tons), large coal-fired facilities (1.3 million tons), and residential wood stoves and fireplaces (0.6 million tons). These three sources contribute 80 percent of the fuel combustion derived particulate matter emissions, versus only 1 percent for all natural gas combustion.[10]

National Estimate of Particulate Emissions by Source
(Millions of Tons per Year)

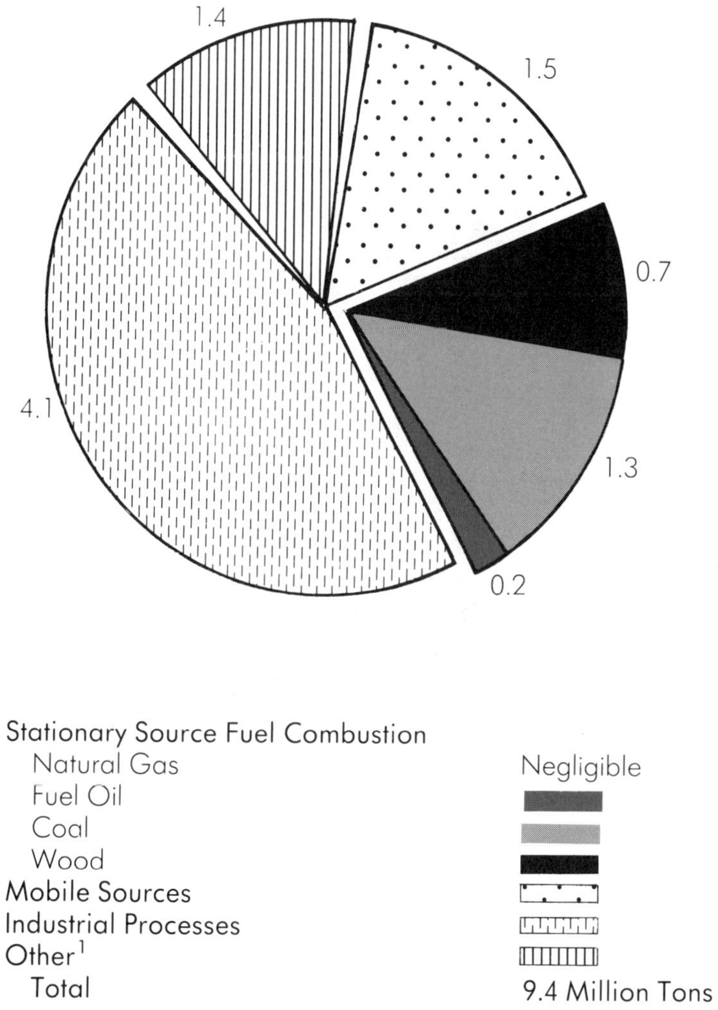

[1]Includes solid waste disposal, forest fires, open burning and other miscellaneous sources.

Source: U.S. Environmental Protection Agency, Office of Air Quality Planning and Standards, *National Air Pollution Emission Estimates,* (Research Triangle Park, N.C.: September, 1982.

Particulate Emissions per Million Btu of Fuel Consumed in Large Industrial or Utility Boilers

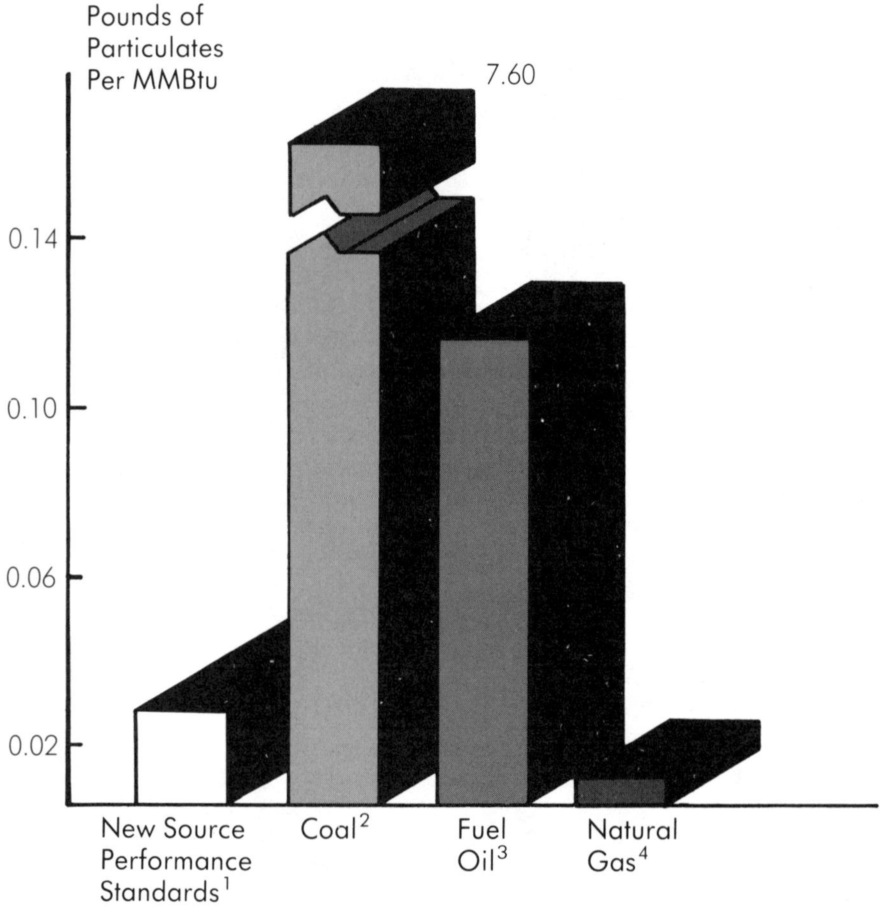

[1]Apply to *new* coal-fired electric utility powerplants; 70 percent reduction required for new oil-fired units.

[2]Bituminous coal, 10 percent ash content, uncontrolled.

[3]Residual fuel oil, 1.5 percent sulfur content, uncontrolled.

[4]Average, may be as low as 0.005 pounds per MMBtu.

Source: U.S. Environmental Protection Agency, Office of Air and Waste Management, *Compilation of Air Pollutant Emission Factors*, (Research Triangle Park, N.C.: August, 1977).

Natural gas combustion emits almost no particulate matter.

The burning of natural gas results in particulate emissions averaging only 0.01 pounds per MMBtu of fuel consumed. In contrast, the combustion of 1.5 percent sulfur oil emits 0.12 pounds of particulate matter per MMBtu, and Eastern coal with a 10 percent ash content emits 7.60 pounds of particulates per MMBtu (uncontrolled).[11] Various control devices are used to limit particulate emissions from large coal-burning facilities. For example, electrostatic precipitators can reduce the particulate emissions at a large (500 megawatt) powerplant by over 98 percent at a capital cost in the range of $50 million.

Carbon Monoxide

High concentration levels of carbon monoxide (CO) can cause serious illness or, in extreme instances, death.

The incomplete combustion of carbon-based fuels produces carbon monoxide, a colorless and odorless gas. The concentration of CO in the atmosphere is dependent on time of day, location, weather and human activities. CO levels tend to be highest in areas of heavy vehicular traffic. When concentrations exceed 100 parts per million, nausea, headache and dizziness are common.

Nearly three-fourths of all carbon monoxide emissions stem from the operation of highway vehicles.

Some 99.6 million tons of carbon monoxide are emitted annually nationwide. Of the total 99.6 million tons, 68.5 million tons are attributable to highway vehicles, and only 6.9 million tons are due to the combustion of fuel at stationary facilities.[12]

The use of natural gas as a vehicular fuel can significantly reduce (and in many cases virtually eliminate) carbon monoxide emissions.

Although nearly all vehicles in this country are currently powered by gasoline or diesel fuel, it is both technically and economically feasible to operate vehicles on either compressed or liquefied natural gas. In fact, there are hundreds of thousands of cars and trucks throughout the world that are currently fueled by natural gas.

In April of 1981, the U.S. Department of Energy (DOE) and the Environmental Protection Agency (EPA) tested the performance of two late-model vehicles (a 1979 Chevrolet Impala and a 1980 Dodge Diplomat) operating on gasoline and then natural gas. Prior to their conversion to natural gas, the Impala emitted 9.2 grams of CO per mile, and the CO emissions of the Diplomat were 7.3 grams per mile.

National Estimate of Carbon Monoxide Emissions by Source
(Millions of Tons per Year)

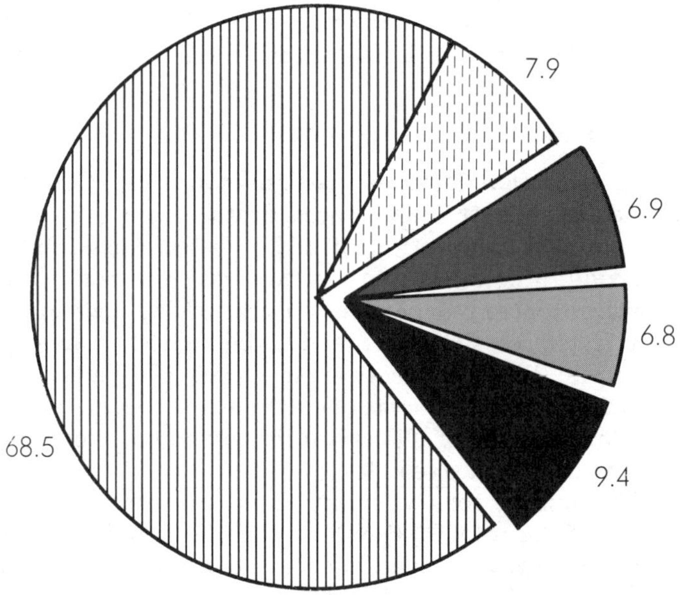

Stationary Source Fuel Combustion[1]
Mobile Sources
  Highway Vehicles
  Aircraft, Railroads and Other
Industrial Processes
Other
  Total — 99.6 million Tons

[1]Breakout by fuel type not available.

Source: U.S. Environmental Protection Agency, Office of Air Quality Planning and Standards, *National Air Pollution Emission Estimates*, (Research Triangle Park, N.C.: September, 1982).

Carbon Monoxide Emissions per Mile of Automotive Travel

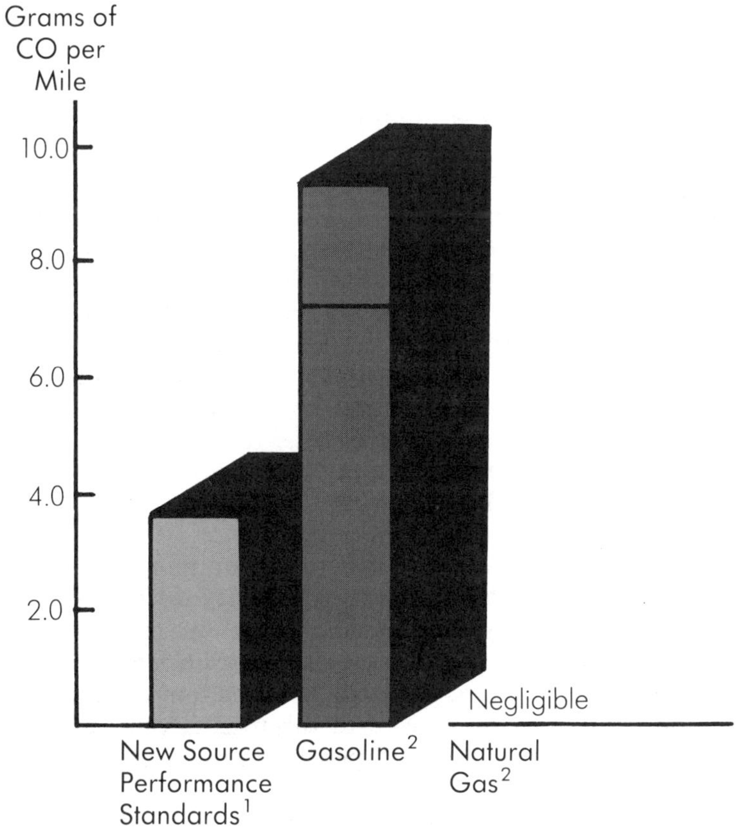

[1]Apply to post-1980 automobiles.

[2]Range for gasoline and natural gas based on testing of two vehicles by the Environmental Protection Agency, a 1979 Chevrolet Impala and a 1980 Dodge Diplomat.

Source: U.S. Department of Energy, Assistant Secretary for Conservation and Renewable Energy, *Assessment of Methane-Related Fuels for Automotive Fleet Vehicles*, (Washington, D.C.: February, 1982).

The CO emissions of the two vehicles were reduced dramatically when converted to natural gas—to 0.06 grams of CO per mile for the Impala and 0.003 grams per mile for the Diplomat.[13]

Hydrocarbons

Photochemically reactive hydrocarbons (HC) contribute to the formation of smog.

Some hydrocarbons react in sunlight with other air pollutants to form smog which can irritate the nose and eyes, reduce visibility and lung function, aggravate respiratory diseases, and corrode and soil fabric. The primary component (over 90 percent) of natural gas is methane, a non-reactive hydrocarbon which does *not* contribute to smog formation. The U.S. EPA has acknowledged the non-reactive nature of methane, and a non-methane hydrocarbon emission standard for automobiles was *proposed* by the EPA in December of 1981.[14]

Gasoline- and diesel-powered transportation vehicles and various petroleum operations are responsible for the majority of man-made hydrocarbon emissions.

Hydrocarbon emissions are similar to carbon monoxide emissions in that automobiles are their primary source. Hydrocarbons may also be emitted by stationary source fuel combustion, chemical processing, and petroleum refining, storage and distribution. The U.S. Department of Energy estimates 1985 hydrocarbon emissions at 10 million tons, with over 50 percent of this total attributable to transportation vehicles, and nearly 20 percent due to various petroleum industry operations.[15]

The conversion of a gasoline-powered vehicle to natural gas can reduce reactive hydrocarbon emissions by as much as 56 percent.

The two vehicles examined at the Motor Vehicle Emissions Laboratory of the EPA in the tests cited in the previous section showed significant reductions in the emission of non-methane hydrocarbons when operating on natural gas. These emissions were reduced in one vehicle from 0.34 grams per mile when operating on gasoline to 0.15 grams per mile when converted to natural gas—a 56 percent decrease. The reduction in the second vehicle was 37 percent—from 0.38 grams per mile to 0.24 grams per mile.

National Estimate of Hydrocarbon Emissions By Source
(Millions of Tons per Year)

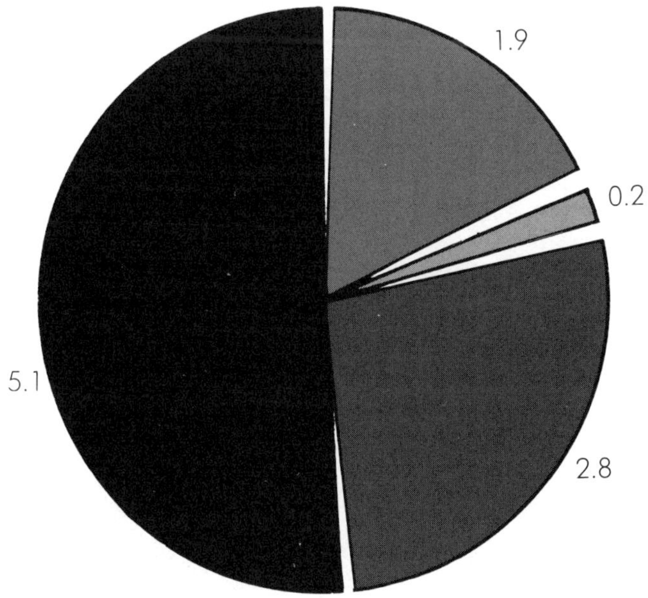

Mobile Sources	
Petroleum Industry Operations[1]	
Municipal Sewage	
Other	
Total	10.0

[1] Petroleum refining, distribution, storage and combustion.

Source: U.S. Department of Energy, Office of Environmental Assessment, *National Environmental Impact Projection II*, (Washington, D.C.: March 1980.)

Non-Methane Hydrocarbon Emissions per Mile of Automotive Travel

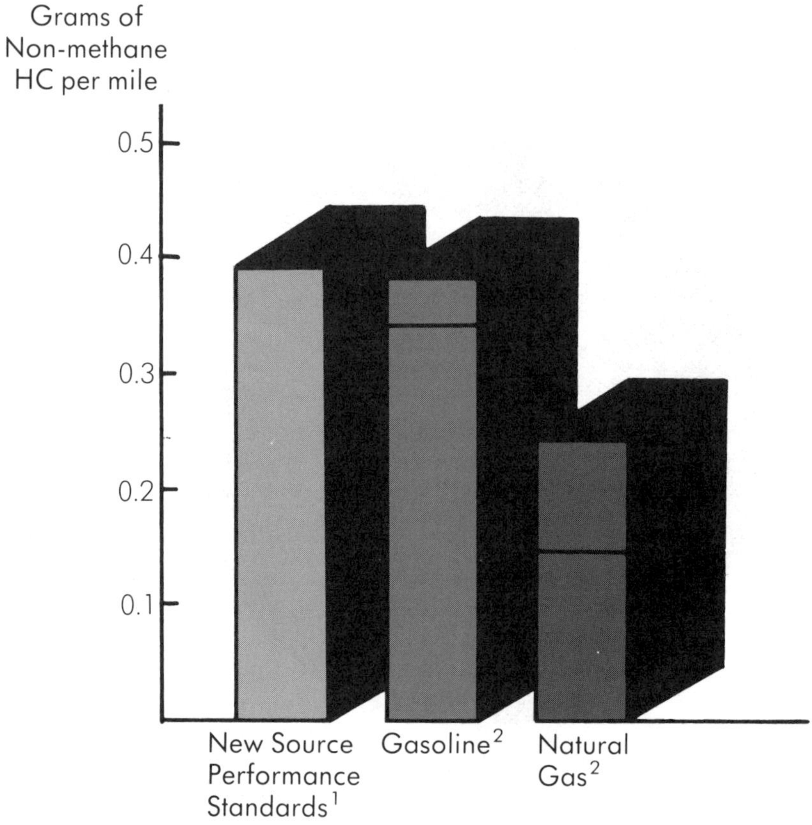

[1]Proposed standard.

[2]Range for gasoline and natural gas based on testing of two vehicles by the Environmental Protection Agency, a 1979 Chevrolet Impala and a 1980 Dodge Diplomat.

Source: U.S. Department of Energy, Assistant Secretary for Conservation and Renewable Energy, *Assessment of Methane-Related Fuels for Automotive Fleet Vehicles*, (Washington, D.C.: February 1982).

References

1. U.S. Department of Energy, Office of Environmental Assessment, *National Environmental Impact Projection II*, (Washington, DC: March 1980) pp. 3-17–3-24.
2. Comptroller General of the U.S., *The Debate Over Acid Precipitation*, (Washington, DC: September 1981) p. 16.
3. National Academy of Sciences, *Acid Deposition—Atmospheric Processes in Eastern North America*, (Washington, DC: 1983).
4. U.S. Environmental Protection Agency, Office of Air Quality Planning and Standards, *National Air Pollutant Emission Estimates, 1970-1981*, (Research Triangle Park, North Carolina: September 1982) p. 6.
5. U.S. Environmental Protection Agency, Office of Air and Waste Management, *Compilation of Air Pollutant Emissions Factors*, (Research Triangle Park, North Carolina: August 1977) pp. 1.1-3, 1.3-2, 1.4-2.
6. U.S. Environmental Protection Agency, Office of Air Quality Planning and Standards, *Op. Cit.*, p. 7.
7. U.S. Department of Energy, Assistant Secretary for Conservation and Renewable Energy, *Assessment of Methane-Related Fuels for Automotive Vehicles*, (Washington, DC: February 1982), Vol. II, pp.. 3-10.
8. U.S. Environmental Protection Agency, Office of Waste Management, *Op. Cit.*, pp. 1.1-3, 1.3-2, 1.4-2.
9. U.S. Department of Energy, Office of Environmental Assessment, *Op. Cit.*, p. 3-12.
10. U.S. Environmental Protection Agency, Office of Air Quality Planning and Standards, *Op. Cit.*, p. 5.
11. U.S. Environmental Protection Agency, Office of Air and Waste Management, *Op. Cit.*, pp. 1.1-3, 1.3-2, 1.4-2.
12. U.S. Environmental Protection Agency, Office of Air Quality Planning and Standards, *Op. Cit.*, p. 9.
13. U.S. Department of Energy, Assistant Secretary for Conservation and Renewable Energy, *Op. Cit.*, Appendix B-25.
14. U.S. Environmental Protection Agency, *Non-Methane Hydrocarbon Standards for Mobile Sources*, Federal Register Vol. 46, No. 246, (Washington, DC: December 23, 1981), p. 62366.
15. U.S. Department of Energy, Office of Environmental Assessment, *Op. Cit.*, p. 3-44.

Chapter 4

Air Pollution Laws, Rules, and Regulations

DAVID J. BARDIN
Arent, Fox, Kintner, Plotkin & Kahn
Formerly, Administrator,
U.S. Economic Regulatory Administration, and
Commissioner, New Jersey Department of Environmental Protection

DR. BENJAMIN SCHLESINGER and
JAMES J. NEWCOMB
Benjamin Schlesinger and Associates

The country is divided into *Air Quality Control Regions* (or AQCRs). Emitters operate one or more "sources," that is sources of air pollution emissions. The demand for natural gas in the select use mode will depend on a variety of emissions restrictions, for example:

- Limitations on sulfur content of fuels (e.g., by percent),
- Compliance deadlines,
- Overall regional sulfur emission restrictions (e.g., New York and other states' acid rain controls),
- Whether the emitter wants to expand process or production capacity,
- Whether the emitter is interested in such things as "emissions trading," "banking," or "bubbling" (which we will define below).

Basic Air Regulatory Approaches and Terminology

The following summarizes basic regulatory approaches and terminology:

Air Quality Control Regions and Sources

A gas service area may include two or more AQCRS, or it may be only a part of one AQCR. There are well over 200 AQCRs in the nation. In some cases, other systems of regionality are important, rather than AQCR's, and must be understood to know the basis of enforcement that affects the gas user. (Such areas of designation are codified in 40 CFR Part 81.) The State air quality control agency or EPA Regional office can readily identify the boundaries of AQCRs or other areas of interest.

The various air quality agencies can not only help identify local enforcement boundaries, but they also help identify all major sources —probably down to a size that is too small to be useful for air quality control purposes. Incidentally, there has been considerable controversy over whether each smokestack is a separate "source" or whether several stacks at one site can be considered a single source for the purpose of environmental regulation. The trend tends to favor a combined approach.

Ambient Air Quality

Testing of the ambient air in each AQCR determines whether the air is clean enough to satisfy *National Ambient Air Quality Standards* (NAAQS). Ambient air is the surrounding outdoor air (i.e., "external to buildings, to which the general public has access" according to EPA), not the air at the top of a stack, or in a home or workplace. The NAAQS are goals based primarily on human health criteria. Each NAAQS sets a limit for pollution of the surrounding air. Separate NAAQS are set for each of the several defined "criteria" pollutants. The criteria air pollutants include:

- Sulfur oxides (sulfur dioxide),
- Particulate matter, or total suspended particulates (TSP),
- Carbon monoxide,
- Nitrogen dioxide,

- Ozone, related to unburned, non-methane, hydrocarbons (HCs), volatile organic compounds (VOCs), photochemical oxidents,
- Lead.

Objectives of clean air regulation include:

- Attainment by each AQCR of compliance with air quality standards for each of the criteria pollutants.
- "Maintenance" of such compliance and "prevention of significant deterioration" of air quality, including in some especially clean air areas, protection of that better-than-minimal purity.
- Control of emissions of some additional toxic (or otherwise harmful) chemicals for which an ambient air quality standard has not yet been set.

Sometimes several ambient air quality standards must be met for the same pollutant, e.g., an annual standard and a daily standard. Moreover, some States have imposed restrictions that go beyond the federal EPA standards. Also, Congress and EPA change standards from time to time.

Geographical regions, such as AQCRs, which are in violation of a particular air quality standard for a pollutant are said to be in "non-attainment" for that pollutant. For example, the AQCR consisting of City X is in non-attainment for sulfur oxides if sulfur oxide concentrations averaged over a particular period of time exceed air quality standards. Conversely, areas whose air quality complies with standards are said to be attainment areas. A few areas are unclassifiable.

Regulatory Agencies

Implementation of the clean air objectives (i.e., achieving attainment of standards) is largely in the hands of State air quality control agencies, supervised by EPA Regional Offices and, to a lesser extent, the Washington, D.C. office of EPA. These State agencies adopt *State Implementation Plans* (SIPs), approved by the federal EPA. Each SIP regulates individual sources so as to achieve or maintain compliance with the overall NAAQS.

State agencies enforce the SIPs. Some States have delegated part of the job to district agencies. Emitters will want to learn what federal, State and local agencies are involved in clean air implementation for their service areas.

Ordinarily, each of these agencies has at least two offices or groups that will be of interest: One group approves plans, and issues permits to sources, such permits determine what one or more sources may do in the future. Another group monitors and enforces compliance by individual sources.

However, the boundaries may be fuzzy, and both groups may not be distinct nor separate in some State agencies. Moreover, the compliance and enforcement group may enter into consent orders with a non-complying source and set a special timetable for equipment or process changes to satisfy that consent order that differs from the timetable or requirements that a permit would normally contain.

Regulation of Sources of Emissions

Implementation of the clean air objectives, pollutant by pollutant, involves restrictions on emissions by individual sources or categories of sources. Some sources are subject to one (or two) source review programs, discussed below, before construction (or expansion or modification). Most sources are subject to the State Implementation Plan (SIP) all the time.

State Implementation Plan

The SIP is usually a set of rules setting specific emissions or operational limits. For example, the SIP may have rules for power plants, cement plants, incinerators, etc. Alternatively, the SIP may prescribe maximum sulfur content for each fuel (e.g., 0.5 percent sulfur fuel oil), or it may categorically address all industrial boilers larger than some fuel input rate (e.g., greater than 250 MMBtu per hour for all sources on a single site, dropping to 100 MMBtu per hour after a certain date).

The SIP is changed from time to time. Each change involves a State agency decision followed by EPA review and approval. SIPs usually differ from State to State and SIP provisions often differ for different AQCRs in one State. A SIP governs on-going operations of a source, not just new operations.

Source-Review Programs

Under the Clean Air Act, EPA requires that certain major emission sources comply with at least one of three programs before they may begin to operate. The three programs apply nationwide to attainment areas and to non-attainment areas, as follows:

- As to certain specified categories of facilities EPA has issued New Source Performance Standards, called "NSPS." For example, there are NSPS for petroleum refineries, kraft pulp mills, glass furnaces, all industrial boilers over 250 MMBtu per hour (Note: to be lowered to 100 MMBtu per hour), and electric utility steam generating units. New sources for which NSPS have been issued must comply no matter where located and regardless of the "attainment" status of the area. The NSPS may deal with one or more of the six criteria pollutants or with another pollutant. NSPS do not involve a permit (unless called for by the SIP).
- In an attainment area major new facilities and modifications (even if not subject to NSPS) are subject to a permitting program designed to prevent significant deterioration (PSD) of air quality. Note, for this reason, attainment areas are also referred to as PSD areas.
- In non-attainment areas there is a permitting program for major new facilities designed to help achieve the air quality goals.

A particular facility may be subject to the standards of more than one program. For example, the facility might be subject to PSD permitting standards with respect to one pollutant and to non-attainment permitting standards with respect to a second. It may also be subject to a NSPS. (It would, in any event, be subject to SIP emission standards, but the permitting programs standards are generally stricter.)

EPA source-review requirements generally apply also to "major modifications" of existing sources as well as new sources. In addition to EPA's three programs, an EPA-approved SIP for a given state may impose a permitting requirement with respect to other facilities, including existing facilities although no modification is proposed. A single source may require more than one permit.

Each permitting program can involve quite onerous procedures. Moreover, case-by-case permitting (in contrast to set performance standards) has frustrating tendencies because of uncertainty, delay, and possible regulatory impetus to squeeze out a little extra "cleanup" commitment in that case.

In various circumstances, clean air regulators require that facilities control emissions of a given pollutant by—

- Best available control technology (BACT),
- Reasonably available control technology (RACT), or
- Lowest achievable emissions rate (LAER).

Offsets, Bubbles, Trading, Banking

In some cases, extra emissions reductions by one, complying source may "offset" emissions from another to achieve regulatory compliance for that other source. For example, "bubbles" may be allowed to combine all the stacks at a single plant. The decision to "bubble," however, may require case-by-case regulatory review. These concepts include outright emissions "trading" between two unrelated companies, in which one company pays another to reduce emissions from a nearby source, and the "banking" of emissions reductions for future use.

Expansion Plans for Sources

When plant expansions cause an increase in total pollutants emitted, an existing facility might face a change in regulatory category so that tougher requirements apply. (In other words, expansion of the plant might bring about a form of "bracket creep" under clean air regulations.) However, by including some conversion to select use of gas in the expansion plan, a user may be able to decrease instead of increasing total emissions after the plant modification, and thereby to avoid the application of new requirements. Although this approach will not always work, it has real promise for many sources.

Intermittent Compliance Problems

Some sources have only rare problems complying with the SIP. It is sometimes possible and cost-effective to solve the problem of occasional violations with select use, either under the existing SIP or with a small change in the SIP. If SIP emission standards must be amended, however, they will not satisfy Clean Air Act standards if geared directly to changing pollution conditions in the ambient air.

This limit on "intermittent" solutions in the SIP to achieve NAAQS, which appears in Section 123 of the Clean Air Act, does not apply to strategies addressed only to State standards, or to intermittent solutions based on factors other than the atmospheric air quality (such as, reduced gas price, or condition of pollution-control equipment, or manufacturing changes).

For example, if a State ambient standard turns on a 24-hour integrated measurement, a facility might find increased gas use an advantageous and permissible way to meet emission standards whenever the facility went from one 8-hour shift to two or three shifts.

For further information on this point, Chapter 12 discusses when seasonal or intermittent select use is acceptable and when it is not.

Enforcement

Regardless of the kind of select gas use project that is developed, success in obtaining the requisite environmental approvals will depend on the enforceability of the arrangement. Each select use application must be defined in a way that assures air regulators that it will continue to function as developed, with no unforeseen increase in air emissions.

Enforcement issues with regard to select use are likely to include agreement to stick with the plan, what to do if the plan can't work temporarily for some reason, default conditions, and the like. In general, the simpler the select use arrangement, the easier it will be to convince air regulators of its enforceability.

Moving Target of Regulatory Change

Air quality regulation in the U.S. rarely stands still. Instead, it keeps changing with new technical findings; the changing focus, direction, and intensity of public pressure; court decisions; and of course, politics. Here are some current examples of this ever-changing air regulatory picture:

Proposed EPA "Tall Stacks" Regulations

On July 8, 1985 the EPA promulgated regulations regarding air quality that could create new applications for select gas use. The EPA action will roll back significantly the credit previously given to dispersion of air pollutants by use of tall smokestacks in past SIP approvals.

Tall stacks disperse emissions from the immediate plant vicinity and, so, tend to prevent violation of NAAQS. However, amendment of the Clean Air Act in 1977 limited the credit to be given for tall smokestacks as a recognized means of air pollution abatement. Congress was responding to a call to reduce the overall amount of pollutants "injected" into the air, regardless of the height of the stack and the actual compliance with NAAQS. In 1984, the U.S. Court of Appeals sustained an appeal by the Sierra Club contending that EPA had not gone far enough to carry out the 1977 Amendment.

In response to these concerns, the EPA's new rule and its implementation by SIP revisions in affected States may require some 400 industries and electricity generating plants with tall stacks to act to reduce air pollution. If so, they will have to install air pollution controls, switch to lower sulfur and ash fuels, or derate (i.e., partially or completely shut down). For some such facilities, select gas use could prove the cost-effective alternative.

State Acid Rain Legislation

Several states in the Northeast and Midwest are moving to enact acid rain legislation in the absence of Congressional action on the subject. Some mandatory versions of these state laws set deadlines for adoption of programs for reductions in air emissions of sulfur oxides (e.g., New York State's requirement of an interim program to be adopted by January 1986 and submission of a more comprehensive program for 1991, intending a 12 percent sulfur emission reduction in 1988 and 30 percent reduction in the 1990s). States now having or considering explicit acid rain controls, such as New York's, include:

- Massachusetts
- Wisconsin.

Other states have enacted a milder form of acid rain legislation, calling upon the federal government to act on the issue.

As is the case with EPA's "tall stacks" proposal, state sulfur removal requirements can make select gas use economically attractive, as against costlier solutions, if the law gives the fuel user discretion on the most cost-effective mode of compliance. For example, the New York statute expressly allows the option of switching to some (or all) gas fuel and has no bias in favor of scrubbers.

U.S. Supreme Court Decision on Bubbling

Environmental law is constantly evolving through judicial interpretation as well as legislative and regulatory actions.

For example, a U.S. Supreme Court Decision in 1984 confirmed the validity of EPA's "bubble policy" as a means of complying with the Clean Air Act in non-attainment areas. That decision has opened the door to a significant increase in select use of natural gas in expanded or modified facilities.

Chapter 5

"Emissions Trading" for Air Pollution Control-- A View from EPA

MICHAEL H. LEVIN
Director, Regulatory Reform Staff,
U.S. Environmental Protection Agency

"Emissions trading" (ET) can allow firms to comply with the Clean Air Act (CAA) in the most cost-effective manner rather than by meeting uniform requirements. It offers a chance to control more of your environmental destiny instead of having centralized agencies tell you the "how" as well as the "what" of compliance.

ET can allow *state agencies* to encourage sources to do more than the minimum required, and to avoid hydra-headed EPA review of every permit change.

ET can benefit the *EPA* by providing faster compliance and better air quality planning data. With trading, the Agency can assume the role Congress initially envisioned—a role in which states are fundamentally responsible for pollution control because they're closer to the problems, and EPA is a manager setting the framework and auditing, not running, state programs.

ET can also help *environmental groups* reach clean air goals because it represents opportunities for more real reductions, better enforcement, improved regulatory ability to respond to new knowledge, and incentives for sources to reveal (rather than conceal) emission sources and better ways to control them.

The CAA has worked, has produced substantial environmental progress by almost any measure since its enactment in 1970. *The price of air pollution control is now generally seen as a normal cost of doing business.* But the easy shots have all been taken; there are few large uncontrolled sources left. We're faced with cranking down further on big well-controlled ones, or regulating smaller and smaller ones for smaller unit returns.

That's the traditional direct regulation approach. It *will* produce further progress, but is increasingly seen as carrying a heavy price:

- in terms of *EPA/state resources*, spread too thin.
- in terms of steeply rising *control costs* and resulting delayed compliance.
- in terms of the *legitimacy of the whole air pollution control system*, because it becomes very difficult to tell a plant manager why he must get further reductions at $2000 or $5000/ton, when he *knows* he could get the same tons for $100 each and is facing dozens of other regulatory requirements, as well as the need to make a profit. He will see those as unnecessary costs, and will begin to wonder what's wrong with a system that operates this way.

There's another way to get equal or better environmental results at far less cost, without disrupting the regulatory framework and procedures we've all learned to live with, and without compromising environmental goals.

Hundreds of firms have used or are using the bubble or other emissions trades in over two dozen states, realizing control cost savings of nearly a billion dollars through existing-source bubbles alone. Firms are putting their money where their mouth is, and staking capital—even venture capital—on ET's continued growth.

Nevertheless, we are at a crossroads in air pollution control. EPA will shortly issue a final ET policy, replacing the 1982 interim policy. That policy is designed to provide a constitution for ET, a set of principles that will be good into the indefinite future.

There are many issues that the final ET policy addresses, but two of them are fundamental:

1. *What is a "surplus" reduction* available for credit in bubble or

other trades, where there's no attainment demonstration and 'you don't know what's needed to attain?' Is it

- beyond what's currently required?
- beyond what's needed to attain?
- beyond the best anyone can do?

This "baseline" issue is important for NSPS trades as well as trades by existing sources under State Implementation Plans (SIPs).

2. Are incentive-based approaches like the bubble and ET an essential tool for getting further progress, or a privilege to be withheld until "progress" is assured through traditional means?

Beneath these questions lies the issue of whether meaningful innovation is possible under the CAA. If you take the position that CAA works perfectly and actually produces everything the statute says will be produced, it's easy to see every innovative departure as a retrogression, a "sellout."

If, on the other hand, you believe that there are real CAA problems which these approaches can help cure, these approaches are easy to see and paint as producing more real world progress, a stronger, more rational, and more predictable CAA.

A good deal of EPA's effort over the last three years has been aimed at assuring, based on past experience, that trades do produce real world progress, and that they do so in ways which provide a more *flexible*, *simple*, *reliable*, and *stable* regulatory environment for us all.

What Is ET?

CAA generates uniform state- or nation-wide emission limits for regulated processes (e.g. >0.3 lb/pollutant per pound of of product). It does so, for example, through the SIPs, for existing sources, and the NSPS, for new facilities.

But often it costs 100 times more to remove the same pound of pollutant from one stack or vent as from an adjacent process. Often it costs far more to go from 70% to 90% control on a regulated stack than to get that 20% from a nearby uncontrolled one. Removing that second pound for $1 instead of $100 can save the firm (and society) $99, with same environmental results.

ET consists of four basic components:

The *bubble* lets managers of existing plants (or groups of plants) treat all their stacks or vents as though under a giant dome and rearrange emission limits, controlling less where control costs are high in exchange for extra, compensating control where costs are low.

Netting exempts in-plant modernizations, expansions, or other "modifications" from federally-mandated new source review (NSR), so long as plant-wide emissions don't increase by significant amounts.

Offsets allow new sources that can't "net out" to locate in nonattainment (NA), and in some cases PSD areas, by securing reductions greater than the new source will add.

Banking allows firms to store extra reductions in a legally-protected manner for future use or sale.

All four components

- provide balance-sheet reasons (large cost savings or actual profits) for better-than-required performance.
- focus on environmental results rather than uniform limits, so long as results consistent with the Act are secured.
- recognize that central agencies can't foresee, or afford to acquire knowledge of, site-specific chances for cheaper equivalent control.
- involve the same common currency, the emission reduction credit (ERC)—a reduction which is surplus, permanent, quantifiable, and enforceable, and may therefore be used to meet control requirements elsewhere, consistent with CAA.
- let firms and states act like good environmental consumers, substituting cheap surplus reductions for costly required ones—exactly what you do when you comparison-shop for a car or go to the supermarket.

EPA's 1982 ET policy superceded the 1979 bubble. It removed many constraints which experience had shown were not necessary to protect air quality. It used the concept of ERC to integrate and streamline all these components.

And it took two important additional steps: it authorized generic rules (no more site-specific SIP revisions). It extended use of bubbles to nonattainment (NA) areas without approved demonstrations of attainment, using the concept of a RACT baseline (one reflecting Reasonably Available Control Technology, the Act's most stringent general requirement for existing sources) to define what was "surplus" there.

As a result, by January 1985 EPA had approved 40 bubbles as SIP revisions, with another 7 proposed. States had approved 33 more under generic rules. 130 additional bubbles were under development or review. All told, there were over 200 bubbles we knew about in 29 states, with total estimated savings of over $800 million, averaging over $3 million each, from existing-source bubbles alone. 10 generic rules were approved, with another 12 under development. EPA had approved or proposed to approve 8 banking programs, with banks adopted or being developed by at least 16 more states or areas. By January 1985, 3000 offset transactions, and hundreds of "nets," had been successfully implemented. And a major GAO report had projected savings of billions of dollars per year with positive air quality effects, and called for expanded trading.

More important, ET had shown incentives can work to produce better environmental results, as well as large cost savings. Numerous documented cases now show bubbles producing attainment, not just SIP-equivalent reductions, as well as better emission factors, reduced PSD threats, better inventory data, and faster compliance.

ET had begun to change corporate behavior, shifting pollution control decisions from cost centers which limit damage, to influential profit centers which decide where to invest company funds for the greatest return.

ET had begun to change EPA's behavior, converting it to a manager auditing state programs rather than a policeman involved in each permit change at every regulated plant.

ET had spread far beyond the 1982 EPA policy, to shape pending acid rain bills, the pending stack heights "bubble" policy for compliance with EPA's new tall stacks regulations, the single-plant water bubble EPA has issued for plant-wide best-available technology (BAT) compliance by the steel industry, and interplant trades to meet water quality standards "beyond BAT."

Two further developments have also occurred. First, a year ago in its Chevron decision, the Supreme Court unanimously affirmed the validity of "netting" in NA areas, indicating that the Act's NSPS language "itself implies a bubble of sorts" and sending a strong signal that, at least where Congress has not explicitly addressed the issue, EPA is not required to wring the last possible pound of reduction out of every regulatory transaction where it reasonably believes such quests for perfection to be counter-productive.

Second, in January 1985 EPA proposed to approve its first NSPS compliance bubble, for the Central Illinois Public Service Company (CIPS). That bubble involves two new 600 MW boilers, each subject to the 1.2 lb SO_2 NSPS. Under the bubble, one boiler will emit at 0.6 and the other will meet a 1.8 pound requirement to achieve a combined 1.1 lb limit. The bubble will save CIPS half a billion dollars over the facilities' useful life—with 3000 tons per year less SO_2, better enforcement and the creation of local mining jobs.

If everything's so rosy, why do we still have problems, and why aren't the final ET policy or NSPS bubble policy out yet? First, there were some "bad" bubbles, including some that relied on shutdowns made years before any trading proposal, and others that relied on reductions many times more than RACT, which happened to be produced by routine installation of standard required equipment. This gave rise to charges of "paper trades."

Second, and more important, was a question of basic philosophy. Though many of these bubbles were in attainment areas or NA areas with approved demonstrations in which "paper trades" are irrelevant, they raised a difficult conceptual issue: "Extra" reductions don't matter where all sources meet uniform requirements; they're just a bonus for faster progress. (Though of course there's little reason for sources to try to produce them.) But where sources can comply in the aggregate through bubbles, overall emissions can *increase* if reductions "would have happened anyway," and are used to avoid controls elsewhere.

Put differently, for *existing sources*, how can you say any reduction is "surplus" when you don't know how much reduction is needed to attain, and all reductions are therefore "needed"? Further, for *new sources* subject to technology-based limits, how can you say any reduction is "surplus" where installation of the required technology "might have produced that reduction anyway," without the source doing anything different?

These kinds of questions struck at the heart of incentives. They implied that reductions had to be made without any economic motive in order to be credited; and they threatened to turn voluntary reductions into new regulatory requirements. Moreover, they would have:

- disallowed voluntary reductions produced by extra add-on controls or more efficient, less-polluting process changes (actions EPA should encourage).

- barred trades structured to produce net reductions rather than "equivalence"—the only real chance of interim progress for states unable to identify further sources of reductions or get costly traditional regulations passed.
- mired the Agency in endless subjective inquiries into whether particular non-required reductions "would have happened anyway."

These questions led directly to contentions that no existing-source bubbles should be allowed until a fully-approved demonstration was made, and no netting or NSPS trades should be allowed *ever.*

If the answers don't seem to make sense, you should look to see if you've asked the right questions. That's what EPA did; and it turned out the question wasn't "might reductions have happened anyway." The question was how best to assure continued ambient progress and constantly improving programs in these areas.

The answer given last spring was "progress" bubbles—trades reasonably certain to produce extra real reductions, based on objective factors which will neither tax bubbles out of existence, nor require detailed scrutiny of the corporate motives behind every trade.

Under the draft final ET policy:

1. *In attainment areas and nonattainment areas with approved unquestioned demonstrations*, bubbles can use SIP limits and assumptions as their baseline. Any reductions below that baseline are surplus and can be used in trades, whether produced by shutdown/curtailment, extra add-on controls, better O&M, or less polluting process changes, so long as they are consistent with the SIP. Where a state has shown it will timely attain ambient standards, it's free to reallocate limits under the umbrella of that demonstration. The CAA requires no more.

2. *In primary NA areas which lack an adequate demonstration, "progress" bubbles are specifically encouraged*, and will generally be approved if they

- use as their baseline the actual, SIP allowable or RACT allowable emissions rate, whichever is lower, for all sources in trade, determined as of application to bank or trade (whichever is earlier).

- use actual capacity utilization and hours of operation to compute the baseline, based on the two years before application for credit.
- meet general ambient equivalence tests and produce a net air quality benefit, which shall consist at a minimum of a 20% reduction in emissions remaining after application of appropriate baselines.

3. These tests represent a significant *tightening* of bubble requirements in these areas. They generally disallow pre-application reductions, on the theory such reductions have nothing to do with the opportunity to trade. But they are objective, clear, and designed to assure real interim progress in these areas, in ways which also enhance a state's ability to develop a complete SIP, while crediting reductions associated with trades. Because of them, past "bad bubbles" will not be repeated, and all bubbles should be more predictable to obtain.

4. We've also taken one further step to assure real reductions and legal defensibility. Under the draft policy, Regional Administrators are authorized to conduct a limited number of further "special reviews" of bubbles which meet these threshold tests, where they can conclude, based on objective facts, that a particular bubble would not produce the reductions required—for instance, that a post application shutdown was announced 5 years ago, but is only now being implemented. "Special reviews" recognize that general objective tests may not provide a complete answer in all cases. But they are meant to be limited in scope and number, and to focus only on trades which may have substantial adverse impacts if further scrutiny is not conducted.

Under the draft NSPS compliance bubble policy,

- bubbles between different new facilities each subject to an NSPS are specifically encouraged, and will generally be approved if they produce reductions as great or greater than source-specific application of uniform NSPS, and if they insure equivalent source utilization and enforceability.
- the baseline for defining "surplus" reductions will be actual or expected performance of required "best demonstrated control technology" (BDT), except where sources show they can and

would "dial controls" to meet the nominal NSPS allowable limit (in which case the allowable limit will likely be the baseline).

- NSPS bubbles between facilities subject to different NSPS, or located in different plants, will be considered, provided they meet the policy's other tests.
- fuel switch bubbles between operating new facilities will also be considered, but not those for sources which have not yet commenced operation. It was just too difficult and resource-intensive to get into what fuel an unbuilt source would have used "but for" the bubble, at this early stage.

What Does This Have to Do With Gas and Select Use?

The answer should be obvious. One of the first bubbles EPA approved, in January 1981, was a select use bubble for New England Power's Narragansett subsidiary. Narragansett had two 100 Btu boilers within 1/4 mile of each other in downtown Providence, Rhode Island, both using 1% sulfur imported oil. Under the bubble, one switched to natural gas and the other to 2.2% sulfur domestic oil, saving $1.5 million per year in fuel costs, with significant additional SO_2 reductions.

There is no general reason under CAA why the same approach couldn't be used for other existing-source bubbles or for NSPS compliance bubbles, so long as the general tests of those respective policies are met.

Select use bubbles can be used either by switching one source completely to gas, or by switching it to a lower-emitting mixture of coal and gas, in order to allow an equivalent increase in emissions at another source. That's especially true in attainment areas, which most parts of the country are for SO_2.

One major concern is that the increasing source not trigger new source requirements which could wipe out the benefits of the bubble. There are several fairly straightforward ways to avoid this result, all as legitimate as use of exemptions under the tax code:

- If sources are within the same plant, a bubble represents a "netting out" which should not trigger preconstruction review because there is no plant-wide increase in emissions.

- If the increasing source is below 250 MMBtu per hour heat input (the NSPS cutoff), NSPS don't apply.
- Where the increasing source could burn coal on December 21, 1976 and is switching to it.
- Where the increasing source converts to coal through a conversion order issued under ESECA or Section 125 of CAA—orders in which state governors play a major role.
- More generally, if the increasing source does not increase because of a substantial capital expenditure or a 50% reconstruction, new source requirements should generally not apply. Under EPA regulations, you need not just a significant emissions increase, but an increase due to particular types of actions. Actions which do not involve capital expenditures generally do not trigger new source requirements, because an increase was never meant to convert an existing source into a new one by virtue of that fact alone.

Gas can also be used for mobile-source fleet conversions. Such conversions can create emission reduction credits that may be used to meet stationary-source control requirements. The draft final ET policy specifically authorizes such mobile/stationary source trades. Moreover, unlike transportation measures, such "hardware" conversions are among the easiest to qualify as surplus, permanent and enforceable.

A word of caution. What we've described are pending policies, which could change in some respects before they're finally issued. And even then, EPA regulations can get complicated. You should contact states or EPA regions to work out bubbles or "combined combustion" approaches well in advance of formal application. We want building blocks for the future—applications which are easy to approve and will encourage others to take similar steps, not "test cases" that drag on for years. EPA wants to help develop bubble applications—whether for select use or otherwise—that are as environmentally and economically sound as they can be.

Chapter 6

"Emissions Trading" to Meet Clean-Air Laws

MALCOLM WEISS and
JOHN PALMISANO, *President, AER *X Division, Residuals Management Technology Inc.*

Recent regulatory reform undertaken by the EPA is not an attempt to do away with the Clean Air Act, as some opponents claim, but rather to allow for the significant differences in pollution-control costs and strategies across the spectrum of an industry. Emissions trading is one of these reforms. It comprises the offset policy, bubble policy, netting, and emissions banking. These concepts allow firms to substitute relatively cheap, surplus emissions reductions for expensive reductions at other emission points—overcontrolling where the marginal cost of control is low in lieu of having controls where the marginal cost is high. Trading is accomplished under well-defined conditions that are designed to improve local air quality, while allowing industry flexibility in meeting environmental responsibilities.

These reform initiatives are intended to improve the environmental efficiency of each dollar spent on air-pollution-control activities. Environmental efficiency is improved when the cost of control declines while environmental results remain constant or are improved, or when regulatory strategies improve health or aesthetic values, or show other measurable benefits at relatively constant costs.

The first environmental programs in the U.S. were based on so-called command and control modes of regulation. The basic assumption was that regulators could obtain enough familiarity with industrial processes to develop reasonably cost-effective control strategies for industrial sectors. Regulators' knowledge and good intentions would, it was thought, assure equitable and economically efficient pollution-control programs across industries.

Command and control regulation was appropriate for taking the first steps to combat the nation's most serious environmental problems. However, as targets for pollution reductions have become less obvious, this approach has become less dependable. Soon, regulators and those they regulated were painfully aware of the limitations inherent in traditional regulatory strategies. In today's complex industrial world, it is unlikely that a regulator in Washington, DC can specify with accuracy a control strategy for an entire industrial sector that is either equally cost-effective across an industry or results in improvements in health or ecology for similar populations.

To address these limitations, an effort was made to develop opportunities, within current statutes, to refine existing regulations and promote cost-effective pollution-control strategies. Information on the shortcomings of regulatory programs has been used by EPA to develop newer, more efficient approaches to achieving congressionally established environmental goals. The result, among others, was emissions trading.

EPA's emissions-trading program gives plant managers the flexibility to take advantage of control alternatives to find the most cost-effective and even profitable ways to control air pollution. Because the cost of controlling a pound of the same pollutant from adjacent processes in plants can vary by as much as 100 to 1, by allowing one firm to compensate another for extra pollution controls, this new flexibility can save plant managers millions of dollars in control expenditures, with no adverse effects on air quality.

Creating ERCs

Emissions-reduction credits (ERCs) are the common currency of emissions trading. Credits can be created by product, process, or pollution-control changes, such as decreasing air emissions below current legal requirements. Reductions are encouraged where they are least expensive. Facility shutdown or curtailing operations can also create ERCs.

An ERC is an official certification that excess emissions reductions are:

- Surplus. The reductions are beyond applicable permit or state-implementation-plan (SIP) requirements.
- Permanent. The company will maintain the reduced emission level.
- Enforceable. The reduction is legally binding.
- Real. It is an actual reduction, rather than on paper.

Fig. 1. demonstrates a method of creating ERCs and how they are used.

Trading Emissions

Emissions trading is made up of four similar concepts:

- Offsets.
- Bubbles.
- Netting.
- Banking.

Offsets. EPA's emissions-trading policies began in 1976 with the adoption of the offset policy. The requirement for offsetting major *new and modified sources* of emissions was subsequently incorporated into the 1977 Clean Air Act amendments. Offsets allowed new industrial siting and major expansions in areas not meeting the National Ambient Air Quality Standards, the so-called nonattainment areas.

Existing firms were encouraged and allowed to reduce emissions below permit levels, thereby creating emissions reductions that could qualify as ERCs. ERCs can be bought, leased, or donated to new sources of air emissions within local air sheds. But total emissions reductions must be greater than emissions increases, to ensure that total emissions in nonattainment areas will decline. The farther away the ERC-generating firm is from the ERC-using firm, the greater the tradeoff ratio must be. Almost all states require new sources to obtain offsets.

In clean-air areas, a firm may need offsets to ensure that new emissions do not degrade air-quality conditions and move the air shed from attainment to nonattainment status for one or more pollutants.

Bubbles. The bubble policy allows *existing* sources (plants or groups of plants) to develop alternative emissions-control strategies.

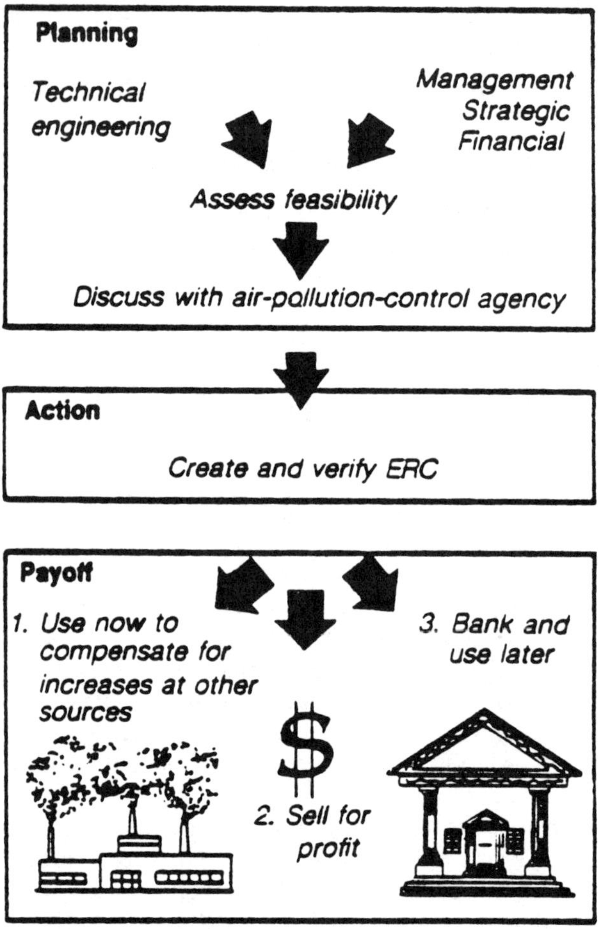

Fig. 1. Emissions-reduction credits are used, sold, or banked for later use in trading.

Firms are allowed the flexibility to develop a wide variety of customized control strategies so long as there is no degradation in the ambient-air quality of an industrial area. Alternative control strategies can focus on fuel switching, equipment shutdowns or curtailments of production, use of more effective control technologies on specific emission points, changes in operations, mobile-source trades, and area source and fugitive credits.

A bubble is geographically limited by the area affected by emission decreases and increases. The pollution from *all* area stacks is combined. Resulting ambient-air-quality impacts are the binding constraint on bubble trades.

For some pollutants, like volatile organic compounds and NO_X, the area of impact resulting from emissions increases and decreases may be very large; localized impacts, however, may be slight. For other pollutants—particulates or SO_2—both local and area-wide impacts must be evaluated prior to bubble approval.

A bubble can include emissions increases and decreases within one or more plants; multi-plant bubbles, however, are uncommon, though their future use is likely to increase. Cross-firm particulate or SO_2 bubbles are so constrained by ambient-air-quality requirements that only firms situated nearby or on the same site can benefit from bubbling.

Bubble programs can be delegated to states after EPA reviews and approves the states' individual so-called "generic rules." These rules preapprove certain classes of emissions trades. Many states have developed generic rules covering one or more pollutants.

Netting. Netting refers to sources using emissions credits to avoid new-source-review (NSR) requirements under both the prevention-of-significant-deterioration and nonattainment sections of the Clean Air Act. Since the definition of source includes an entire plant, a wide range of point-specific emissions increases and decreases can take place without having to meet strict NSR requirements, so long as the *net* effect is insignificant. For netting purposes, only ERCs that are generated from *on-site* sources can be used to compensate for *on-site* emissions increases.

Prior to June 1984, netting was permitted only in attainment areas. Last June, the Supreme Court recognized EPA's authority to extend the netting concept to nonattainment areas. The effects of this decision on utilities may be significant since expensive state-of-the-art lowest-achievable-emissions-rate and best-available-control-technology requirements can be avoided. Facilities may also be exempted from obtaining preconstruction permits, offset requirements, and other rules if offsetting emissions decreases can be secured from within a plant. New-source performance standards (NSPS), however, must still be met by any new source.

Emissions banking. Emissions banking is the creation, certification, and storage of ERCs for future use in a bubble or in offset transaction, or for "netting-out" of a new-source review. Once banked, credits are transferrable to third parties.

The rules governing emissions banking are defined by each SIP. A few states and localities have formally developed emissions-banking systems, while some support informal banks. Informal banking, however, offers little protection against state confiscation of banked credits.

Fig. 2 shows how banking strengthens and expands offset, bubble, or netting opportunities.

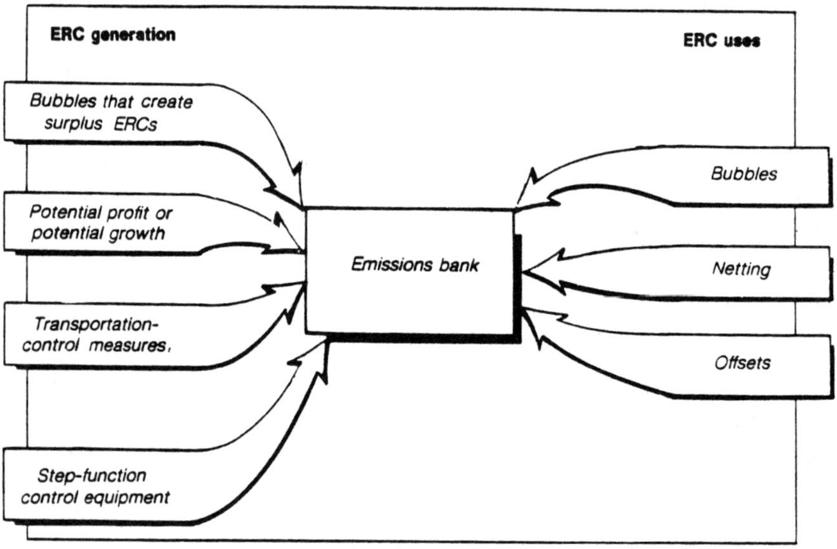

Fig. 2. An emission bank accepts deposits of credits which can be withdrawn for use in offset, bubble, or netting transactions.

Utility Trading Experience

EPA, which does not track offsets, estimates that over 2000 offset transactions have occurred. The vast majority of these have been internal offsets (within a single corporation). Of these 2000 cases, only seven involved electric utilities.

Four utilities initiated and then cancelled permitting actions that would have required offsets; two utilities received approval for offset transactions. The City of Columbus, Ohio, found that its coal- and refuse-fired powerplant would contribute toward creating an SO_2-nonattainment area. By raising the stack of a nearby source and converting a second source to natural gas, the area influenced by the new control strategy remained in attainment for SO_2. In the other case, Nevada Power Co. paved a mile of highway three miles from the plant site to control fugitive dust and created needed total-suspended-particulate (TSP) credits.

One offset transaction was still pending in January 1984. The City of Cleveland has built a cogeneration plant, intending to sell electricity to its Div of Power & Light. The steam will be sold to another party. Although the plant has yet to receive its permit, offsets in the area are expected to be easy to come by, since there is a surplus of credits resulting from local plant closures.

The bubble policy has been in effect for almost six years. As of October, 1984 approximately 200 bubbles were approved, proposed, or under development in 29 states. The 50 currently in operation have saved over $700-million. Only five of the bubbles are for electric utilities. The five operating bubbles are New England Power Co., Burlington (Vt) Electric Dept., Tampa Electric Co., and Kentucky Utilities Co., all for SO_2; and Toledo Edison Co., for TSP.

Bubble-like transactions are not always designated as bubbles. Other electric utilities may have processed (or may be processing) SIP revisions that are, in effect, bubbles. SIP-processed trades cannot be easily identified, though perhaps two or three electric-utility trades fall into this category.

A review of the list of offsets and bubbles processed by electric utilities understates that industry's familiarity with emissions trading. An informal survey of 23 electric utilities by Roger Raufer of the University of Pennsylvania found that:

- Some analysis of trading options had been done by 21.
- Five had been approached by other firms regarding an ERC sale.
- Three had had discussions with state agencies regarding possible trades.
- Two had developed bubbles. But whether or not these were accounted for in EPA's tracking system is unknown.
- One had approached other firms to sell ERCs.

Netting transactions, by their very nature, cannot be counted or tracked. Netting avoids a portion of the regulatory process and leaves almost no administrative trail.

Three state-wide banking rules have been approved by EPA. Three other state rules have been proposed to the agency. Ten other states or localities have adopted banking rules. Eight others are working to develop or are considering adopting such rules.

There is no evidence of electric utilities placing ERCs into formal emissions banks. Utilities may have unprotected credits in informal banks, but only a state-by-state or city-by-city search could identify such instances.

One utility recently generated nearly 1000 tons of NO_X and 175 tons of SO_2 units. The local air-regulatory agency proposed to discount these credits by 50%. This is not the only case of ERC discounting or partial confiscation. But with upfront planning, these impediments can be mitigated. Regulators may discount banked ERCs to provide a growth allowance or to meet the "reasonable further progress" requirements in the Clean Air Act.

Despite the still developing utility interest in emissions trading, there is no doubt that, when the opportunity arises, utilities pursue trading-based control strategies. Furthermore, they will seek out these strategies even more aggressively in the future as the marginal cost for higher levels of emissions control escalates.

Some utilities have advocated extending emissions trading in new directions, such as NSPS "compliance bubbles." In addition, evidence shows that trading could contribute to the resolution of the economic issues involved in the acid-rain debate. More sulfur can be removed from fossil fuels, if the cost to remove a marginal ton is decreased.

How to Use an ERC

Since all emissions trading is based on ERCs, individuals and firms can participate only after creating emissions reductions. As noted earlier, all emissions reductions are not equal—not all can be credited. However, those that can be credited become an asset like other intangible corporate holdings that have value. Creating ERCs has been shown to be profitable.

After certifying emissions reductions, the next step is determining how the ERCs can be used to obtain their greatest value. Answering this question is difficult—the solution depends on many variables.

Some firms will want to maintain a personal inventory of ERCs for internal offsets, bubbling, and netting purposes. Other firms, particularly those closing or discontinuing certain operations, will realize they do not need to hold their ERCs and will sell them on the open market.

If they are not used internally, ERCs can be sold, leased, optioned, given away, or otherwise conveyed to specific third parties under valid contractual agreements. Before deciding whether to trade or use an ERC, these five factors must be evaluated.

(1) How many credits are there? Quantifying credits would appear to be a simple measurement exercise. However, uncertainties do exist. The key factor is determining the baseline against which reductions are to be measured. For example, if the ERCs are in a nonattainment area, a new SIP may be required that will allocate new emissions reductions to existing sources. If a firm's current emissions limit is reduced by a regulator, then the quantity of excess reductions also decreases.

(2) How can the credits be used? Not all emissions reductions can qualify as ERCs. Certain control measures can be used to generate ERCs, other measures cannot. In some states, shutdown-derived credits or curtailments cannot be used for bubbles or offsets. In other states, fuel switches may not be used to yield credits.

Once credits are generated and banked, their use may not reflect the quantity of credits created because of the greater-than-1:1 tradeoff-ratio requirements.

(3) What are the opportunity costs related to the credit-generating activities? Opportunity costs are an economic concept referring to the costs of not doing something—a foregone opportunity. If, for example, a firm creates credits by shutting down, the cost to generate credits equals zero. Opportunity costs are absorbed if nonshutdown credits are sold, and later a firm must purchase external credits from another firm to permit an expansion.

(4) What is the best way to handle the various regulatory and normal uncertainties and delays? Uncertainty can affect a transaction from both the seller's and buyer's viewpoints. It can encourage ERC sellers to quickly dispose of credits for fear that their value may be reduced by confiscation, a moratorium on their use, or an unfavorable change in the offset ratio. Uncertainty can also serve to inhibit

ERC transactions because sellers are generally not knowledgeable about pricing, selling, or conveying credits.

The buyer is also plagued by uncertainty that can affect the price it is willing to pay. Perceived uncertainties, along with delay, are very costly to buyers and can be the cause for abandoning an otherwise-successful venture.

(5) What are the estimated transaction costs? Transaction costs are the costs—measured in both dollars and delay—incurred during the process of searching for ERCs, locating potential sellers, negotiating the ERC price, and conveying ERCs. They can be very significant and have a profound effect on ERC prices. These costs probably dominate other trading-related costs, including the cash costs for an ERC.

These factors affect buyers, sellers, and users of ERCs in many different ways. They represent ERC market forces. Thorough analysis of these factors reduces uncertainty. The less uncertainty, the more trading. More trading promotes lower transaction costs, and—in turn—reduces uncertainty.

Experience Vital

The individual firm may not care about these market forces, for the firm is only an occasional buyer and seller. Just like other occasional investors or occasional purchasers, the occasional ERC trader is more likely to err than the professional.

Approaching trading in an ad hoc fashion ensures that there will be problems. Firms can deprive themselves of over 50% of the financial rewards of trading if they have insufficient information on past, current, and future ERC-trading activities. Using the right skills and methodologies means a successfully completed trade and substantial profits. Knowing how to play the market to receive the best price (or at least a fair one) for credits is complicated and requires special skills.

Many professional disciplines may encompass the skills and interests necessary to assist with ERC transactions. Individuals who can assist in transactions include:

- Consulting engineers.
- Lawyers.
- Economists.
- Environmental consultants.

- Regulatory analysts.
- Planners.
- Vendors of pollution-control equipment.
- Brokers of ERCs.

Each profession harbors some of the skills required for a successful transaction, and input from them is necessary. The skills can be combined in an emissions brokerage firm that can:

- Appraise the market value of ERCs in specified locations.
- Assess the cost of producing ERCs.
- Provide advice on market facts and trends.
- Act as liaison with the regulatory agency.
- Provide engineering analyses as necessary, such as quantifying and qualifying ERCs.
- Prepare, modify, or obtain operating permits.
- Assist clients in obtaining financing.
- Negotiate the deals.
- Assist conveying credits between parties.
- Evaluate tax consequences of trades.

In addition, brokers can bring interested parties together, reduce transaction costs, and help develop an efficient market for ERCs.

Outlook

Emissions trading is not a static program. It has evolved over time and will probably become more rather than less flexible over the next few years. Of course, individual states can choose not to endorse some trading options, but the long-term trend appears to show a broadening of the applicability of trading concepts.

Before becoming an automatic part of pollution-control strategy, however, firms must recognize that they must conduct extensive research to identify and determine local and state rules and their applicability to particular circumstances. In addition, their current and future pollution-control activities should be mapped out to provide accurate estimates of their current and future control costs in order to plan for upcoming trading activities.

Section III
"Select Use" of Natural Gas for Environmental Purposes

DAVID J. BARDIN
Arent, Fox, Kintner, Plotkin & Kahn
Formerly, Administrator, U.S. Economic Regulatory Administration and Commissioner, New Jersey Department of Environmental Protection

DR. BENJAMIN SCHLESINGER
JAMES J. NEWCOMB
Benjamin Schlesinger and Associates

Chapter 7

What Is "Select Use?"

Select use of natural gas is an environmental quality control method in which natural gas is burned, alone or as part of a mixture of fuels, to achieve compliance with air quality regulatory requirements or to improve ambient air quality.

Select use includes both full-time and part-time applications. Select use may involve either gas used by itself or with other fuels. It offers cost-effective solutions to many environmental compliance and fuel combustion problems, especially for reducing air pollution, because of its clean-burning attributes.

Concept and Application of "Select Use"

Clean-burning attributes make natural gas an ideal companion for other, less environmentally desirable fuels such as coal, high-sulfur fuel oil and wastes, in a variety of combustion situations. Consequently, clean air requirements may sometimes be met by "select use" of gas.

Add-On Technology versus Process Modifications

There are two categories of pollution control or reduction methods: "Add-on technology" and "modified processes." Familiar add-on pollution removal devices include wet or dry scrubbers to remove sulfur dioxide from the flue gases and precipitators to remove fly ash.

Equally important, but less standardized, are modified processes, i.e., environmental quality control methods based on changes in the production process itself.

Regarding modified processes, EPA has explained:

> Selection of the type of fuel burned in furnaces using modified processes is an integral component of modified process techniques. [48 F.R. 50674, November 2, 1983.]

Under this definition, modified processes include select use. In many cases, select use may also be added on to the previous process (e.g., afterburning or over-firing with gas).

Pollutants Controlled by Select Use

Select use of natural gas can reduce sulfur oxides, particulate emissions, opacity, oxides of nitrogen, unburned hydrocarbons that contribute to photochemical smog, and combinations of the foregoing. Select use can also control emissions of other chemical pollutants. For example, natural gas may be used to increase combustion temperatures so as to break down certain toxic chemicals, and thereby help dispose of them.

Nitrogen oxides are the major pollutants of concern when burning natural gas. Nitrogen oxide emissions are functions of combustion chamber temperature and combustion product cooling rate. Emission levels vary considerably with the type and size of unit and with operating conditions.

Current Applications

Applications of select use in the United States now include the following:

- Gas and high sulfur fuel oil burned simultaneously in separate boilers, e.g., in a number of northeastern locations.
- Gas and coal burned in combination in separate boilers.
- Gas and industrial or municipal wastes burned in combination in the same boiler (e.g., wood wastes of poor quality).
- Periodic or seasonal uses of gas, in varying combinations with other fuels, to the extent required or allowed by environmental regulatory standards (e.g., emergency episode plans; emissions trading).

- Gas and coal burned together in the same boiler (e.g., including pulverized coal, washed coal, coal-water mixtures), making possible both dilution of emissions and gains in efficiency.

During 1981-84, EPA approved select gas use as an air quality control method at nearly two dozen major U.S. fuel burning facilities, including industrial and electric utility plants.

More widespread application of select use has been limited by a number of factors. Perhaps the principal impediment has been the impression, spawned by curtailments in the 1970s, that natural gas is in short supply and that its use is supposedly only a temporary air quality control measure. Furthermore, the argument goes, gas supplies will eventually run out, and become prohibitively expensive along the way to extinction.

Recent gas developments and events in North America, however, have disproved all the imminent extinction theories where natural gas is concerned. Gas supplies are ample and in surplus. The only serious barrier to new field developments appears to be the lack of sufficient demand in new markets. Appendix A and other available materials provide evidence to potential select users of gas and air quality regulators how abundant a resource gas is and seems likely to remain. That information can lay the foundation for vigorous marketing of select use.

Special Advantages of Natural Gas as an Emissions Control Technology

The select use of natural gas as an air emissions control technology offers a number of special advantages in comparison with other cleanup technologies customers ordinarily turn to.

- Select gas use offers the flexibility to respond quickly and economically to changes in air quality standards. Whereas capital intensive control technologies such as scrubbers or baghouses may be well-designed to meet a given emissions standard, they may have to be redesigned or substantially modified if standards change. In contrast, co-combustion of natural gas allows industries to respond readily to newly imposed regulations by altering the fuel mix. This flexibility may be a substantial economic advantage in an uncertain or unstable regulatory environment.

- In many applications, select gas use offers a way of simultaneously controlling emissions of several different pollutants—SO_X, NO_X, particulates, and hydrocarbons—thereby reducing the likelihood that other emissions control equipment will have to be installed in the future if standards become tighter.
- Unlike many other emissions control technologies, select gas use entails minimal de-rating or efficiency losses in industrial boilers. In fact, in some cases, co-combustion of natural gas will *improve* unit rating and efficiency. Alternatively (and perhaps more importantly) select use can also allow the power plant or boiler to operate de-rated when less than full output is desired, and then use more gas when full power is needed.
- Combined usage of gas and coal can reduce fly and/or bottom ash disposal problems, decrease consumption of limestone and other sulfur-removal materials, and lower costs of scrubber-sludge disposal. Some owners of facilities that implemented select use have reported lowered maintenance costs for emissions control equipment resulting from the reduced load on that equipment.
- In addition, select gas use reduces the likelihood that units will have to be totally shut down for maintenance of emissions control equipment. If necessary, boilers can be operated on 100 percent natural gas while repairs on scrubbers or baghouses are completed, avoiding costly idling of plant capacity.

These special benefits of natural gas as an emissions control technology come in addition to the usual advantages of natural gas use, such as the ease and economy of fuel-handling compared with other fuels and lack of storage costs.

The advantages of select gas use translate directly into lower costs for users seeking to meet their environmental requirements. Specific dollar benefits arise to the extent that the need to spend capital and/or O&M dollars on air pollution abatement equipment can be reduced by involving modest (i.e., select) volumes of natural gas in overall fuel combustion. This, along with gas's other advantages, means that gas will be especially valuable when users:

(1) Are not in compliance with EPA emission standards,

(2) Have had compliance problems in the recent past,

(3) Anticipate spending substantial capital in the near future for air pollution controls,

(4) Are now seeking to meet an emissions reduction schedule that will bring them into compliance, with EPA standards, or

(5) Are subject to particularly stringent emissions standards.

Typically, they may avoid intensive capital investment requirements for emissions control equipment by adopting select gas use as an alternative.

Chapter 8

Select Gas Use With Fuel Oil

Select gas use with fuel oil can help oil users control emissions of any number of air contaminants—including sulfur oxides, particulates, opacity, or other air pollutants resulting from oil burning. However, in virtually all the current select gas use applications involving fuel oil, gas is used to control emissions of sulfur oxides.

Generally, this form of select gas use works in the following way:

- Under the existing SIP, a given group of combustors in the area, industrial boilers and electric powerplants alike, must burn a higher cost, low sulfur grade of fuel oil (e.g., 0.5 percent sulfur No. 6 residual fuel oil).
- Under EPA's "bubble policy," the facility is allowed to burn a mix of high-sulfur fuel oil and natural gas that, all told, emits no more sulfur from the site than if low-sulfur oil is burned.
- Therefore, under a select gas use plan, some of the facility's boilers switch to high-sulfur oil, while its other boilers convert to natural gas; note it is important that the net emissions from the facility do not increase.
- If gas must be diverted to other customers from the industrial user (e.g., as typically occurs on a few days in the winter for interruptible gas customers), then the whole facility temporarily switches back to low-sulfur fuel oil.
- The customer saves fuel operating expenses at no net emissions increase.

Clearly, other forms of select gas use with fuel oil are possible. For the sake of uniformity, however, the remainder of this chapter presumes the above select use concept, in which gas is used selectively to control sulfur emissions from oil-fired boilers.

Note also that if the facility is now using oil, adding or substituting gas raises no new FUA issues because 1981 amendments took out its anti-gas discrimination sections.

The following four conditions should be present for select gas use with fuel oil to warrant consideration:

- Significant volumes of oil are used for fuel in the facility (e.g., No. 6 residual).
- Air emission rules that apply in the region preclude combustion of available fuels (e.g., 2 percent sulfur oil cannot be burned because of a one-half percent sulfur requirement).
- The lower sulfur fuel oil sells at a premium price, versus lower cost higher sulfur oil.
- Natural gas service is available in the region, regardless of whether or not the facility is already a gas customer.

The following information must be known for both electricity generating stations and industrial boilers, before a select gas use plan can be developed:

- *Size*—rated and actual boiler or powerplant capacity, for scale purposes. (Note: Inventory should separate and you should give lower priority to facilities below 10 MW and/or 100,000 pounds of steam per hour, which are often exempt from air quality requirements.)
- *Location*—preferably related to prevailing air quality control regions (AQCRs), as designated by state or county air quality officials (e.g., zones, cities, districts, etc.)
- *Current and projected fuel use*—including fuel oil types and grade (e.g., 0.5 percent sulfur No. 6 residual, 1.0 percent sulfur No. 4 fuel oil, natural gas, etc.); and if different, volumes of fuel use projected in the short term (e.g., imminent coal conversion plans, expansions, shutdowns, etc.)
- *Duration of facility*—age, remaining useful life, degree of capitalization

- *Fuel prices*—costs of fuels burned and available, including required versus higher sulfur content, expressed in $ per MMBtu, cents per gallon, etc.
  - Current
  - Projected in the mid-term (1-5 years)
- *Fuel contractual considerations*—any special clauses which affect the facility's price or volume terms, e.g., take-or-pay, volume discounts, special rates, etc.
- *Natural gas service*—whether or not a gas meter is set at the facility; adequacy rating (optional)
- *Prevailing air quality designation*—PSD or nonattainment, as classified in the State Air Implementation Plan (the "SIP"), from
  - SO_X
  - NO_X
  - Particulates
  - HC

  as well as SIP requirements for this unit and any pending proposed revisions of classification. Note, these designations relate to location (see above).
- *Other*—special political, regulatory, or other fuel use considerations, e.g., target of coal conversion efforts, community resistance to coal, etc.

This inventory can be used to help set priorities among facilities for initial focus.

For example, priorities among facilities within the inventory may be based on such criteria as size of facility, average cost of fuels, and/or overall volume of fuel use. The inventory may also be segmented with respect to such factors as air quality control region (PSD versus nonattainment), or gas service versus none.

Performing the Benefit/Cost Analysis

One of the most important criteria for evaluating a select use plan (i.e., the facilities contained in the inventory described above) is how much money the user will save if he decides on select gas use, versus his next best air quality control strategy (e.g., burning low sulfur fuel oil throughout his facility).

In the most basic example, if an oil customer burns gas in the select use mode, and thereby saves the added cost of purchasing low-sulfur oil, then the benefit of gas relates to the direct cost savings to the user. Exhibit 8-1 presents a formula for calculating this economic benefit based solely on relative fuel costs; i.e., it ignores all the other benefits of select use.

The benefit/cost formula will change as a result of specific conditions at the user's site, grades of oil used (i.e., different fuel blends on either side of the equation), or other considerations. Therefore, the formula should be only a guide to the relative attractiveness of select gas use.

Facilities using gas in the select use mode report other cost savings as well, such as:

- Reduced maintenance expenses
- Improved efficiency of combustion (but NOTE: This could work both ways, depending on the age of the boiler, etc.)
- Reduced oil inventory carrying expenses

These and all other applicable benefits of gas use should be considered in a benefit-cost analysis of select use.

Securing Regulatory Approvals

There are several steps required to secure regulatory approvals for select gas use in oil-burning facilities. The detailed procedures necessary to obtain approval for a particular select use application will depend on state and county regulations, and on specifics of the prospective select use application. Given the wide range of such possibilities, the purpose of this section is to:

- Provide information on how to learn more about the regulations and procedures relevant to a particular select use application, and
- Overview the steps necessary to obtain approval in some typical select use cases.

Exhibit 8-2 summarizes the steps described below.

Exhibit 8-1

Calculating the Economic Benefit of Select Use with Fuel Oil

An understanding of the economic benefit of select use—i.e., gas's true value over oil can help show whether or not select use is warranted, and if so, by how much. Here's how it might work, based solely on relative fuel prices:

If gas costs less than low-sulfur oil, then the user should be burning gas anyway and he probably already is as long as he's gas capable. In any event, he will not need select use to justify gas.

But if gas costs more than low-sulfur oil, then some extra benefit is necessary to justify gas use.

In other words, the total cost of select gas use—including the gas, plus the high-sulfur oil, less any other savings arising out of gas use (e.g., reduced need for oil storage, etc.)—should be less than the cost of using low-sulfur oil throughout the facility. This relationship can be written as follows, again based strictly on relative fuel prices:

$$\text{Cost of using low-sulfur fuel oil} >= \text{Cost of using high-sulfur fuel oil} + \text{Cost of using gas}$$

The above equation may be used to derive the benefit of select gas use, i.e., how much more than low-sulfur oil the gas is worth to this select gas user:

$$\text{Benefit of gas in select use (\$/MMBtu over low-S oil cost)} = \frac{L - H}{R - 1}$$

Where:

R = Ratio of sulfur content in high-sulfur oil: low-sulfur oil

H = Price of high-sulfur oil that a user will be able to burn with select gas use (in \$ per MMBtu)

L = Price of low-sulfur oil a user must burn without select use (\$ per MMBtu)

For example, if a user can obtain 1.5 percent sulfur No. 6 residual fuel oil for \$4.00 per MMBtu, but he's obliged to burn 0.5 percent sulfur oil costing \$4.50, then the select gas use benefit works out to \$0.25 per MMBtu *over the price of low-sulfur oil.*

A worksheet is provided for user wishing to apply this limited formula.

Exhibit 8-1 (continued)

WORKSHEET

Computation of the Economic Benefit of Select Gas Use with Fuel Oil

Complete the following information for each facility:

1.	Grade of low-sulfur fuel oil required (% sulfur)	________
2.	Price of above ($ per MMBtu)	$ ________
3.	Grade of high-sulfur fuel oil that could be used (% sulfur) . . .	________
4.	Price of above (% per MMBtu)	$ ________
5.	Subtract line (4) from line (2), and place the result here	$ ________
6.	Divide line (1) into line (3)	________
7.	Subtract 1 from the number in line (6) above	________
8.	Divide line (7) into line (5)	________

Line (8) above is the select gas use benefit, expressed as an add-on to the cost of low-sulfur fuel oil, in $ per MMBtu.

Emissions Standards

The first step is to determine what air pollution emissions standards apply to the facility. Information on emissions standards can be obtained by contacting local air pollution control authorities (see Appendix D for a directory of governmental air pollution control authorities). If a facility is located within a city or county that has its own air pollution control authority, then the local authorities listed in the directory in Appendix D should be contacted to determine the applicable standards. Otherwise, the state air pollution control office (or, in some states, a district office) will provide this information.

Next, an assessment must be made of air pollutant emissions under the proposed select use configuration. Select use configurations may include:

- *Cofiring*, i.e., burning natural gas together with high-sulfur fuel oil in a single boiler, or
- *Multiple boilers*, i.e., burning natural gas in one or more boilers in order to allow the burning of high-sulfur fuel oil in other boilers.

Exhibit 8-2

General Regulatory Approval Procedures for Select Gas Use

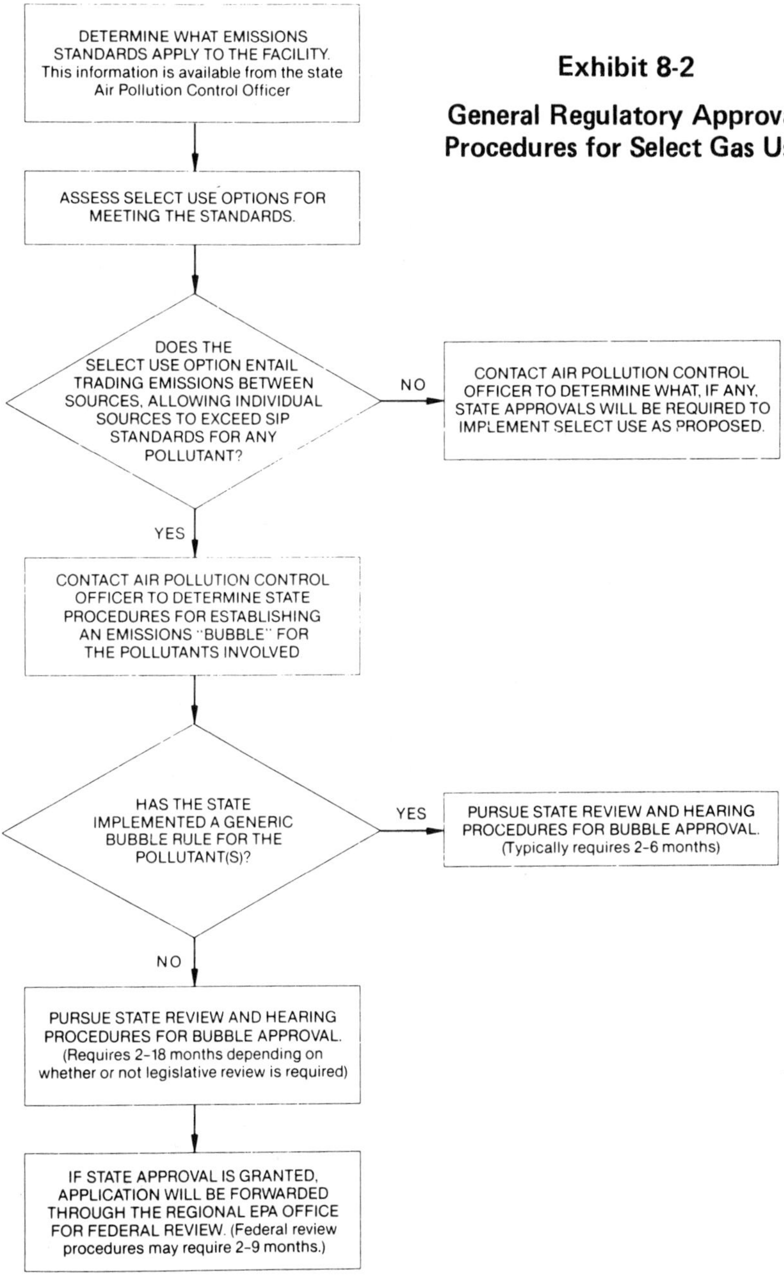

These two alternative select gas use configurations require quite different types of regulatory approvals.

Select gas use options that entail co-firing natural gas and oil in a single boiler require only that the proposed select gas use project offer a means of meeting existing emissions standards on a continuous basis. In these instances, obtaining regulatory approval for implementing select gas use is generally quite simple.

Where a proposed select gas use project entails trading emissions between air pollution sources—allowing any point source to exceed emissions standards for any pollutant—the facility must obtain state approval for an emissions bubble. By establishing an emissions bubble, several point sources within a facility can be considered a single "source" from a regulatory perspective and trading of emissions between individual point sources can take place.

State procedures for obtaining approval to establish a bubble vary widely. Some States have had extensive experience with emissions trading and have streamlined their procedures in order to shorten the lead-times for regulatory approvals. The most important factor in determining how readily a bubble application can be approved in a given State is whether or not the State has obtained EPA approval for a generic bubble rule.

A generic bubble rule avoids case-by-case review of bubble applications by the regional EPA offices. States that have implemented a generic bubble rule typically process bubble applications in two to six months. Appendix E provides a listing of those States that have implemented or proposed generic bubble rules.

The time required for obtaining approval for select use varies widely from State to State. In States with generic bubble rules, applications may be approved in two to six months. In other States, the complete approval may take more than a year because federal review of each application is required.

The Application Process: An Example

Suppose that Company A must obtain approval to burn a high-sulfur fuel oil in boiler No. 1 in exchange for burning natural gas in boiler No. 2. The trading of emissions between the two sources requires that the facility be "bubbled." How can the company obtain approval?

First, an application must be prepared which includes all documentation required under State regulations as required by the Clean Air Act. Such an application will typically require documentation of sulfur emission rates from each point and assurance that the proposed emissions control technology can provide continuous emissions control.

In some cases, air quality dispersion modeling may be required. Modeling requirements are frequently waived for small sources or for cases in which the sources to be included in the bubble are quite close to one another.

The application typically must be submitted to a State environmental quality agency which will review the application, and if appropriate, forward it to the bureau of air quality control which issues a public announcement of the application, accepts written comments on the proposal and, where requested, schedules public hearings. After evaluating written comments and hearing testimony from interested parties, that bureau will either approve or disapprove the request. For States with a generic bubble rule, no further approvals are required.

States without a generic bubble rule must submit the State-approved application to the regional EPA office which prepares an initial draft of a Federal Register notice describing the State action. This notice is submitted to EPA Headquarters in Washington, D.C. for approval and issuance.

After a thirty day period during which written comments are received and testimony heard on the proposed rule change, the regional EPA office re-evaluates the application in light of public comments and prepares a final Federal Register notice. Final approval of this notice by EPA Headquarters completes the process.

Implementing the Project

Implementing select gas use with oil is likely to be the simplest of the various select use applications. Many of the select gas use projects in this category are likely to be existing or former gas customers who will therefore be both physically equipped to buy gas and familiar with typical contractual and financial arrangements.

Nonetheless, implementation will require that final agreement be reached on terms of a gas purchase contract suitable to the needs of the select use customer. The design and approval of such an agreement, of course, will require the involvement of gas company personnel for rates, operations, gas supply and/or other functions.

In some cases, construction of new gas lines or metering facilities may be required. Gas users should be informed as to the lead-times and planning requirements for such construction.

Chapter 9

Select Gas Use With Coal

As Exhibit 9-1 illustrates the application of select gas use with coal falls into the following categories, for purposes of dealing with the regulations, especially the Clean Air Act and the Fuel Use Act:

- Large boilers:
  - Pre-1971 boilers, including both industrial and electric utility boilers, which are not subject to a NSPS and may apply select gas use to meet SIP standards
  - Post-1971 industrial boilers and 1971-1978 electric utility boilers, which are subject to a NSPS that allows select gas use, but not bubbling
  - Post-1978 electric utility boilers, which are subject to a NSPS that restricts select use
- Small boilers
- Other non-boiler applications.

Where a NSPS applies, bubbling may not be used to meet the NSPS itself. However, it is allowable to develop offsets for a number of other purposes, such as trading with another source.

SIP provisions vary widely. All large boilers (over 250 million Btu per hour) constructed after August 1971, are subject to a uniform new source performance standard (NSPS) set by EPA. However, large electric utility boilers, constructed after September 1978, are subject to an even more stringent NSPS which limits select-use opportunities.

Exhibit 9-1

Segmentation of Select Use Markets Under Clean Air Act and Fuel Use Act

<table>
<tr><th>CLEAN AIR ACT</th><th>Large Industrial boiler</th><th>Large Electric utility boiler</th><th>Other Combustors</th></tr>
<tr><td>Pre-1971</td><td>No NSPS
Select use encouraged</td><td>No NSPS
Select use encouraged</td><td rowspan="3">No NSPS
(proposal pending)
Select use encouraged</td></tr>
<tr><td>Post-1971</td><td rowspan="2">1971 NSPS
Select use encouraged</td><td>1971 NSPS
Select use encouraged</td></tr>
<tr><td>Post-1978</td><td>1978 NSPS
Select use restricted</td></tr>
</table>

<table>
<tr><th>FUEL USE ACT</th><th>Large Industrial boiler</th><th>Large Electric utility boiler and combined cycle</th><th>Other Combustors</th></tr>
<tr><td>Pre-1978</td><td>No limit on gas
Select use encouraged</td><td>No limit on gas
Select use encouraged</td><td rowspan="2">No limit on gas
Select use encouraged</td></tr>
<tr><td>Post-1978</td><td>Some limit on gas
Select use limited</td><td>Some limit on gas
Select use limited</td></tr>
</table>

(Only for a few units did construction commence after September 18, 1978, the controlling date. FUA restrictions apply only to power plants or large industrial boilers where construction commenced after November 1978 plus a few older power plants specifically ordered, under FUA or the previous law, not to burn gas.)

Small boilers, not yet subject to NSPS, like non-boiler applications, are generally subject to SIP emission standards only.

Select gas use with coal offers the possibility of "blending" two very different fuels in a way that captures the benefits of each: the economy of coal and the premium fuel-handling and environmental characteristics of natural gas. Briefly, the technologies for select gas use with coal include:

- Cofired applications, in which coal and gas are burned simultaneously in the same boiler; and
- Boilers fired in parallel, where coal is fired in one boiler and gas in a second, with both boilers operated in a coordinated fashion.

In either case, select gas use with coal may be an attractive option for a number of different types of facilities:

- Some of the facilities now burning low-sulfur coal in order to meet environmental regulations may be able to burn cheaper, high-sulfur coal in some boilers in exchange for burning natural gas in other boilers if the facility can be placed under an EPA bubble.
- Select gas use may be the only alternative to installing scrubbers for coal-fired facilities that are required to burn local, high-sulfur coal, but still must comply with strict sulfur emissions regulations. Electric utility industry experts estimate the costs of retrofitting scrubbers to existing electricity generating stations to be $250-$500 per kilowatt of capacity. Investments necessary for cofiring natural gas with coal are roughly $150-$250 per kilowatt.
- Oil-burning facilities may be able to convert to burning coal-gas mixtures or, in the case of multiple oil-fired boilers, establishing a bubble around the facility may allow gas to be burned in some boilers and coal in others.
- Gas customers that are converting their facilities to burn coal may find select gas use to be an attractive way to enhance their fuel flexibility and comply with emissions standards.

In all, select gas use offers an economically competitive means of reducing air pollution emissions from coal-fired utility and industrial boilers. The approaches above can provide for reductions of emissions of oxides of sulfur and nitrogen as well as particulates.

Select gas use also reduces operating costs related to fuel-handling and to collection and disposal of ash and sulfur-removal materials. Moreover, select use requires relatively small capital investments compared with alternative emissions control equipment.

In the future, attempts to control acid precipitation through the imposition of stricter sulfur emission standards could dramatically increase the market for select gas use. Should acid rain legislation be passed by the Congress, many coal-burning industrial and utility facilities would be forced either to retrofit existing boilers with scrubbers or to implement select use. Recent legislative proposals specifically permit the select use of gas by large powerplants to achieve environmental compliance by offsetting some or all of the SO_2 emissions that would otherwise be associated with direct coal combustion.

In general, almost every sizable industrial coal-burning facility should consider select gas use. Many coal-fired facilities that presently comply with emissions standards may be dissatisfied with their existing emissions control equipment. In many cases, installation of scrubbers and/or baghouses has caused a substantial derating of boiler capacity. Furthermore, the problems of ash collection and disposal and the handling and disposal of scrubber sludge may spur plant operators to consider alternative emissions control methods.

Performing the Benefit/Cost Analysis

Economic analyses comparing select gas use with other emissions control technologies will entail estimation of net present values of capital, financial, and operating costs of the alternatives. These analyses are likely to be quite sensitive to assumptions about future prices of gas and coal, as well as to financial considerations. In general, the more detailed and comprehensive the benefit/cost analysis is, the more the economic advantages of select use will be revealed.

The combined use of gas and coal can produce benefits such as:

- Reduced capital cost of scrubbers, and/or other air pollution control equipment
- Reduced emissions of SO_2 and NO_X

- Increased flexibility in siting new facilities in or near nonattainment areas
- Increase in thermal efficiency of the boiler by reducing the percentage of air in the combustion mix
- Better turn-down ratio and improved capability to accommodate load swings
- Dramatically reduced emissions of suspended particulate matter (TSP) through use of gas burners in an overfiring mode
- Increased fuel flexibility during unplanned interruptions in coal supply (e.g., caused by freeze-ups, etc.)
- Reduced costs of fly and bottom ash collection and disposal
- Decreased consumption of limestone and other sulfur-capture materials
- Decreased costs of land required for scrubber sludge disposal
- Increased flexibility in responding to changes in emissions control regulations
- Reduced derating of boilers.

These benefits must be weighed against the capital costs of facility conversion and operating costs of burning natural gas.

A generic benefit cost formula cannot be offered in this chapter because each coal-fired facility may entail a separate set of alternatives for implementing select gas use. In general, however, a benefit-cost formula and strategy should be prepared that:

- Develops an understanding of the next best alternative to select gas use, and
- Ensures that the cost of operation with select gas use does not exceed the cost under that next best alternative.

The formula contained in Exhibit 9-2 provides one approach toward calculating the select use benefit of gas in the context of coal use.

Securing Regulatory Approvals

The first step toward securing the regulatory approvals necessary to implement select gas use with coal is to determine what air pollution emissions standards apply to the facility. Information on emissions standards can be obtained by contacting local air pollution control authorities as listed in the directory provided in Appendix D.

Exhibit 9-2
Calculating the Economic Benefit of Select Use with Coal

Select use with coal can benefit in a number of ways—it can save the expense of installing and operating (or upgrading, etc.) air pollution equipment, reduce operating costs by enabling burning of high-sulfur coal, or some combination of these two along with efficiency gains. Thus, since the cost savings can accrue in any of several ways of combinations, no universal worksheet is available. The following provides a generic approach to developing a formula, however, so as to serve as a starting point in assessing the benefits and costs of select gas use with coal.

If select gas use with coal is to have a net benefit, then a basic formulation would be as follows, with all cost estimates expressed in terms of Net Present Value:

Cost of plant operations & maintenance with coal	+	Capital cost of air pollution control equipment	>=	Cost of using coal and gas in combination	+	Capital cost of select use

Each of the costs in this type of formulation will need to be evaluated, taking into account capital expenses involved together with realistic financing mechanisms, grades of coal on either side of the equation, and the exact configuration of select gas use that will be involved in the project. Note the results of applying this formulation will differ depending upon which air contaminant (e.g., sulfur, particulates, etc.) select gas is in place to control.

Where a proposed select gas use project entails trading emissions between air pollution sources, the facility must obtain state approval for an emissions bubble. By establishing an emissions bubble, several point sources within a facility can be considered a single "source" from a regulatory perspective and trading of emissions between individual point sources can take place.

State procedures for obtaining approval to establish a bubble vary widely. Some States have had extensive experience with emissions trading and have streamlined their procedures in order to shorten the lead-times for regulatory approvals. The most important factor in determining how readily a bubble application can be approved in a given State is whether or not the State has obtained EPA approval for a generic bubble rule.

A generic bubble rule avoids case-by-case review of bubble applications by the regional EPA offices. States that have implemented a generic bubble rule typically process bubble applications in two to six months. Appendix E provides a listing of those States that have implemented or proposed generic bubble rules.

Implementing the Project

Select gas use with coal can take several forms:

- *Parallel-firing* gas and coal, with each fuel burned in separate boilers
- *Cofiring* gas and coal in the same boiler, utilizing either:
  - *Comingling*, i.e., entailing injection of pulverized coal and natural gas using the same burners
  - *Over-firing*, i.e., involving two-stage combustion brought about by sequential injection of coal, followed by natural gas.

Furthermore, some applications of select gas use with coal may implement the concept in "coal conversion" boilers (i.e., currently burning oil), while others might introduce select gas use into units that are presently coal-fired. The remainder of this implementation discussion characterizes some of the kinds of engineering issues involved with each of the foregoing approaches to select gas use with coal.

Parallel-firing. Implementing select gas use in parallel coal- and gas-fired boilers should require only minimal investment in equipment. Facilities with boilers that already have the capability to fire either coal or gas, or those capable of burning oil, will need to make little or no changes to their plant.

For boilers that are capable of burning only coal, select gas use in parallel fired boilers may require installation of a gas line (if such does not already exist), as well as gas burners, air registers, control equipment, and a gas meter.

In some cases, but clearly not all, modifications of fireboxes may be required in order to enable optimal combustion and heat transfer conditions. Modification of control equipment in conjunction with parallel-firing (i.e., in separate boilers) offers additional advantages insofar as the gas-fired units, which have a much shorter response time, can be used to meet fluctuations in steam requirements.

Comingling. Implementation of select gas use in a way that involves cofiring gas with coal in the same boiler, using the same burners, may require burner retrofit and/or modification of the firebox. Burners designed to combust gas and powdered coal have been in use for decades, particularly in steel mills, and are available from a number of manufacturers at modest costs. Other firms have developed burners capable of burning mixtures of gas and coal from 0 to 100 percent gas.

Over-firing. Some innovative designs for coal-gas burning employ two-stage firing systems. Initial engineering experiments involving select gas use cofiring with coal indicate the performance superiority of over-firing configurations, versus direct comingling. (For example, see reference to A. E. Green in the Bibliography, Appendix B.) Many of the benefits of combined use of gas and coal cited above—reduced particulate emissions, increased thermal efficiency, etc.—can be obtained in the over-firing mode.

Coal conversions. Some of the special facility modifications that may be required for converting an oil-fired boiler to select gas use with coal are listed below:

- Burners for coal and/or gas must be installed and the wind-box must be modified for burner installation.
- Furnace bottom hoppers must be modified and an ash removal system must be installed.
- Convective passes may require modification through addition of plattens.
- A particulate control device and flyash handling system must be installed.
- Fuel handling, storage and preparation equipment must be installed.
- Forced-draft and induced-draft fans may have to be replaced or upgraded.

- Instrumentation and controls will require modification.

Typically, conversion of an oil-fired unit to coal requires a substantial derating of the unit (up to 60 percent). The combined firing of gas and coal, on the other hand, especially over-firing, negates the need for boiler derating and, at the same time, increases the efficiency of coal combustion by reducing the percentage of air in the combustion mix.

Finally, the instrumentation and controls for a coal-fired boiler converted to select gas use may also need to be modified. These modifications, however, are likely to substantially increase operational control flexibility of the boilers. This is because (1) natural gas can be used as a "swing fuel," responding quickly to changes in the energy demand of the system, and (2) unlike most coal-fired boilers, boilers that are co-fired with coal and gas can be "turned down" to very low levels during idle periods; natural gas enables a precise energy input to hold the combustion chamber to any low temperature desired.

Chapter 10

Select Gas Use to Enhance Incineration of Wastes

Because of the breadth and diversity of waste materials generated in the nation, no single select gas use concept is articulated in this chapter. In general, the use of natural gas in a cofiring (or "over-firing") mode can enhance disposal of a variety of solid and liquid waste materials, hazardous and otherwise.

In general, select gas use for enhanced waste incineration entails making use of two of the "premium" fuel qualities of natural gas: (1) its high-temperature flame characteristics, and (2) its low emissions characteristics. Moreover, the use of one Btu of natural gas in this way provides more than one Btu of fuel; waste disposal costs are reduced as waste becomes fuel.

Select gas use in combination with combustion of wastes could potentially find application in any of a large number of industrial settings.

The important possibilities may include:

- Municipal incinerators, especially in conjunction with solid waste trash-to-steam plants
- Wood or paper products industries
- Furniture manufacturing companies
- Chemical processing plants
- Synthetic rubber manufacturing plants
- Superfund sites.

Special attention should be paid to possibilities for cogeneration. Some paper and wood products industries in northeastern and midwestern States have found it profitable to use waste paper, cardboard, or woodchips, together with natural gas to fire boilers for cogeneration of steam and electricity. The use of natural gas:

- Provides flame stabilization and improved burning efficiency
- Allows the facility to meet particulate emissions standards
- Can make up energy deficiencies in the boiler.

In some cases, burning liquid wastes requires select use of natural gas. A facility in Florida blends virgin residual oil with plant byproduct waste oils. The blend is fired in their boiler in combination with natural gas in the ratio needed to meet the sulfur dioxide emission limit. The sulfur content of the virgin oil is low enough that when gas supplies are interrupted, 100 percent virgin oil may be fired. In another instance, a chemical processing plant is able to burn what would otherwise be a toxic waste product through select use of natural gas.

EPA has released a priority list of existing hazardous waste disposal sites that have been earmarked for immediate clean-up on a priority basis, under the Superfund Program. The National Priorities List (NPL) is available from EPA's Public Information Office, Washington, D.C. Each of these NPL sites should consider select gas use.

The EPA's Compliance Data System lists other potential select use customers in a given service area. In particular, facilities that are burning municipal wastes, wood, or paper products and are experiencing difficulties meeting particulate emissions standards are likely candidates for select use, and are likely also to be included in the CDS report.

Much of the opportunity for select gas use in combination with waste products is likely to be hidden. Many manufacturers are simply unaware of the possibilities for select use in their industry. Many assume that the costs associated with the disposal of their waste products are unavoidable.

These manufacturers need to become aware of the availability of natural gas, and aware that many industries are now discovering new methods of solid and liquid waste disposal through the select use of natural gas.

Performing the Benefit/Cost Analysis

Benefit-cost analysis in the case of select gas use for incineration of waste products cannot be set out in general because of the differences in individual applications. This section aims, instead, at presenting a checklist of items that should not be overlooked in developing such a benefit-cost analysis.

Evaluation of the benefits of such an application should include assessments of:

- Avoided costs of waste disposal
- Avoided costs of fuel and/or electricity purchases
- Value of increased fuel flexibility through the use of natural gas as a back-up fuel.

The costs of implementing select gas use entail:

- Capital costs of plant or of modifications in order to allow select use
- Costs of natural gas necessary to operate the system
- Potential costs of bringing natural gas service to the site.

Securing Regulatory Approvals

Since cofiring natural gas with waste products does not involve the creation of an EPA bubble, the procedures necessary to implement select use are likely to be relatively simple. Cofiring natural gas with waste wood products in order to meet particulate emissions standards, for example, may require no advance approvals whatsoever.

A facility can, in some cases, simply begin burning natural gas, subsequently demonstrating that it is in compliance with emissions regulations. New sources, however, could be subject to more stringent regulation through either (1) a permitting system or (2) the application of New Source Performance Standards (NSPS). However, only a NSPS with respect to particulate matter exists for waste incinerators at this time.

Construction of resource recovery plants in nonattainment areas may present special needs for emission offsets. Here, coordination will be necessary by the planners of the resource recovery plant, with an existing fuel user who could reduce emissions (e.g., of non-methane hydrocarbons or particulates) by using more gas, and with the regulatory agency.

Implementing the Project

Some facilities implementing select gas use to enhance incineration of wastes may be new gas customers, unfamiliar with standard industry practices in purchasing natural gas. The implementation phase of such an application will therefore require cooperation between the gas company and the new customer to assure that final gas purchase arrangements are made smoothly.

In some cases, installation of a new gas line and gas metering facilities will be required. Potential users should be informed as to the lead-times and planning requirements for such construction.

Chapter 11

Select Gas Use With Cogeneration

Cogeneration facilities now vary from 300 megawatt industrial plants to 60 kilowatt commercial package units. Most applications of cogeneration must be individually engineered because of differing energy balance, fuel, physical and meterological conditions at each site. This chapter makes no attempt to address each of these factors, but, instead, is limited to describing how select gas use in cogeneration projects might be helpful to their overall success.

Cogeneration systems produce both electrical or mechanical energy and thermal energy from the same primary energy source. Cogeneration captures otherwise wasted thermal energy, usually from a heat engine producing electric power, and uses it for applications such as space conditioning, industrial process needs, or water heating.

Over the last several years, cogeneration has evolved as one of the most important potential new markets for natural gas. Some potential cogeneration applications, however, have been impeded by air pollution regulations. In these cases, select gas use may facilitate the implementation of cogeneration projects in a number of localized situations. For example,

- In some nonattainment areas, proposed cogenerators may have to arrange for offsetting reductions in air pollutant emissions from other sources in the area in order to implement the project; select gas use can provide for such an offset.
- In areas subject to PSD standards or other stringent emissions limitations, burning natural gas as a part of the fuel mix may allow the cogenerator to meet emission standards.

Examples of cogeneration technologies include:

- Steam turbines
- Open-cycle combustion turbines
- Combined cycle systems
- Diesels
- Steam Rankine bottoming cycles.

The economic attractiveness of cogeneration depends on the costs the cogenerator would incur for electricity purchases from a utility and thermal energy production, versus the cost of cogeneration. In addition, the higher the rates at which the local electric utility will buy electricity from industrial and commercial cogenerators, the greater the incentives for cogeneration.

Hence a key step toward assessing cogeneration and select use is determination of local electric buy-back rates. Some projects turn on selling electricity; others involve only on-site use of the cogenerated electricity. In some cases, gas companies finance or own cogeneration facilities.

Opportunities for cogeneration exist in a number of industries as well as in commercial and residential buildings. According to the U.S. Office of Technology Assessment:

The *pulp and paper industry* represents a sizable market for cogeneration because of the large amounts of burnable process wastes that can supply energy needed for plant requirements. Integrated pulp and paper plants dispose of wood wastes and processing fluids and recover process chemicals in furnaces that typically supply more than half of a plant's energy needs. Electric power production through cogeneration has become an integral part of the manufacturing process in this industry.

Select gas use in this context can offer advantages such as:

- Improved combustion properties
- Increased control over energy input
- Lower air pollution emissions.

The *chemical industry* is a major steam-raising industry that represents a substantial potential market for topping-cycle cogeneration. The *steel industry*, another energy intensive industry, also represents substantial potential for additional cogeneration. In both cases, select gas use may allow some plants to implement cogeneration without exceeding emissions standards.

The *food processing industry* represents a potentially large-scale cogeneration candidate.

Opportunities also exist for cogeneration in *commercial buildings* (such as office buildings, hospitals and hotels). Although commercial enterprises typically use thermal energy only for space heating and water heating, the economics of cogeneration systems for commercial buildings are favorable under many circumstances. Select gas use is likely to play a particularly important role in the context of commercial building cogeneration because such facilities are likely to be located in urban areas where emissions controls are toughest.

Finally, through emissions trading, select gas use may make cogeneration possible in many marginal contexts. For example, a new gas-fired cogeneration plant installed in conjunction with a plant expansion may replace separate generation of steam and of electricity. It may thereby make possible economically an expansion with no "net" increase in emissions. Alternatively, it may prove possible to sell a reduction of the previous emissions.

In summary, key factors shaping select use and cogeneration include

- State Public Utilities Commission support for cogeneration
  - California and Texas, for example, have strongly supported cogeneration, speeding its growth
- "Avoided cost" of electricity generation (the rate that electric utilities may pay cogenerators for power sold to the grid)
- Availability or legality of "self-service" wheeling from one location of the cogenerator to another
- Air pollution regulations

Performing the Benefit/Cost Analysis

Benefit/cost analysis of cogeneration systems must include assessments of:

- Fuel cost and availability
- Ownership and financing
- Tax incentives
- Utility rates for purchases of cogenerated power or opportuni-

ties for "self-service" wheeling where avoided utility retail charges exceed utility purchase rates
- Availability of outside financing, as by a gas company or a limited partnership
- Investment risks.

The investor must weigh the cost of cogeneration against the costs of purchasing electric power and producing thermal power. The federal requirement that electric utilities purchase power produced by cogenerators at a rate equal to the full avoided cost of electricity generation amounts to a guarantee of a market for cogenerated electricity. Avoided cost rates, however, vary substantially across the country, so that the economic attractiveness of cogeneration varies according to the price of utility-produced power.

In some service areas, avoided cost rates are lower than the standard retail electricity rates. In such areas, cogenerators may want to investigate the prospects for "self-service" wheeling to a second site, if they can generate more electricity than they can use on-site. Conversely, they may want to attract a second consumer to the generation site.

The benefits and costs of cogeneration systems in a generic sense were compared in the energy analysis entitled *An Economic and Environmental Comparison of Natural Gas and Coal Use for Large-Scale Industrial Cogeneration* (American Gas Association, October 1984).

The methodology employed in that study set forth options for configuring benefit-cost calculations for cogeneration in commercial building applications by comparing such alternative combinations as:

- Gas boiler/electric chiller
- Electric resistance heat/electric chiller
- Gas-fired cogenerator/double stage absorption chiller

in a variety of regions, using a hospital facility example. For industrial applications, the analysis compared gas-fired cogenerators with oil and natural gas boilers, using a food processing plant example.

In addition to the items in the 1981 A.G.A. study, benefit-cost analyses of cogeneration projects may also have to account for expenses associated with stand-by power from the local electric utility, as well as costs of equipment required to interconnect with the local electricity network.

The methodology employed in that report offers a guide to evaluations of cogeneration options against conventional energy systems. Where measurable air quality offsets can be gained through the application of select gas use in conjunction with cogeneration projects, select use can be added to the equation by reducing the capital expenses associated with the gas-fired cogeneration options by the dollar value of the savings provided by select use.

Some special considerations must be made in cases involving select gas use. In particular, the costs of implementing select use must be weighed against the costs of alternative air pollution emissions control strategies, if any such alternatives exist. In order to make such an assessment, the potential cogenerator must know the rates to be paid for gas as well as the long-term outlook for gas availability and price.

In addition, potential users should be aware of the concomitant advantages of burning gas, specifically,

- Ease of fuel handling (especially important for urban cogenerators)
- Operational flexibility
- Increased reliability through use of gas as a back-up fuel.

Securing Regulatory Approvals

Regulatory approval may be necessary in order to implement select gas use with cogeneration. If the cogenerator is located in a nonattainment area, any increase in air pollution emissions will have to be offset by reductions in emissions from another source in a manner accepted by state regulatory authorities.

State procedures for obtaining approval for such a trading of emissions between sources—even if they are part of the same facility—vary substantially from state to state. Such emissions trading will be relatively simple in states that have implemented a generic bubble rule for the pollutant involved. Appendix E indicates that status of state emissions trading programs.

Where a generic bubble rule has not been implemented, the procedures for emissions trading are likely to be more time consuming, although many states have moved to streamline the review process. Further information on your state's requirements can be obtained from state air pollution control officers, as listed in the directory provided in Appendix D.

Cogenerators subject to Prevention of Significant Deterioration (PSD) standards must obtain a permit from state air pollution control authorities in order to implement a cogeneration project. Although once again each state has its own procedures, in many cases the combination firing of gas and some other fuel can be used as an emissions control procedure in order to meet the strict PSD standards. Approval of PSD permits typically requires two or more months.

Implementing the Project

Implementing select gas use with cogeneration will entail a coordinated effort to respond to any technical or economic problems related to gas supply that could delay or impede the start-up of the project. Agreement must be reached with the user on terms of a gas purchase contract which is suitable to the requirements of the cogeneration facility.

In some cases, construction of new gas lines or gas metering facilities will be required. Potential new customers should be informed by gas companies as to the lead-times and planning requirements for such construction in order to coordinate the construction of gas facilities and cogeneration units.

It should be emphasized that "building-in" equipment to burn natural gas is likely to be substantially more economical than retrofitting such equipment at some later date. Given the possibility that additional federal and/or state actions will be taken to control acid rain, the flexibility to control emissions through select use strategy may be particularly attractive. Moreover, the cogeneration unit equipped for select gas use may have additional advantages associated with dual-fuel capability.

Chapter 12

Overcoming Obstacles to Select Use: A Question and Answer Review

Even though select gas use offers a logical and cost-effective solution to a variety of air quality compliance problems, potential users will often be surprised to find that some law, policy or perception seems to stand in the way.

The following "question-and-answer" review deals with the kinds of issues that are most likely to be raised about select gas use:

Fuel Use Act

Question

Isn't the use of natural gas as a boiler fuel forbidden by federal law; e.g., doesn't the Fuel Use Act prevent gas use in electricity generating plants where it hasn't been used before?

Answer

Not necessarily. In most cases, federal law does not now restrict select use of gas at all.

The Powerplant and Industrial Fuel Use Act of 1978 placed restrictions on some uses of oil and natural gas by electric utilities or industries to fuel some large boilers and other combustors. This Act is sometimes referred to as PIFUA or FUA. The purpose of FUA is to restrict the use of oil and natural gas in large:

- Electric utility powerplants, which include electric utility boilers, gas turbines, and combined cycle units which produce electricity for sale and cogeneration units selling over half their output to electric utilities.
- Industrial facilities (referred to in the law as "major fuel burning installations," or MFBIs), including industrial cogeneration units that sell less than half their output. The industrial facility (MFBI) restrictions are currently applied only to boilers, although they could be extended by rule to new gas turbines, combined cycle units and internal combustion engines as well.

Congress repealed some of FUA's key restrictions in 1981. Regulatory changes since then have limited or softened its few remaining restrictions, although it may be necessary for the user to apply to the Economic Regulatory Administration in the U.S. Department of Energy for an exemption or file a certificate with that agency in order to take advantage of those changes.

The key to understanding FUA is that it treats old and new facilities very differently from each other. In fact, FUA does not apply at all in the following situations:

- FUA *does not prohibit* any amount of gas use in electric powerplants existing or under construction before November 1978, unless the government has issued a plant-specific order under FUA (or its predecessor statute) prohibiting gas use at that plant.
- FUA *does not prohibit* any amount of gas use in industrial facilities (i.e., at major fuel burning installations) installed or under construction before November 1978.

Even as to new electric powerplants or industrial installations:

- FUA *does not apply* at all to boilers rated below 100 MMBtu per hour input.

Note, however, that while the small facilities rated below 100 MMBtu per hour are exempt from FUA, they may frequently be exempt from Clean Air Act regulations as well.

Even if FUA does apply, the rules and exemptions allow substantial leeway. For example, a new electric powerplant boiler subject to FUA may use up to 25% gas with no advance approval required. In the case of a non-powerplant industrial boiler, the owner may be allowed to raise gas burning up to 43% by merely filing a certification of a qualifying fuel mixture (such as gas and coal). In other cases, the owner may have to follow more cumbersome DOE procedures in order to try to secure gas exemption.

Gas Supply Concerns

Question

Is there really enough natural gas around for select use? Specifically, does my local utility have sufficient supplies to enable select gas use at my facility?

Answer

Yes. U.S. gas supplies are more than ample for select gas use, and for any other new markets the gas industry foresees at this point. The current gas surplus of some 2–3 trillion cubic feet is about 10–15 percent over present sales and looks like it will last into the 1990s, especially with the arrival of reasonably priced Canadian gas. Furthermore, discoveries of new gas supplies in the U.S. lower 48 States has generally kept pace with gas sales during the 1980s. The market is now setting new gas prices and determining the amount of gas available. The ultimate gas resource base may well exceed the coal resource base.

Appendix A gives further details.

Question

Is there enough gas for select use and other uses? Won't select use simply deprive other uses of the limited supply, so that no net clean-up of the air will ever happen?

Answer

There's plenty of gas to go around.

Gas pipelines have unused capacity to deliver clean-burning gas as a result of conservation and of the permanent reduction in "smokestack" industries. Gas producers have unused capability to deliver gas to pipelines; they could be drilling more new wells and producing more gas.

The gas industry could produce, transport and consume much

more gas, even if gas markets grow back to 1981 levels. Successful marketing of select use will increase total gas consumption, improve utilization rates, and stimulate more drilling for gas that would otherwise stay in the ground for want of a market. From a national perspective, other uses, such as residential uses, demand only a fraction of the gas that producers can extract and the pipeline network can transport. To forego select use opportunities would leave gas unused in the ground.

Question

If EPA relies on select use of gas and a dirty fuel at the same time, what will we do next time we have a gas shortage?

Answer

We do not expect a repeat of sustained gas curtailments, because of changes in the basic rules governing U.S. natural gas supplies—namely, field price decontrol of new gas, as allowed by the Natural Gas Policy Act of 1978. A vastly improved gas supply situation has resulted from this gradual decontrol. New gas prices in the field now represent bargained-for market prices, not government set prices.

Environmental Laws, Regulations and Policies

Question

Does EPA allow us to switch to gas use on days of intense air pollution only?

Answer

EPA does not prohibit any use of gas. Fuel users may use gas on any day they like. Some gas suppliers offer special rates for such "air quality episode" use. Moreover, air quality regulators may require or approve plans for fuel switching. For example, the Los Angeles Basin of Southern California has a standby plan in place to switch to gas in case of emergencies involving public health.

Although States generally may, and often do, provide for switching to gas (or other clean fuels) on days when ambient air becomes polluted, the issue remains whether readiness to use gas during periods of dirty air gets full credit from EPA as meeting clean air standards. It does not.

For example, in a nonattainment area, a State cannot satisfy federal ambient air quality standards by a legally-binding commitment to switch to a clean fuel if and when the air becomes too dirty. However, even in that situation, switching back and forth on other grounds may satisfy federal standards. (In the case of State standards which exceed federal standards, the Clean Air Act disqualification of such switching does not even apply.)

The issue arises only in "modelling" projections and estimates of what will be necessary to meet federal ambient standards. States "model" the anticipated performance of each major source plus the aggregate of minor sources in order to predict ambient air quality conditions. For that purpose, planning to switch on the basis of pollution conditions in the atmosphere may not count.

A State must submit "modelled" projections to justify EPA approval of its SIP and revisions to the SIP. Also, a State must perform extensive modelling in order to change designation of a region from nonattainment to attainment status.

Nonetheless, it is interesting that switching in fact may avoid a monitored violation of Federal ambient air standards in an attainment area (thereby helping preserve that status), even though planning to switch does not count to prove that the State is doing enough to reduce pollution.

Finally, a State could get full credit from EPA for plans that tie switching to gas to operating factors, such as total fuel consumption (e.g., the more fuel is burned, the higher the percentage of gas)—rather than to ambient air conditions.

Question

Since the Clean Air Act prohibits the use of "intermittent" emissions control technologies, doesn't it also prohibit the select use of natural gas where gas is purchased under an interruptible contract?

Answer

Actually, the Clean Air Act does not prohibit individual sources from using "intermittent" emission controls. Indeed, approved SIPs provide for "intermittent" control technologies. (The "intermittent" control issue under the Clean Air Act directly concerns area-wide planning credit for certain controls, as discussed in the previous answer.) One may expect, however, that regulatory authorities will demand an emission-control solution that will work all the time.

It is true that select gas users buying gas on interruptible contracts must be able to demonstrate that they can meet emissions standards even if gas deliveries are interrupted. Across the country, however, many select gas users have, as part of their bubble approvals, simple and economical ways of complying with emissions standards during gas delivery interruptions (which typically occur in winter).

A select gas user in Connecticut, for example, burns high sulfur fuel oil in one of its boilers and natural gas in another under an EPA-approved "bubble." When gas deliveries are interrupted each winter, the plant fires both boilers with low-sulfur No. 2 fuel—stocked as "ignition fuel"—for the short period until its stock of high-sulfur No. 6 fuel oil can be replaced with low-sulfur residual oil.

Emission standards vary as to time period. Some apply hour-by-hour, others to much longer periods. Some allow one or two exceptions a year. Moreover, emissions trading arrangements (in which the trading source cuts into emissions below the minimum) may have their own unique periodic constraints.

Advantages and Disadvantages of Select Use

Question

Some industrial users say they have "taken a look" at select gas use, but don't think that it is economically viable. What's the answer to this?

Answer

Have these users made a realistic assessment of all the benefits of select gas use? In many cases, "back-of-the-envelope" assessments omit many of the economic advantages of select gas use. Have potential users attached any value to the economic and operational flexibility that being able to burn natural gas affords them?

If they are burning coal, have they considered the reduced costs of operating emissions control equipment, the avoided "down-time" for maintenance of scrubbers or baghouses, or the ease of boiler start-up made possible by select gas use?

Have they considered possible interruptions of coal deliveries? Have they considered the possibility that acid rain legislation will be passed by the Congress that will force them to reduce emissions beyond present levels?

Question

If EPA and local air pollution control authorities are lax in enforcement of air pollution laws, won't that eliminate select use?

Answer

Inadequate enforcement of the Clean Air Act certainly could weaken the incentives for select gas use. EPA's budget for enforcement has declined by more than 30 percent since 1981, so the resources for enforcement at the federal level are already stretched pretty thinly.

On the other hand, the EPA and State regulators are managing the results of field inspections, self-reporting and the computerized database of air pollution sources to identify significant violators, track their compliance progress and assure that enforcement actions are taken against noncompliance. Select gas use may help to solve some of the enforcement problems faced by both violators and regulators.

Some regulators feel pressed to "go easy" on emission violations where they know that marginally profitable industries in their region cannot afford the substantial capital investments necessary to meet present standards. Select use may offer a solution for both regulators and emissions violators insofar as it represents an economical, low capital-cost emissions control technology.

For these reasons, it is important that local air pollution authorities be informed about the possibilities for select use. In nonattainment areas, select gas users may find that they can create *marketable* emissions offsets; local air pollution control authorities will be able to play a role in finding customers for such offsets.

Question

Our local air pollution regulations prohibit the use of high sulfur fuel oil. Doesn't this make select gas use with oil impossible?

Answer

No. In most select gas use applications involving high-sulfur fuel oil combined with natural gas, users obtained variances from fuel use regulations. They established, in ways that satisfied local air pollution control authorities, that their emissions through select use would be less than or equal to emissions from burning low sulfur fuel oil.

Section IV
State Regulator's Perspectives on Natural Gas Use for Environmental Purposes

Selected Authors

Chapter 13

Select Use of Gas: Regulator's Perspective

RUTH K. KRETSCHMER, *Commissioner*
and
PAUL A. BALLONOFF, Ph.D., *Senior Economist*
Illinois Commerce Commission

Select use of gas covers certain forms of mixed use of natural gas in conjunction with other fuels. It arises largely from the combined effects of environmental law, the economics and physics of natural gas combustion, and the existing availability of natural gas for uses beyond those given high priority by law and regulation.

The viewpoint of a regulator must be an anomaly. Regulators must be skeptics, by experience if not by statute and nature. Regulators can, of course, advocate, but what they can affect is limited by law. What they may influence is limited only by individual interests and philosophies, but may be constrained by the interests they implicitly or explicitly represent.

Thus, to a state public utility regulator with a particular philosophy, the world looks very different than it must to a federal energy regulator, an environmental regulator at any governmental level, or a member of whatever legislative body.

The opinions and views expressed in this paper are those of the authors and do not necessarily represent the views of the Illinois Commerce Commission.

Regulatory officials do not, of course, make law; they can only act within the body of law established by the Congress or the state legislature. Apart from the environmental laws that may affect energy use, there are two Federal laws that are of interest to regulatory philosophy as applied to select use of gas.

These are the Fuel Use Act of 1978 and the Natural Gas Policy Act of 1978. The apparent legislative intent in these acts was to shape end use of particular fuels according to certain noneconomic concepts of what the ideal mix of fuels should be, while at the same time attempting to shape the economic reality to conform to that preconceived ideal. The specific sections of the Fuel Use Act of interest are those which limit use of natural gas in new boilers particularly for electric generation, while the interesting sections of the Natural Gas Policy Act relate to incremental pricing of natural gas to particular end users.

Both acts have failed in both areas. The acts may have helped reduce the use of natural gas, but on a wrong premise. Natural gas is not, and except for bad law and perhaps poor federal regulation, never was, a resource in absolute scarcity, unavailable at any price. Thus the real effect of these types of energy legislation has been to distort the economic decisions of all users away from whatever economic efficiency would otherwise dictate.

Instead, it creates the "need" to have such a strange concept as "select use" of natural gas, which intent is to sell a potentially efficient solution to an energy use problem, in spite of law which does its best to make economic solutions impossible.

Thus in an area where at best a state regulator has only influence but not direct authority, it is clear that influence should be brought to the repeal of the Fuel Use Act, and certainly the elimination of those sections of the NGPA which relate to incremental pricing of natural gas to certain end users. The best solution to the economic problem of deciding which fuel to use for a particular purpose is to allow open competition among those fuels and among providers of those fuels.

This pro-market philosophy unfortunately does not currently fully extend to environmental regulation. Thus, until there is a politically acceptable and economically feasible way to cause those who "create pollution" to also carry the cost of that pollution in the price of their product, some form of direct environmental regulation is probably necessary. At present that regulation affects the amount of sulfur emissions from combustion of fuels. As Illinois knows all too well, use of a large portion of U.S. coal reserves is thwarted by its high level of sulfur which becomes airborne waste when burned.

For good reason, current and probable future law limit the amount of emission and, explicitly or otherwise, require either the installation of capital intensive cleaning devices or decreased use of the polluting fuel, or both. Future acid rain legislation will probably reinforce these trends.

Thus, if select use contains a way to allow use of high sulfur fuels while at the same time provide economically efficient production of energy, we should all cheer. It would certainly aid the economy of states like Illinois which otherwise may have difficulty in marketing one of their prime resources.

More generally, it could aid the economy of the nation by reducing the dependence on foreign fuels. If foreign sellers see that we are unwilling to use our domestic energy resources for fear of environmental damage, their prices will be correspondingly higher. We can not rely on the goodwill of any nation to protect the interests of our own. The best way to develop our own energy resources is, therefore, to allow the greatest development of domestic supplies in the context of economically efficient decisions.

In a free market, with the least legal constraint, the natural advantages of location and abundance would produce the desirable result of U.S. dependence primarily on domestic energy resources. For international producers to enter our markets would then require that they offer still lower prices for their own products.

Thus, in the energy field, there is a clear correlation between the maximization of competition and free markets and the national goal of energy self-sufficiency. If select use can allow use of environmentally undesirable resources in an economic way that does not cause environmental harm, then by all means select use must be encouraged. If select use also helps the balance of payments, so much the better.

Not all of the means of encouraging this result lie within the realm of state regulatory influence or authority. For example, if the particular boiler being considered for conversion to select use does not belong to a state regulated utility, then the state public utility authority has no direct control at all over that use. Our strongest influence would only be to express opinions to the responsible agency, such as to the State or Federal E.P.A. on pollution questions. End use customers, from the view of the state utility regulator, have the full legal and economic right to make their own decisions as they see fit.

However, the state regulator may be able to affect the structure of the tariffs from state regulated distribution utilities. This does not mean that a state regulatory commission could or would on its own initiative create special discount tariffs for so-called select use customers. In the first place, just as public policy can create undue restrictions on economic decisions, it can also create distortions of incentive.

If select use is really as economically desirable technically as is claimed, then it will require no unjustified "special breaks" to implement. Second, regulators are restricted by the record of a contested rate proceeding to the actions they may take on creating or changing tariffs.

Thus, a good place for regulated utilities and their customers to obtain the needed tariff terms to make select use possible is with the state regulated tariffs and tariff proceedings. If the local utility does not have tariff terms suitable to the cost characteristics of delivering gas for select use (or any use) then the proper procedure is to obtain such tariffs through the regulatory procedure.

The customer may start by direct discussion with the local utility, followed by a regulatory submission; direct application by the end users to the Commission; intervention by the end user in a rate case of the involved distributor; or some combination of all of these.

On the other hand, if the select user is a state regulated utility, then the state utility authority has a direct interest in the decision to implement. Depending on the size of installation or retrofit required, the interest of the state authority may arise from any of several sources.

If the installation is big enough or new enough, it may provoke a proceeding under the siting authority of the Commission. If the installation is costly enough to be considered an "investment" as opposed to an expense, then it will affect the rate base of the utility in a rate proceeding. And even if the installation "only" affects expenses on a current basis, it still may affect the computation of expenses in the rate case cost of service or the allowance of expenses through an energy cost adjustment clause.

Very likely both of these matters will arise in any rate case in which rate base due to boiler conversion is involved. Finally, any such decision involving a large change in utility operations may be subject to a prudence decision of the regulatory authority.

None of this ought to be particularly unexpected by the utility in question. The utility likely deals with such concerns on a regular basis, generally quite successfully; however, it does mean that the utility should pay particular attention to the details of its regulatory framework at the same time as it plans its new energy installation.

If, for example, the energy cost adjustment clause of the utility contains a restriction to a particular mix of fuels as a "base mix" and has limits which exclude fuel mixes that deviate from that base mix, then clearly the time to consider revision of the fuel clause is at the time that the change in fuel mix is first considered. States will differ on how they may treat this situation. Some may exercise the universal safety valve—don't act until necessary.

Nonetheless, an early involvement or at least notification of the state staff may prevent or conceivably solve cost recovery problems before they occur. As a general rule, state commissions are happy to support anything that saves money for ratepayers. The problem in dealing with a state staff over a select use installation is not an adversary one but is in the realm of avoiding pitfalls while reaching a common goal.

Ways do exist, however, to convert a common interest into regulatory adversity. Two prominent examples are the notorious regulatory "successes": the Alaska Natural Gas Transportation System (acronym ANGTS, but pronounced "ANGST" for good reason) and imported liquefied natural gas or LNG.

The ANGTS project was to construct a project to bring Alaskan gas to the "lower 48" states. Following passage in 1974 of the Alaska Natural Gas Transportation Act, a complex series of negotiations, political discussions, and studies led to a decision by 1980 to construct a project. As late as 1981, private companies, such as Exxon, Arco, and Soho, which had previously been excluded were allowed to buy into the project.

Over $500 million in private (at risk) funds were spent on engineering studies alone. The annual Federal cost for oversight ran as much as $20 million. Yet by the end of 1982 the project was essentially suspended, if not abandoned. Already completed portions, such as Northern Border pipeline, went into rates but at less than full volumes.

For example, in March 1984 the portion chargeable to Panhandle Eastern Pipeline volumes through Northern Border was $3.18/MCF for transport alone, the gas cost even more. In their March, 1985

PGA, the Panhandle share of Northern Border was down "only" to $1.70/MCF for transported cost, still a substantial amount.[1]

The history of LNG projects in the same period suffered a similar fate. Pipelines including Southern Natural and Trunkline Gas Co. (actually, their affiliate, Trunkline LNG) constructed significant capital projects to import LNG. Yet by the time the projects were put into service, in 1981 through 1983, economic conditions were very different from those which had been presumed would occur. Neither project is presently processing LNG.[2]

To a large extent, both of the LNG and ANGTS projects resulted from use of faulty and overly optimistic (or, as it were, pessimistic) forecasts. If select use is indeed economically efficient, it should not have to depend on sales pitch numbers or outdated forecasts. Each installation should stand on its own merits, and certainly in any state siting or rate case review it would have to do so.

There remains a possibility, however, that some of the advocacy material for select use lead to the same pitfalls as happened to the ANGTS or LNG projects. For example, one of the more detailed studies of comparative economics of select use was that of the American Gas Association funded simulation study by a Florida academic institution,[3] comparing select use options to each other and to a base case without select use.

---

[1] Based on costs filed in Panhandle Eastern Pipeline Company purchased gas adjustment filings before the Federal Energy Regulatory Commission Dockets TA 84-1-28 and TA 85-1-28.

[2] Both mentioned LNG projects have extensive administrative records. The Trunkline LNG Company proceedings are generally part of F.E.R.C. consolidated dockets with lead number CP 74-138. The Southern Natural project is reviewed in F.E.R.C. Opinion No 83 at 10 FERC 61,287 (1980).

[3] See A.E.S. Green, Editor: An Alternative to Oil: Burning Coal with Gas. University Presses of Florida. 1981.

A study by the American Gas Association, Policy Evaluation and Analysis Group, (dated March 13, 1981 and published as an Appendix[1] to the *Proceedings* of the 1982 International Conference on Combined Combustion of Coal and Gas, by the A.G.A.) also appears to rely on 1980 vintage assumptions on energy price.

A more recently dated study (November, 1983) jointly by the G.R.I. and A.G.A. (Evaluation of the Environmental and Other Benefits of the Select Use of Natural Gas) was evidently based partly on prices of the January to March, 1983 period. This period however was one of the most volatile single quarters in recent energy pricing history and, therefore, probably not a good example of what a current cost evaluation of a particular current project might be. The remaining prices and price indices used by the study, however, are apparently based on 1981 reports and are even more dated. If recent history of fuel

The study was published in December, 1981 and embodied what was at that time an honest assessment of the horrors of the energy future: oil at $30 a barrel, coal at $40 a ton, and natural gas at $2.50 per MCF. With these assumptions, the study showed the cost comparison of oil generation at 7.2 cents per KWH to select use of coal/ natural gas at 3.4 cents, less than half the cost.

Yet today there are places in the U.S. where natural gas is delivered for industrial use at $4.00/MCF,[4] coal runs around $25 to $30/ton,[5] and oil can be reasonably bought at $26/barrel and near term forecasted at $22/barrel.[6] These are precisely the kinds of economic reversals which caused havoc for the ANGTS and LNG projects.

Select use decisions and the ANGTS or LNG projects have one major difference. Select use is not a single large project requiring a huge one time investment and corresponding risk. Or, at least, to date it has not been represented as such.

Instead, select use would result in many separate installations, each of which must stand on its own merits. Unlike the mass scale projects, the particular institution or business fully responsible for the success of each installation is completely identifiable. It should be expected, therefore, that each such installation would be reviewed on the particular costs and benefits of that installation and will not depend, as did certain other projects, on academic and potentially misleading forecasts.

---

prices demonstrates anything, it should demonstrate that simple (or even technically fancy) trending of fuel prices is a very unreliable index of future pricing. Site-specific studies are what is required.

The same 1982 A.G.A. conference volume noted above also contains a number of realistic assessments (such as the article by Richard J. Drummond, "Combination Fuel Burning at the James River Plant" in Springfield, MO) of the relative benefits of combined fuel use which do not depend on macroeconomic assumptions or simulations. Thus, in spite of skepticism on the price projects, there seems good ground for believing real benefits might exist.

[4] The March, 1985 issue of the Department of Energy's *Natural Gas Monthly* quotes average price of industrial sales by pipelines at $4.19/MCF and of gas purchased by electric plants at $3.77/MCF, for January, 1985 (table 10, page 23). Prices were so volatile that the $2.50 estimate used in the study was actually out of date before the study was published.

[5] *Energy Users News,* Vol. 10 No. 24 of Monday, June 17, 1985 on page 17 reports January, 1984 coal prices from as low as $22.42/ton to as high as $47.80/ton depending on location and coal quality.

[6] As of this writing "oil price" ranges from $24 to $28/bbl depending on which newspaper report one picks. Prices in the entire range can be documented daily. Price decreases or reports of "expected" decreases are common.

One must, therefore, understand that regulators react cautiously to a new development which is presented with great enthusiasm based on forecasts. When asked for specific decisions and authorizations, the regulatory body must require and receive much harder, better, numbers in the siting application than have been provided in the past.

Finally, note that the support for select use is not limited to the kind of mixed boiler fuel technology that has currently come under that term. Good economic policy should encourage economically efficient technology, particularly one that does not aggravate environmental problems.

For example, fuel cells seem to be a very promising advance which may be environmentally cleaner and economically more efficient than select use boiler mixtures. If present indications hold up, fuel cells may also have a far more widespread application since their location of use would not depend on boiler applications nor have as stringent zoning restrictions.

The Gas Research Institute, funded by a surcharge on FERC jurisdictional gas sales, is actively engaged in research on many such technologies. In residential and commercial application these include efficient natural gas absorption air conditioning and improved efficiency furnaces and cooking devices. In industrial applications the combined forces of legal incentives (through the Public Utility Regulatory Policy Act cogeneration sections), GRI research and plain old competition, are leading to increased efficiencies in gas-driven applications as well as expanded markets for natural gas.

These applications, particularly air conditioning, not only benefit the gas transport industry, they also benefit energy efficiency in general. To a large extent, electric plant construction has been designed to accommodate peak load growth due in large measure to summer air conditioning loads. Further, as a new electric plant is constructed, older and less efficient units often become peaking or cycling/peaking generators.

Thus, expanded use of natural gas for summer air conditioning has a triple benefit: gas pipelines may be fuller in the summer, when their present loads are lowest, thus helping reduce unit costs on the pipeline without incurring additional capital expense; the need for future electric plant construction to meet peak loads is reduced; and use of the least fuel-efficient electric plant on peak is also reduced.

One does not need to advocate any particular technology to recognize that expanded natural gas markets can lead to lowered average unit costs in interstate and local utility transport and distribution margins, and thus may be a net benefit to customers generally. Select use, from the utility regulator's view, is but one segment of a much larger market. If it can solve an environmental problem while also benefiting, or at least not making more difficult, the job of utility regulation, then it is certainly worth encouraging.

We should encourage such good things as may be done now through select use to increase economic efficiency. But we should also be careful that changes made in current laws do not apply to only one energy technology. It would be a temporary dubious benefit to replace existing and uneconomic legal restrictions that inhibit select use with other restrictions that encourage select use now but prohibit other technologies later.

To the extent consistent with environmental quality, regulatory policy must encourage the fullest development of open and competitive markets in all areas of energy use.

Chapter 14

Use of Natural Gas as an Air Pollution Control Measure in California

ALAN GOODLEY, *Chief*
Strategy Assessment Branch
Stationary Source Division
California Air Resources Board

As most people are aware, the Los Angeles Metropolitan area, known as the South Coast Air Basin, has the most severe air pollution problem in the nation. Other areas of California also have severe air pollution problems. For the last twenty years in California, the most effective way in which air pollution has been reduced from stationary sources has been the use of natural gas.

All of the power plants in California, with one minor exception, are equipped to burn natural gas. With one exception, Scattergood Unit 3 of the Los Angeles Department of Water and Power, all of these power plants are also equipped to burn fuel oil. With minor exceptions, natural gas is used exclusively in industry, commercial and governmental establishments and in private residences. Unlike many other parts of the nation, heating oil is rarely used in private residences in California.

One of the ways in which the use of natural gas has been encouraged in California by air pollution control districts has been to limit the allowable sulfur content of liquid fuels to 0.5 percent. In the Los Angeles area, the sulfur content of liquid fuels burned by electrical utilities and oil refineries has been limited to 0.25 percent. The result of these limitations on the sulfur content of fuel oil has been to raise the price of the fuel oil to a level somewhat above that of natural gas. As a result, there is generally an economic incentive to burn natural gas.

Exhibit 14-1 shows a comparison of the emissions from a 1000 megawatt power plant at full load when it is burning natural gas and when it is burning 0.5 percent sulfur fuel oil. This table shows that when burning fuel oil, emissions of particulate matter are about 30 times those when natural gas is burned, emissions of oxides of nitrogen are double, and emissions of sulfur dioxide are about 500 times greater than when natural gas is burned. Therefore, it is highly desirable to encourage the burning of natural gas as a way of reducing emissions from stationary sources.

Exhibit 14-1

Comparison of Emissions of Particulate Matter, Sulfur Dioxide and Oxides of Nitrogen from a 1000-Megawatt Power Plant When Operating on Fuel Oil and When Operating on Natural Gas

(Pounds per Hour)

Fuel	*Particulate Matter*	*Oxides of Nitrogen*	*Sulfur Dioxide*
Fuel Oil (0.5% Sulfur)	867	3000	5080
Natural Gas	29	1500	10

Most of the air pollution control districts in California have regulations which require large industries and commercial establishments to submit plans for actions to reduce emissions during air pollution episodes. Among the elements which must be contained in the plans is a plan for the burning of natural gas. In addition, Rule 715 of the South Coast Air Quality Management District requires the burning of natural gas by electrical utilities during air pollution episodes to the extent that natural gas is available.

Generally speaking, air pollution control districts in California avoid directly requiring the burning of natural gas. However, air pollution control districts frequently incorporate into permits emission limitations that effectively require the use of natural gas. For example, permit restrictions on the Los Angeles Department of Water and Power's Scattergood Unit 3 are such that that unit can only burn natural gas. Similarly, permit restrictions on cogeneration units to be installed in the South Coast Air Basin are such that they could only be met if those units burn natural gas.

During the middle and late 1970's, natural gas became less available in California and power plants and other industries were forced to burn substantial quantities of fuel oil. However, since 1979, natural gas has become more available and power plants are now burning natural gas almost exclusively. Exhibit 14-2 shows the trend in emissions of sulfur dioxide from 1978 to the present in the South Coast Air Basin and Ventura County. Note that the SO_2 emissions have declined from 450 tons per day in 1976 to 150 tons per day in 1984.

Natural gas is also helpful in meeting the air quality standard for ozone. In most areas of the nation that are not attainment for ozone, the control of hydrocarbons is the strategy for attainment of the ozone standard. For sources emitting low concentrations of hydrocarbons, natural gas is frequently used to incinerate the hydrocarbons. An example would be a natural gas assisted flare at a refinery or chemical plant.

Furthermore, modeling studies of air quality for Kern County have recently been conducted and show that control of emissions of oxides of nitrogen as well as hydrocarbon emission control may be required to reduce ambient concentrations of ozone in order to achieve the ozone standard in that county.

Kern County is the site of the world's greatest concentration of steam generators used in thermally enhanced oil recovery. At the present time, most of these steam generators burn either residual oil or produced crude oil. There are proposals by natural gas suppliers to replace the oil that is burned with natural gas. If further NO_X reduc-

Exhibit 14-2

Sulfur Dioxide Trends in the South Coast Air Basin and Ventura County

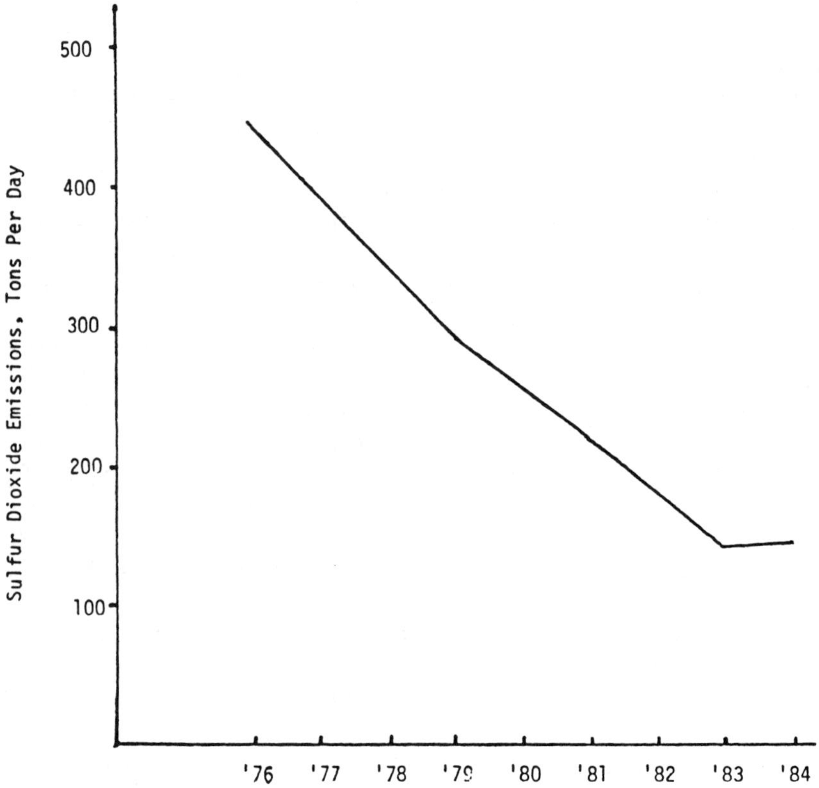

Source: ARB/TSD/SSD

tions are required from oil field steam generators, replacing oil-burning with natural-gas-burning would be helpful in attaining the ozone standard in that county.

Another possible use of natural gas to improve air quality is the production of methanol. Nearly all methanol produced in the world is made from natural gas. In California, the state has a large-scale methanol demonstration program on state-owned vehicles. About 500 state-owned vehicles are equipped to burn methanol fuel.

The analysis of emissions from those vehicles show that they are about 40 percent less photochemically reactive than are emissions from gasoline-powered vehicles. If widely used, this would result in a reduction of ambient concentrations of ozone. The price of methanol at the present time makes it competitive with the use of premium unleaded gasoline.

In addition, several trials are underway to determine the use of methanol in heavy duty diesel engines. The trials are being made in two-engine configurations. In one case, a glow plug is used to start the engine and the engine has controls to maintain the air-fuel ratio within specified limits. In the other case, the diesel engine is modified and equipped with a spark plug. These engines are being used in buses that are operating in normal service.

Emissions of oxides of nitrogen and particulate matter are greatly reduced from these engines. The end objective of the methanol demonstration program is to create a market for methanol produced from coal. Until methanol from coal plants are built, however, natural gas would be the source of the methanol. The major obstacle in the path of methanol-fueled vehicles is economic. However, the emissions reduction potential is attractive.

In summary, natural gas has been for many years the most important strategy for controlling emissions from stationary sources in California. Continued supplies of natural gas will be available at least to the end of the century.

Section V

"Select Use" Case Studies

Selected Authors

Chapter 15

Bellefield Boiler Plant—Simultaneous Coal and Natural Gas Firing for Pollution Control

EDWARD PERMAR, P.E.,
Manager of Design
University of Pittsburgh

This case history report shows how a major district heating boiler plant has complied with current environmental regulations by the simultaneous firing of coal and natural gas. This has been accomplished without abandoning existing equipment.

In 1900 Andrew Carnegie donated a sum of money to expand the Carnegie Library. This new facility contained an interior volume of 14 million cubic feet with a total floor space of over 15 acres, which was quite large for its day.

One of the largest institutional power plants ever constructed at that time provided steam to this new structure to generate light, heat, and power. Electricity for light and power continued to be generated until 1957.

In 1933, Carnegie Library, Carnegie Institute, the University of Pittsburgh, and The Mellon Institute of Industrial Research, agreed to share ownership in the plant for mutual benefit.

Current owners of the plant are as follows:

Owners	*Percent Interest*	*Capacity Pounds Per Hour*
Carnegie Library	3.150	18,427
Carnegie Institute	7.350	43,000
Carnegie Mellon University	19.768	115,641
Children's Hospital	8.115	47,474
City of Pittsburgh	2.195	13,390
Eye and Ear Hospital	2.097	12,265
Presbyterian University Hospital	5.756	33,670
School District of Pittsburgh	1.865	10,907
University of Pittsburgh	51.676	315,226
	100.000	610,000

The plant consists of 6 boilers with an installed capacity of 610,000 pounds per hour, and a firm capcity of 460,000 pounds per hour. The general configuration of the plant and the type of installed equipment is shown in Exhibit 15-1.

The Bellefield Boiler Plant was first subject to environmental regulation in 1960 when the Bureau of Smoke Prevention of the City of Pittsburgh and the Bureau of Smoke Control of Allegheny County were combined under the Allegheny County Department of Health and governed by the "Smoke Control Ordinance of Allegheny County." In 1960 the plant was in compliance with this regulation.

First Consideration of New County Ordinance on Air Pollution

On January 1, 1970, the Allegheny County Health Department released "Rules and Regulations, Air Pollution Control, Article XVII." This second ordinance regulated particulates and sulfur dioxide. If two or more existing fuel burning units were connected to a single flue, their total capacity rating was used to compute the amount of particulate matter that may be emitted.

Sulfur dioxide was limited to an emission of not more than 2.5 pounds per million Btu of the input, and particulates were limited to 0.26 pounds per million Btu. On March 3, 1970, the plant manager filed for a variance from the regulation to allow time for the

Exhibit 15-1
Bellefield Boiler Plant Components

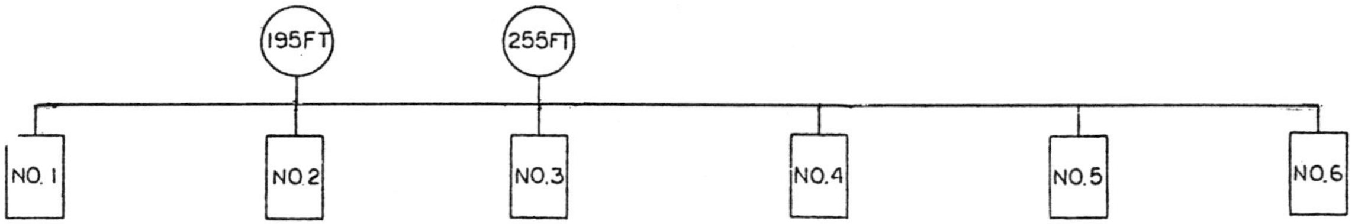

Babcock and Wilcox Co. Stirling Boiler Water Tube Two Drum Installed 1957	Springfield Boiler Cross Drum Water Tube Sectional Header Installed 1946	Zurn Industries Two Drum Water Tube Installed 1977	Springfield Boiler Cross Drum Water Tube Sectional Header Installed 1946	Erie City Iron Works Two Drum Water Tube Installed 1965	Zurn Industries Packaged Boiler Water Tube Scotch Marine Installed 1972
Coal Babcock and Wilcox Co. Chain Grate 100,000 PPH	*Coal* Westinghouse Nine Retort Underfeed Stoker 85,000 PPH	*Coal* Laclede Chain Grate Stoker 100,000 PPH	*Coal* Westinghouse Nine Retort Underfeed Stoker 75,000 PPH	*Coal* Laclede Chain Grate Stoker 100,000 PPH	*No Coal Capability*
Stock Co. Conical Nonsegregating Distributor No Pollution Controls	Stock Co. Conical Nonsegregating Distributor No Pollution Controls	Stock Co. Conical Nonsegregating Distributor Cyclone Dust Collector	Stock Co. Conical Nonsegregating Distributor No Pollution Controls	Stock Co. Conical Nonsegregating Distributor Cyclone Dust Collector	
Gas 57,000 PPH Ring Burner Erie City	*No Gas Capability*	*Gas* 100,000 PPH Erie City	*No Gas Capability*	*Gas* 57,000 PPH Ring Burner Erie City	*Gas* 150,000 PPH Erie City
No Oil Capability	*No Oil Capability*	*Oil* 100,000 PPH	*No Oil Capability*	*No Oil Capability*	*Oil* 150,000 PPH

installation of a gas fired boiler. The Boiler Plant expected some changes in equipment or facilities to meet the new regulations. A sixth boiler was needed to meet new steam demand and natural gas was considered as a fuel to allow the total plant to meet the new regulations.

At that time the Bellefield Boiler Plant was burning coal with the following analysis:

	As Received	*Dry Basis*
Moisture	5.94	—0—
Volatile matter	33.32%	35.42%
Fixed carbon	49.76%	52.91%
Ash	10.98%	11.67%
Sulfur	1.64%	1.74%
Btu per pound	12,391	13,174

On December 4, 1970, the Peter F. Loftus Corporation, Consulting and Design Engineers, delivered a feasibility and cost study for installing gas burners in boilers #1 and #5. The study indicated that since gas has no ash or sulfur content, the installation of gas burners in these two steam generating units would provide substantial dilution of pollutants in the flue gas.

Boilers #1 and #5 were chain grate stoker fired units, had water wall furnaces, and were entirely suitable for either coal firing, for gas firing, or for simultaneous coal and gas firing. Boilers #2, #3, and #4 were brick set units with multiple retort inclined grate stokers and were not readily adaptable to natural gas firing.

The outcome was that one Coen Company intratube gas burner with a capacity of 100,000 pounds per hour was installed in boilers #1 and #5. Unfortunately, in 1979 a past plant manager replaced the 100,000 PPH per boiler and thereby reduced the plant's ability to comply with pollution regulations at maximum demand. The consultant also suggested combination low sulfur and high sulfur coal firing, and low sulfur coal and gas firing.

Multiple fuel burning strategies evolved as shown below:

	Source ↑		*Source ↑*			
Boiler Fuel	1	2	3	4	5	6
Gas	X		X		X	X
Oil			X			X
Coal	X	X	X	X	X	

In 1972 low sulfur coal was relatively expensive compared to gas. At that time gas was burned in combination with 1.6% sulfur coal to meet pollution regulations. Combustion control was arranged so that either fuel could be selected for base loading and load swings would be taken by the other fuel, or either fuel could be burned alone. As the price of gas increased, low sulfur coal became a more desirable fuel.

Current Air Pollution Regulations

On January 1, 1981, the Allegheny County Health Department issued new "Rules and Regulations for Air Pollution Control, Article 20." The new regulation requires that emissions be determined based on actual heat input to equipment rather than design heat input. Low sulfur coal could be burned and would meet SO_2 regulations with all coal firing, but combination firing with natural gas was necessary for dilution to meet particulate regulations.

The Bellefield Boiler Plant is required by EPA to provide the best available control technology for pollution control. Current compliance with the Allegheny County regulations means that the plant also complies with EPA regulations and the Pennsylvania Department of Environmental Resource Regulations. At current capacity the plant can use the following control technology to meet EPA regulations for sulfur dioxide:

1) Burn compliance coal.
2) Simultaneous firing of natural gas and coal.
3) Install a scrubber.
4) Convert to fluidized bed boilers.

Due to physical constraints in the size of the site, management has decided to burn compliance coal in combination with natural gas. Particulates now pose the most serious problem for the plant. Plant capacity cannot be expanded unless a baghouse is installed.

The current coal burned is double screened and washed, 0.8% or less total sulfur, 6% or less ash, and has a heating value of 14,000 Btu per pound. All coal is delivered 3/4 x 3/8 with minimal fines to avoid operating problems with the chain grate stokers and multiple retort underfeed stockers.

Note in Exhibit 15-1 that the boiler plant has no unusual pollution control devices other than multitube cyclone dust collectors on boilers #3 and #5.

Exhibit 15-2

§402. PARTICULATE MASS EMISSIONS

A. **Fuel-Burning or Combustion Equipment.** No person shall operate, or allow to be operated, any fuel-burning or combustion equipment in such manner that emissions of particulate matter exceed the following rates at any time:

1. Where the actual heat input to such equipment is greater than 0.50 million BTUs per hour but less than 50 million BTUs per hour, the rate of 0.40 pounds per million BTUs of actual heat input;

2. Where the actual heat input to such equipment is equal to or greater than 50 million BTUs per hour, but less than 850 million BTUs per hour, the rate determined by the formula:

   $A = 3.5E^{-0.56}$ where A = allowable emissions in pounds per million BTU of actual heat input, and
   E = actual heat input to such equipment; or,

3. Where the actual heat input to such equipment is equal to or greater than 850 million BTUs per hour, the rate of 0.080 pounds per million BTUs of actual heat input.

§403. SULFUR OXIDE EMISSIONS

A. **Fuel-Burning or Combustion Equipment.** No person shall operate, or allow to be operated, any fuel-burning or combustion equipment in such manner that emissions of sulfur oxides, expressed as sulfur dioxide, exceed the following rates at any time:

1. Where the actual heat input to such equipment is greater than 0.50 million BTUs per hour but less than 50 million BTUs per hour, the rate of 1.0 pound per million BTU of actual heat input;

2. Where the actual heat input to such equipment is equal to or greater than 50 million BTUs per hour, but less than 2000 million BTUs per hour, the rate determined by the formula:

   $A = 1.7E^{-0.14}$ where A = allowable emissions in pounds per million BTU of actual heat input, and,

   E = actual heat input in millions of BTUs per hour;

3. Where the actual heat input to such equipment is equal to or greater than 2000 million BTUs per hour, the rate of 0.60 pounds per million BTU actual heat input; or,

4. Notwithstanding the provisions of Subsection B.1 through B.3 above, where the rated heat input to existing fuel-burning or combustion equipment is equal to or greater than 5000 million BTUs per hour, the rate of 2.90 pounds per million BTU of actual heat input.

The Bellefield Boiler Plant submits fuel quantities burned and fuel analyses to the Allegheny County Health Department on a monthly basis to show compliance with the regulations. The County Health Department inspects the plant and makes flue gas measurements on approximately an annual basis. An extract of the applicable regulations for sulfur and particulates is shown in Exhibit 15-2.

Plant Operating Data

The International District Heating and Cooling Association's statistical survey of 1983 listed 22 institutional boiler plants. Of these 22 plants the Bellefield Boiler Plant was the third largest in maximum hourly send-out capability and eighth largest in total steam delivered. Of all 51 commercial and institutional steam plants reporting, including Con Edison of New York, the Bellefield Boiler Plant was the twelfth largest in maximum hourly send out capability and sixteenth largest in total steam delivered.

At the present time the Bellefield Boiler Plant has the following fuel costs:

Fuel	*Cost per Fuel Unit*	*Cost per Million Btu*
Coal	\$61 per ton	\$2.18
Natural Gas	\$5.84 per MCF	\$5.84

During fiscal year 1983, the system members paid \$5.4439 per 1000 lb of steam produced including fuel, labor, maintenance, and capital costs. The judicious use of natural gas has allowed the plant to comply with pollution regulations while still using coal as the primary fuel. This fuel burning strategy produced steam in the most economical manner and did not require pollution control equipment be added to the plant. Simultaneous firing of gas and coal also allows the plant to easily accommodate rapid load swings such as those which occur during morning startup of buildings.

The Plant has burned the following fuel quantities during the last three years.

	Natural Gas			*Coal*		
	MCF	*Btu* $\times$ *10*$^{-6}$	%	*Tons*	*Btu* $\times$ *10*$^{-6}$	%
7/84 to 6/85	289,238	289,238	21.4	37,898	1,061,100	78.6
7/83 to 6/84	227,196	227,196	16.8	40,120	1,123,400	83.2
7/82 to 6/83	229,654	229,654	18.5	36,028	1,008,800	81.5

Conclusion

The nine independent owners of the Bellefield Boiler Plant have cooperated effectively in studying pollution regulations, analyzing alternative means of complying with the regulations, and implementing an economical solution. The plant has retained existing boilers and met the regulations without the addition of pollution control equipment. Pollution control and the economical generation of steam was accomplished by the simultaneous burning of natural gas and coal.

Chapter 16

Narragansett Electric Company—Natural Gas Use in a Bubble

ANDREW AITKEN, *Director of Environmental Affairs*
New England Power Service Company

New England Power Service Company is a subsidiary of the New England Electric System, an electric utility holding company located in Westborough, Massachusetts. Other electric company subsidiaries include New England Power Company, Massachusetts Electric Company, Granite State Electric Company and the Narragansett Electric Company.

The Narragansett Electric Company serves 280,000 retail customers in Rhode Island. It also owns two electric generating stations in Providence, Rhode Island, which operate under the first electric utility air emissions bubble. The bubble allows stack emissions from both plants to be counted as a single source. The result has been to save money for utility customers, and to reduce total emissions.

The two generating plants, the South Street Station and the Manchester Street Station, are located on the Providence River which runs North to South, just upstream of a hurricane flood barrier operated by the Army Corps of Engineers. The plants are 1,200 feet apart, which is important in the bubble concept, and are separated by a complex of retail shops and an office building.

The South Street Station was built in the early 1900's and has undergone many changes since. The most recent major change occurred in 1952 with the installation of a 100 MW unit having a single

325 foot stack. Today that is the only generating capacity in service at South Street—the other stacks are attached to boilers no longer in use.

The Manchester Street Station also was built in the early 1900's, originally to service a direct current electric street railway. It now has three generating units of 50 MW each, the most recent having been installed in 1949. It uses two stacks, each 200 feet high.

Both stations burned coal until the mid-sixties when they were converted to burn residual oil because of fuel costs and environmental concerns. The first oil used had a sulfur content in excess of 2.5%. In 1971, Rhode Island sulfur-in-fuel regulations limited the sulfur content of oil to 1% by weight. Both stations use 1% sulfur oil today except when operating under the bubble.

In 1968, the company first contracted to buy natural gas from the Providence Gas Company for use at the Manchester Street Station. The necessary gas lines and burner modifications were then installed. The supply was interruptible; it was limited to the summer months and its cost was below that of oil. Manchester Street Station was selected for burning gas because it had three generating units and could be more flexible than South Street in adjusting to differing gas supply conditions. Plant modifications were never made to burn gas at South Street.

In the years 1968 through 1971, the average annual gas use was 1.76 million MCF and comprised 21% of the heat input to Manchester Street Station. Gas was not available for purchase in five of the seven years from 1972 through 1978. From 1979 to the present, gas has been purchased every summer. In 1983 and 1984, gas purchases extended into November and, for the first time, annual heat input from gas exceeded that of oil at Manchester Street Station.

The Bubble

The increased use of gas in the past four years has, in part, been due to the implementation of an emissions bubble at the two plants. The bubble concept was developed as a result of 1977 Clean Air Act Amendments. It is one of several EPA policies which come under the heading of controlled trading and are designed to permit sources to place a greater burden of control where marginal costs of control are low and to reduce control requirements where the cost is high (see Exhibit 16-1).

Exhibit 16-1

Bubble Concept

- EPA CONTROLLED TRADING POLICY
- GREATER CONTROL WHEN COST IS LOW
- LESS CONTROL WHEN COST IS HIGH
- NEARBY SOURCES
- DIFFERENT METHODS OF CALCULATION
- AMBIENT AIR QUALITY STANDARDS TO BE MET

Emissions from several nearby sources are combined and regulated as a single source. This allows sources with higher emissions of sulfur dioxide (SO_2), for example, that might exceed regulatory limits to be combined with sources of lower emissions so that the total is within the limits. A bubble might be on a weight basis, such as tons per hour or an emission rate basis, such as lbs/MMBtu heat input. A bubble can only be used where it can be demonstrated that ambient air quality standards in the area will be met.

Narragansett Electric could see in the bubble concept an opportunity for fuel savings beyond those realized by burning gas at Manchester Street Station. With a bubble, a higher sulfur, lower cost fuel oil could be used at the South Street Station. Although the two stations are not truly contiguous, they are close enough to fit the definition of a bubble.

We, therefore, proposed to burn 2.2% sulfur oil at South Street Station while burning gas at Manchester Street Station. The proposal also provided that 2.2% sulfur oil could be burned at South Street Station when Manchester Street Station was not operating which made it a bubble of tons, rather than emission rate. Exhibit 16-2 shows that when using gas with no emissions at Manchester Street Station and 2.2% sulfur oil at South Street Station, the total emissions are less than if both stations were using 1% sulfur oil. This assumes that both stations are operating at full output.

Exhibit 16-2
Bubble Concept, Full Load Operation

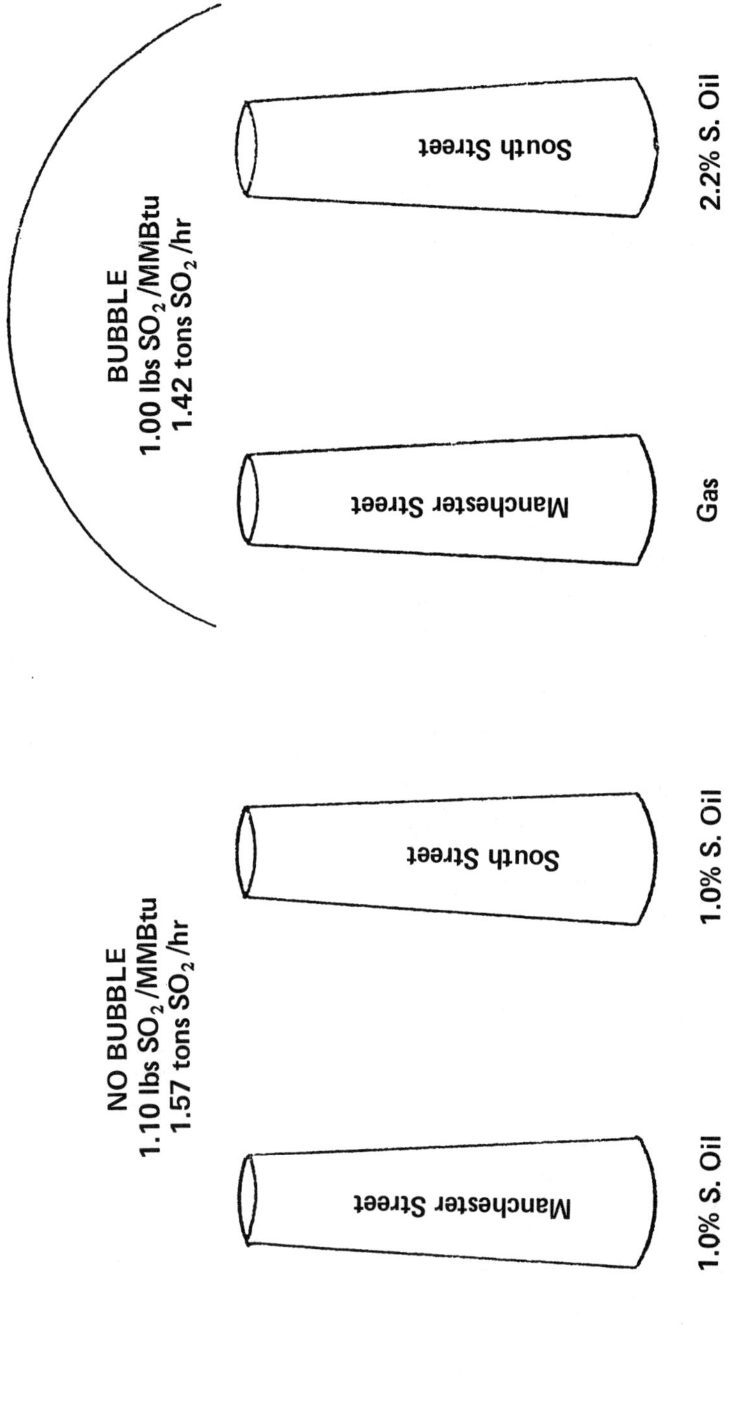

In 1979, we made application, including technical supporting material, to the Rhode Island Department of Environmental Management (DEM) for a variance to the state's sulfur-in-fuel regulations for the bubble. Although Federal law provided for a bubble, there were not yet any corresponding state regulations, so a variance was sought with provision for renewal on an annual basis.

In 1980, the variance was granted. This was approved by EPA as a permanent revision to the Rhode Island State Implementation Plan (SIP) in January 1981. The state variance was renewed in 1982. In 1983, a SIP revision changed the renewal frequency from one to three years.

The use of gas and the bubble have been both an economic and environmental success. Gas burning at Manchester Street Station has displaced the use of foreign oil for generating electricity. Exhibit 16-3 illustrates the barrels of oil displaced by gas on an equivalent Btu basis for the years 1981-1984. An average of 423,000 bbls/year were displaced and the rate of displacement has been increasing.

The customers of Narragansett Electric and other NEES companies have benefitted from the use of lower cost fuel. Exhibit 16-4 shows the dollar savings by year from 1981-1984. The savings are made up in two ways:

- The cost differential between 1% sulfur oil and gas used at Manchester Street Station.
- The cost differential between 1% sulfur oil and 2.2% sulfur oil used at South Street Station.

The total savings since the beginning of the bubble have been over $9 million which have been passed directly to our customers through the fuel portion of their electric bill.

While the emphasis of the bubble has been on dollar savings, it has also resulted in reduced SO_2 emissions to the air. Exhibit 16-5 shows the actual annual SO_2 emissions in tons from the bubble compared to the expected emissions had both plants been using 1% sulfur oil at an equivalent operating rate.

Emissions are on average 700 tons/year less for this example with the bubble. Actual emission reduction is dependent on day-by-day analysis of fuel costs and generating unit dispatch during the summer months.

Exhibit 16-3

Foreign Oil Displaced by the Bubble

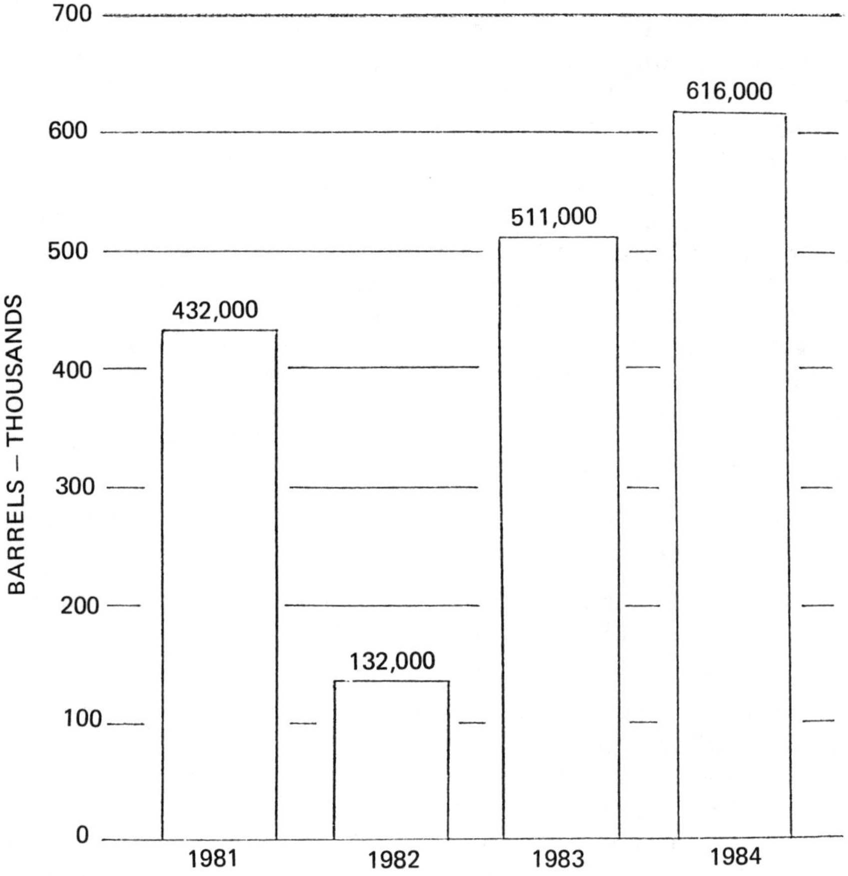

Exhibit 16-4

Savings from Bubble

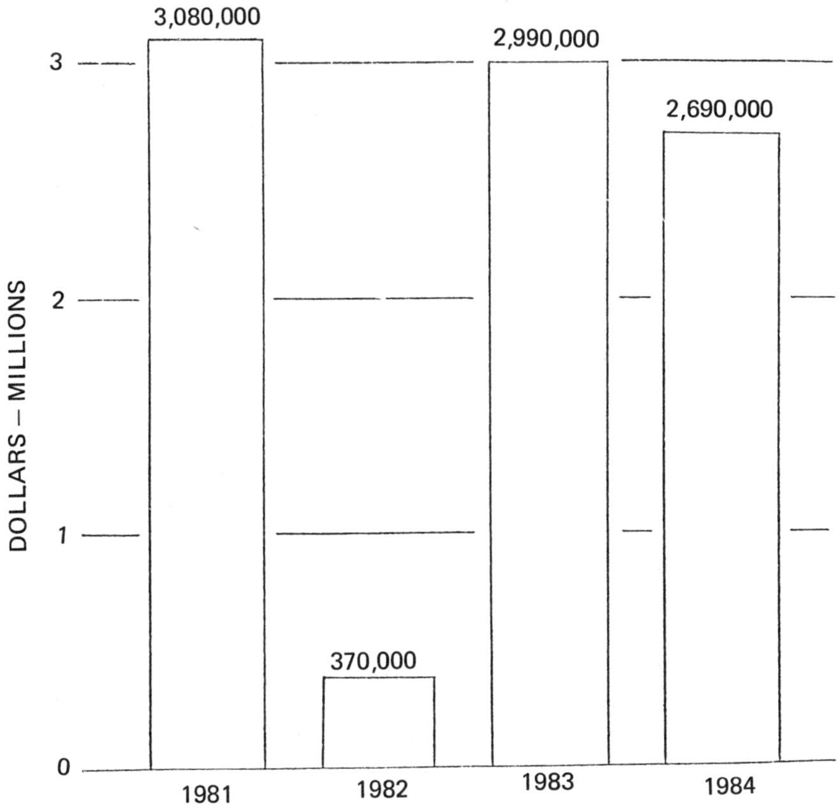

Exhibit 16-5

Sulfur Dioxide Emission Reduction

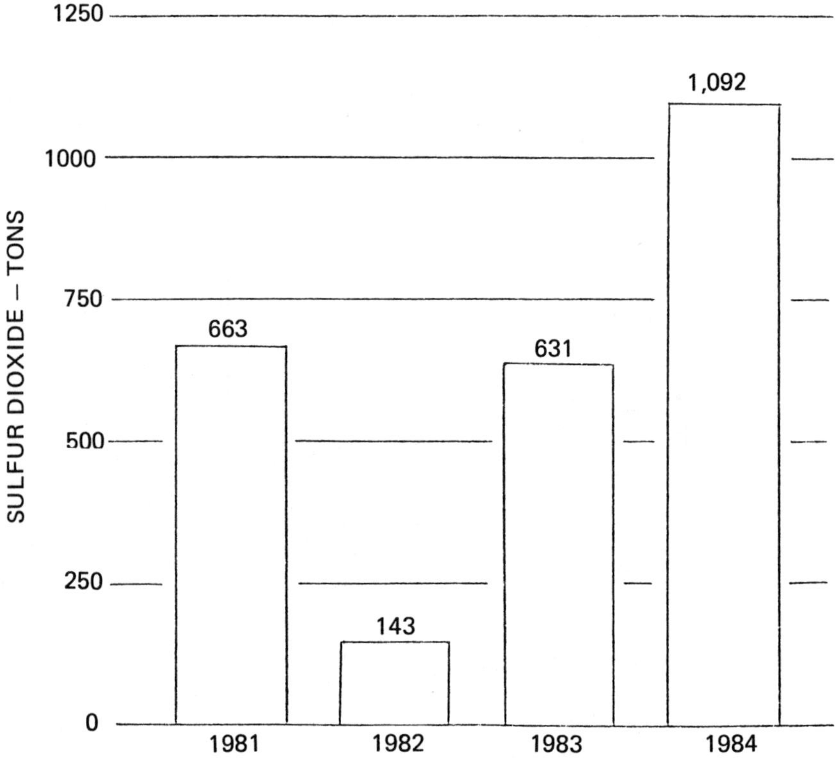

Future Use of Gas

We are now considering the conversion of the South Street Station to natural gas. Preliminary engineering is now underway for the required plant modifications. If converted, we would use interruptible gas, available only for the summer months. Of course, its implementation depends on successful negotiations with Providence Gas on the contract price of gas.

Gas must be priced less than oil in order to offset the cost of conversion and to increase the likelihood that the plant will be called on by the New England Power Pool. New England has excess capacity at this time and South Street Station competes with other 2.2% sulfur oil-fired units. This makes for some tough negotiations considering the presently depressed price of oil.

The Rhode Island DEM has reacted very favorably to the proposed South Street Station gas conversion. A review of the tentative conversion plans resulted in a determination that no approvals or permits would be required, something which rarely occurs in the utility industry. If preliminary engineering work confirms scoping estimates of conversion costs, we are optimistic that South Street Station gas conversion can be completed in time for the 1986 warm weather season.

Assuming that gas remains price competitive and that adequate supply is assured, there are some steps that could be taken to improve the likelihood of future gas conversions. Exhibit 16-6 shows some of them.

Exhibit 16-6

Steps That Will Favor Gas Conversions

- FREEDOM OF CHOICE EMISSION REDUCTION LEGISLATION
- NO NSPS REQUIREMENTS FOR GAS CONVERSIONS
- FINANCING INCENTIVES:
  - CWIP IN RATE BASE
  - POLLUTION CONTROL BONDS

If Clean Air Act amendments are passed which require national SO_2 emission reductions, the natural gas industry should seek to protect its place in the marketplace. The low sulfur coal industry, through its Alliance for Clean Energy, has correctly perceived this issue and opposes any technology-forcing legislation. For example, if legislation is passed which mandates the use of scrubbers for SO_2 emission reduction, natural gas and low sulfur coal could be legislated out of a large part of the marketplace.

Conversions to gas result in quantifiable reduction in emissions of SO_2 and particulates. Other emissions such as NO_X may remain the same. The natural gas industry should work to see that existing sources are not ratcheted into new source performance standards for NO_X when they convert to gas.

Finally, regulatory and financing mechanisms available to other projects should apply to gas conversions. For example, construction work in progress (CWIP) has been allowed in the rate base for utilities converting to coal. This has reduced the overall cost of coal conversion. The same could be done for gas conversions when they are undertaken for environmental reasons. Another example would be the use of tax free pollution control financing for gas conversions if they are done to reduce emissions.

Our experience with the bubble at the Providence plants illustrates that utility projects go forward smoothly only when three interested sectors are benefitted.

These are:

- The customers
- The environmental community
- The company and its stockholders

Natural gas has the advantage over other fuels because it is clean which has direct appeal to environmental interests. If it can be cost competitive also, gas will be in a uniquely favorable position when compared to other costly emission reduction proposals, and should find increasing acceptance as a utility boiler fuel.

Chapter 17

Consolidated Edison— The Role of Select Gas Use in Improving New York City's Air Quality

PETER C. FREUDENTHAL,
Director of Environmental Affairs, Air and Land Use

GORDON A. BEALS,
Manager of Environmental Affairs

DOMINICK J. MORMILE,
Manager of Environmental Affairs

JOHN S. WILSON,
Meteorologist in Environmental Affairs
Consolidated Edison Company of New York, Inc.

Prior to the implementation of air pollution laws beginning in the late 1960's, New York City, like other large urban areas, had poor air quality. Even a few years after the 1970 Clean Air Act Amendments were implemented, the City did not consistently achieve the annual National Ambient Air Quality Standard (NAAQS) for sulfur dioxide (SO_2). It was only after Con Edison implemented a plan for select gas use that the standard was achieved with a margin of safety. Most important, unlike most environmental improvement programs that cost money, this select gas use plan saved electric customers many millions of dollars.

Fuel burning by low level sources is the major cause of air pollution in urban areas. Con Edison's power plants burn residual oil and natural gas, and release their effluents from stacks which are high enough above the ground to minimize the local effect of these emissions on ground-level pollutant concentrations. The emissions from these large sources rise above the stacks because of their buoyancy and are dispersed by the wind.

On the other hand, space heating sources in New York City, which burn the same fuels as Con Edison, release their emissions at or relatively close to the ground, where dispersion is comparatively ineffective. These low level sources are the primary contributors to ambient pollutant concentrations including SO_2, nitrogen oxides (NO_X) and particulate matter.

History of New York City Air Quality

Up to the 1960's, coal was the primary fuel used for power generation and space heating in New York City. Since the combustion of coal produces large amounts of SO_2, annual sulfur dioxide levels in the City had risen above 550 ug/m^3 during the mid-1960's, a level almost seven times higher than the present day ambient air quality standard.

To reduce local sulfur dioxide emissions, the City implemented a rollback strategy in the late 1960's whereby the sulfur content of fuel oil and coal was restricted to 1 percent. With the passage of the Clean Air Act of 1970, an annual sulfur dioxide air quality standard was established at 80 ug/m^3. To meet this federal goal, the City further restricted the maximum sulfur content of fuel oil to 0.3 percent and, with minor exceptions, banned the use of coal.

As a result of these restrictions placed on fuel sulfur content, annual sulfur dioxide levels in New York City improved dramatically, and by 1976, the 80 ug/m^3 air quality standard for SO_2 was marginally achieved. The trend is shown in Exhibit 17-1 which presents the maximum annual SO_2 levels measured in New York City for the years 1960-1979. However, fuel emergencies in 1977 and 1979 forced the temporary use of nonconforming oil in numerous schools, hospitals and dwellings, and contraventions of the standard were again reported in these years in Manhattan.

The highest SO_2 levels in New York City are found in northern and southern Manhattan. The monitor locations and average SO_2

Exhibit 17-1

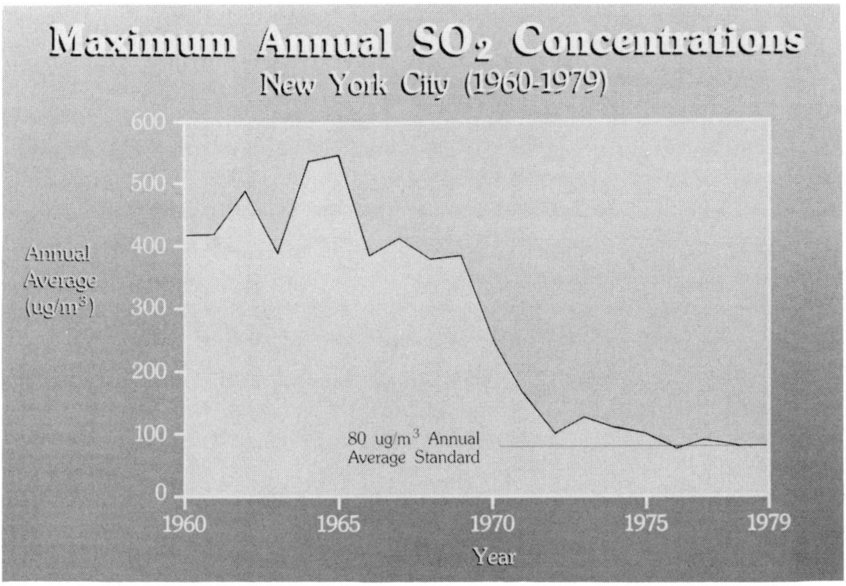

concentrations for the 12-month period ending 1978, prior to the beginning of Con Edison's gas conversions, are shown in Exhibit 17-2.

The Con Edison Affirmative Air Pollution Abatement Program

To alleviate its high electric rates, which had risen by 37% since the 1973–1974 Arab oil embargo, Con Edison proposed, in 1979, to reconvert three of its largest in-city power plants to coal. Conversion of these three units, totaling nearly 1800 Mw, would raise the Company's in-city SO_2 emissions by 67,000 tons per year, which represented a 180% increase.

Although dispersion from the tall stacks would limit the air quality impact of the proposed coal conversion to less than 2.5 ug/ m^3 of SO_2 on an annual basis, Con Edison recognized that even this slight increase might be unacceptable given New York City's marginal compliance with the national ambient air quality standard. The Company therefore proposed to initiate an affirmative air pollution abate-

Exhibit 17-2

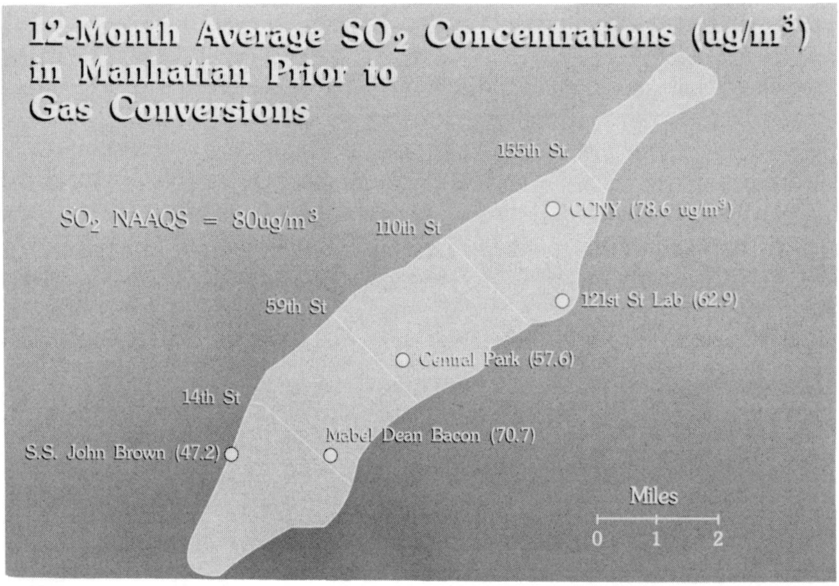

ment program to reduce SO_2 levels in the most polluted parts of the City and to conduct a test burn of oil with a sulfur content equivalent to the 1% sulfur coal being considered for the reconversions.

The affirmative air pollution abatement program consisted of identifying large oil or coal burning boilers in public institutions in the most polluted part of the City, and converting them to natural gas. Fourteen boilers at eleven facilities, each burning between 0.5 to 6 million gallons of oil per year, were identified. After receiving concurrence from the owners of these facilities (the City Housing Authority, two universities, and a hospital), Con Edison obtained Public Service Commission approval to convert these facilities to natural gas and provided technical assistance to expedite the conversions.

Exhibit 17-3 shows the locations of the facilities converted to natural gas under the affirmative air pollution abatement program. Together they emitted about 460 tons of SO_2 per year, about 0.55% of the City's total SO_2 emissions. Although this was a small fraction of the proposed increase in Con Edison's SO_2 emissions, the air quality benefit in the local area was significant.

Exhibit 17-3

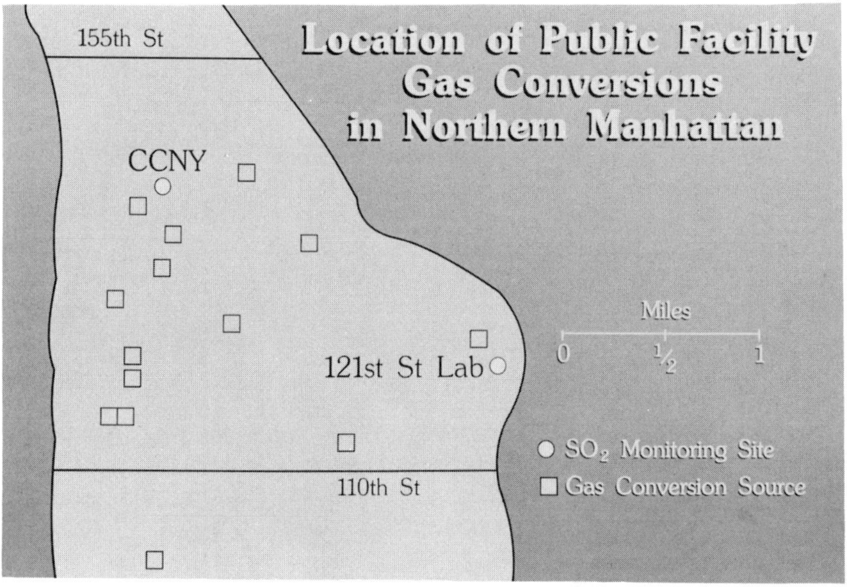

Using a computerized dispersion model that combines meteorological and stack emissions data to calculate ambient pollution concentrations, Con Edison prepared a map showing isopleths of the predicted SO_2 reductions which would result from the gas conversions in northern Manhattan (Exhibit 17-4). This map was used by the environmental regulatory authorities in deciding to permit the Company to conduct a test burn of higher sulfur oil that would produce SO_2 emissions equivalent to the proposed 1% sulfur coals.

About the same time as the Company's affirmative air pollution abatement program was proposed, the Public Service Commission, recognizing the increasing availability of natural gas, rescinded its order prohibiting Con Edison from connecting new gas customers. Subsequently, the Company converted about 1750 low-level buildings from oil-burning to natural gas over roughly a three-year period (February 1979–March 1982).

Most of these buildings were four to ten stories in height and were located in Manhattan. By March 1982, 5.6 BCF of new gas sales had been achieved, displacing the equivalent of about 45 million gallons

Exhibit 17-4

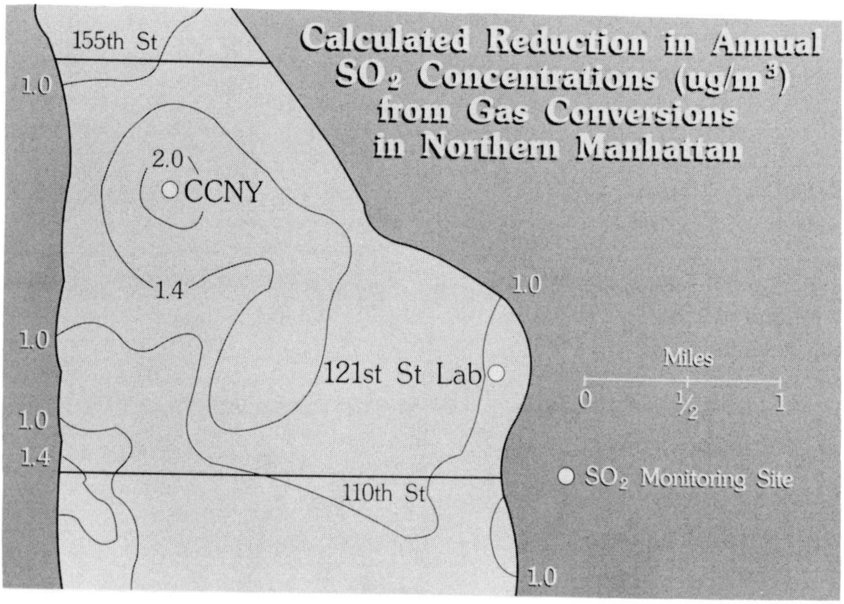

of fuel oil and reducing low level SO_2 emissions by an additional 1010 tons per year, bringing the total SO_2 reduction from all the gas conversions to 1470 tons per year.

In August 1980 Con Edison also began the test burn of higher sulfur oil at the three facilities where use of coal had been proposed. Over a 32-month period lasting through March 1983 these generating units burned fuel with sulfur content up to 1.5%. The average increase in the SO_2 emission rate was 12,000 tons per year, nearly eight times the SO_2 reduction by the low level sources.

Air Quality Impact of the Manhattan Gas Conversions

The decrease in low-level SO_2 emissions in Manhattan which resulted from the gas conversions had a pronounced effect on ground level ambient SO_2 concentrations. The annual SO_2 average at the Mabel Dean Bacon High School monitoring station (MDB), which in 1979 reported the highest SO_2 concentration in the City and marginally exceeded the standard, dropped from 81.2 ug/m^3 in 1979 to 65.0 ug/m^3 in 1982 (a decrease of 16.2 ug/m^3) (Exhibit 17-5).

At CCNY, which in 1979 reported the second highest pollution concentration in the City, the annual SO_2 average fell from 76.8 ug/m^3 in 1979 to 59.5 ug/m^3 in 1982 (a decrease of 17.3 ug/m^3). Further reductions in the annual SO_2 averages at these sites occurred between 1983 and 1984 as additional low-level buildings converted from residual oil burning to natural gas.

Environmentalists opposed to Con Edison's use of higher sulfur fuel questioned whether a 1470 ton per year SO_2 reduction from low-level sources could have been the cause of the substantial improvement in air quality, especially in light of the average 12,000 ton per year SO_2 increase in power plant emissions during the test burn. They argued that the improvement might have been temporary, and resulted from mild winters, energy conservation and the downturn in New York City's economy.

We investigated each of these factors and found that they did not explain the air quality improvement. To evaluate whether mild win-

Exhibit 17-5

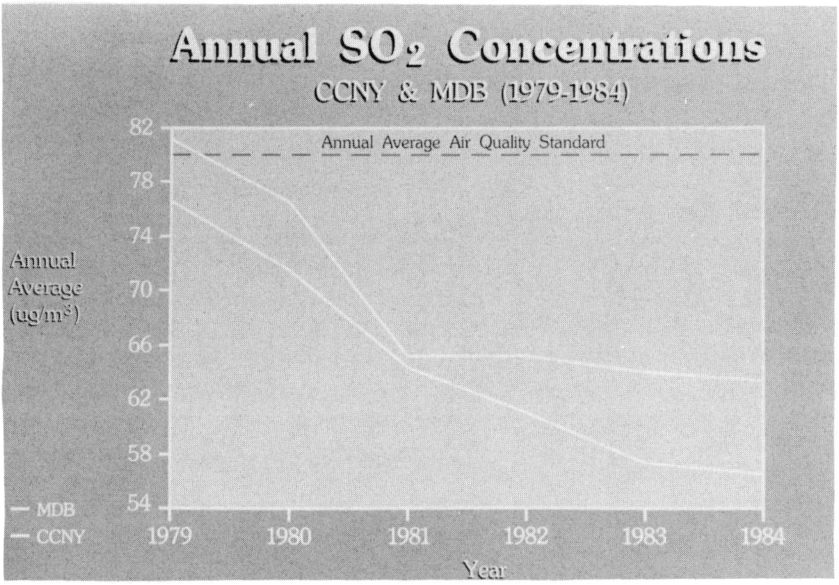

ters could account for the air quality improvement, we analyzed heating degree days, which are directly proportional to oil consumption for space heating. Assuming equivalent meteorological conditions, average ambient SO_2 levels should be nearly proportional to degree days.

Exhibit 17-6 shows both the heating degree days calculated for Central Park and CCNY annual SO_2 averages for the years 1979 through 1984. There were 12% more degree days in 1980 compared with 1979. Hence, one would have expected SO_2 levels to rise by a similar percent in 1980.

In contrast to this expectation, annual average SO_2 levels declined by 7%. The 6% decline in heating degree days between 1980 and 1981 was accompanied by a 10% decline in annual SO_2 concentration, but in the subsequent three years, in spite of nearly constant winter temperature conditions, SO_2 levels continued to decline. These data show that temperature variations, as expressed by heating degree days, do not account for the substantial improvement in annual average SO_2 levels in Manhattan.

Exhibit 17-6

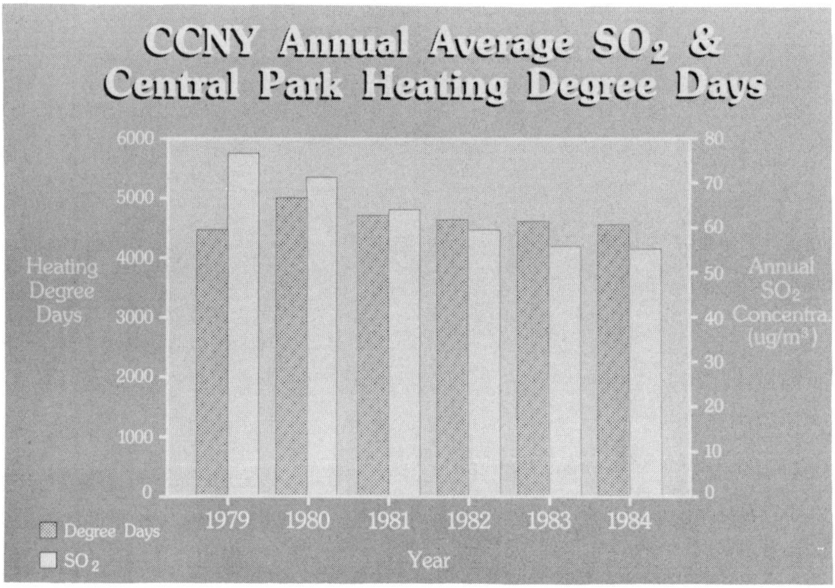

Likewise, the air quality improvement cannot be explained by a decline in economic activity. Although much of the country was in recession in the early part of this period, New York City did not experience this decline. Instead there has been about 1% per year growth in business activity since 1979.

Whereas before 1979 Con Edison could quantify effects of conservation because of reduced gas sales, gas use per customer has actually been increasing since 1979. The inability to detect conservation effects after 1979, the year when energy prices were highest, indicates that conservation effects on total energy consumption have not been significant. Assuming oil users demand for energy is proportional to that by Con Edison gas customers, conservation cannot account for the improvement in air quality.

Because neither mild winters, nor conservation, nor changes in economic activity can account for the air quality improvement, we conclude that it was the gas conversions that were responsible for the substantial measured improvement in air quality.

Applicability of Select Gas Use in Other Urban Areas

The select use of natural gas to replace oil in low level sources has resulted in New York City's firm compliance with the annual NAAQS for SO_2 with a wide margin of safety.

Throughout the period 1980–1984 there has been continued net air quality improvement even though somewhat more than half of that time there was a contemporaneous large increase of SO_2 emissions from the three facilities where coal conversion had been proposed. Environmentally the program was successful because, not only did Con Edison's new gas customers benefit from use of a less costly fuel, but electric customers saved $53 million during the test burn period when the three power plants were permitted to burn nonconforming oil.

Con Edison's select gas use program was based on offsetting air quality impacts rather than emissions. Its success depended on identifying SO_2 sources for emission reduction where the air quality impacts per unit of emissions would be greater in the area of concern than those from the power plants that were proposed for emission increases.

Select gas use based on offsetting air quality impacts by small emission reductions at low level sources to offset large emission increases at major sources may find equally successful application for preventing significant deterioration of air quality. The Clean Air Act sets forth narrow limits by which air quality in areas attaining ambient standards may be allowed to deteriorate, and consumption of available prevention of significant deterioration (PSD) increment is often an issue in the siting of new major facilities.

If a new major source impacts an area where existing low level sources can be converted to natural gas, there would exist a potential for offsetting PSD consumption. By consideration of stack heights and locations in planning these offsets, it may be possible, as in the New York City example, to achieve the equivalent ambient air quality offsets at far lower cost than could be achieved by offsets determined on an equivalent emission basis.

Chapter 18

Ontario Hydro—Using Natural Gas to Clean Up a Coal-Fired Station

D. K. A. GILLIES,
Manager, Environmental Protection
Ontario Hydro

One of the many uses of natural gas is in the upgrading of aged fossil stations who no longer can conform to environmental laws and regulations. Natural gas burns cleanly with no particulate and very little acid gas emissions—at least relative to average coal.

This chapter gives a brief account of the environmental upgrading of a 1200 MW coal-fired thermal plant located along the lakeshore a short distance (2 to 3 miles) from downtown Toronto. The station was designed in the late 1940s and came into service in the early to mid 1950s. The technology of the 1950s was no longer acceptable to the environmental consciousness of the late 1960s.

Some facts dealing with the size and environmental controls are contained in Exhibit 18-1.

Shortcomings of Design and Operation

The stacks of the 4x100 MW and 1x200 MW units were too short to give a plume rise above the downward flow over the station under some conditions. This poor diffusion resulted in large increases in downwind ground level concentrations of SO_2.

Exhibit 18-1

Environmental Design of Fossil Station—R. L. Hearn TGS

No. of Units	4 X 100 MW (1951-1953) 4 X 200 MW (1961-1962)
Fuel	2.2% S, 10% Ash—440 T/hr at full load
Stack Height	5 – 200-foot on 100 MW & 1 – 200 MW unit 3 – 300-foot on 200 MW units
Precipitator (ESP) Efficiency	95% on 100 MW – Design 98% 92% on 200 MW (upgraded to 99% 1967-1968)
Conversion—1971	4 X 100 MW units – all gas 4 X 200 MW units—coal and/or gas 49 BCF = CF 40% 16 BCF – November to April 33 BCF – May to October

The particulate removal was not sufficient to prevent a gray-brown plume from darkening the sky along the water front and over the city centre.

The diffusion of the plume was restricted by the cold body of Lake Ontario which generated ground-based inversions when warm water flowed northward over the cold lake water.

Increasing pressure from air pollution control officials and the general public resulted in a study to recommend improvement.

Pollution Problem and In-House Solutions

Operating at full load with 1.3% sulfur fuel, the maximum hourly concentrations of SO_2 occurring in downtown Toronto were measured around 1.30 ppm. This was over five times the hourly required air quality provided by regulation in 1969.

Hence reductions in ground level concentrations were required to reduce SO_2 readings and particulate emission reductions were needed to meet opacity requirements.

A study carried out by the Corporation cited a number of options for clean up.

Interim Solution	*Comments*
(a) Shut down Hearn Units 1 to 4.	Available immediately but at significant sacrifice to reliability of supply to meet demands until this capacity is replaced.
	Solves particulate problem and reduces significantly, but does eliminate sulfur dioxide emissions.
(b) Keep Hearn Units 1 to 4 in service only to prevent load cuts. Energy replaced by suitable purchases from interconnections and use of combustion turbines whenever available.	Available immediately at significant increase in operating cost. Significantly reduces particulate and sulfur oxide emissions. Requires low sulfur fuel use.
(c) Continue present operating practice of burning low sulfur fuel in all units and of minimizing operation of Hearn Units 1 to 4 by purchasing power at a small premium during onshore winds.	Available immediately. Small increase in operating costs. Reduces sulfur oxide emissions. Unacceptable particulates. Requires supply of low sulfur fuel.

	Long-Term Solutions	*Comments*
(d)	Equip Units 1 to 4 with precipitators of 99.5% efficiency.	Takes two years to complete.
		Will solve particulate problem.
	Burn low sulfur fuel (less than 1.0%) with on-shore winds, including natural gas.	Will reduce sulfur oxide emissions.
		Requires supply of low sulfur fuel (coal or natural gas).
(e)	Equip Units 1 to 4 with precipitators of 99.5% efficiency.	Takes two years to complete precipitators.
	Equip Units with SO_2 removal equipment.	Commercial-sized equipment to remove SO_2 from thermal plants is not proven. Several years would elapse before this method could be operational.
		Solves particulate and sulfur oxide emission problem.
		No low sulfur fuel required.
(f)	Build multi-flue high stack.	Available in less than two years.
		Solves particulate problem.
		Substantially reduces ground level concentrations of SO_2 in Metro Toronto.

The best interim solution is noted by Item (b) which provided substantial improvement in particulate and sulfur oxide emissions.

The best long-term solution is the construction of a multi-flue stack at least 500 feet high into which all eight units would discharge. Extra improvement to SO_2 and particulate emissions would result from burning cleaner fuels.

An analysis of fueling for the next few years recommended diversifying from US coal alone. Hence, residual oil and natural gas, and later Western Canadian coals, were contracted for use in the Corporation with natural gas to be used at R. L. Hearn TGS.

Consultant Study Results

In addition to the recommendations noted above, a consultant was retained. This consultant studied the pollution climate of Metropolitan Toronto and the estimated contribution from R. L. Hearn TGS and reported, in part, as follows:

1. The annual sulfur dioxide and dust levels in Metropolitan Toronto were at least twice as high as permitted.
2. Annual maximum sulfur dioxide effect from R. L. Hearn TGS needed to be reduced by 65%.
3. SO_2 dosages are always higher in winter due to space heating.
4. The new precipitators installed on Units 6, 7, and 8 in 1967 reduced fly ash dosage by at least 40%.

The consultant recommended:

1. A tall stack at least 500-feet tall should be built into which all flues should be combined. This would result in:
   (a) Annual SO_2 dosage reduced by 85%.
   (b) No concentrations of fly ash or SO_2 would exceed the maximum allowable limits.
   (c) Further improvements could be achieved by using stack gas heating to increase plume rise or by increasing stack height.

Decision

The final decision was for a 700-foot stack. At the same time, it was decided to diversify our fuel supply since there were some problems in meeting coal supply targets. It was decided to convert R. L. Hearn TGS to burn natural gas and to build a new residual fuel oil thermal station.

The taller stack and conversion to burn natural gas mixed with coal gave a double improvement in pollution abatement at R. L. Hearn TGS. There were no longer large ground level concentrations in downtown Toronto when the wind blew on-shore, there were no more gray-brown palls of fly ash across the horizon and from 1971 until the station became mothballed in 1983, there were no more complaints about pollution from R. L. Hearn TGS.

Availability of Natural Gas

The delivery of natural gas to the station required special work. No stations were burning natural gas and hence we had no design or operational experience in natural gas firing.

(a) *Time to Convert*

Engineering work on gas conversion began August 1970. Construction work was scheduled to start in May 1971, to be completed with all units discharging into the new stack in November 1971. Boilers were six front-fired (B&W) and two were CE corner-fired.

Ontario Hydro engineered the gas piping and equipment up to the burners, as well as auxiliary services such as air and electrical supplies.

(b) *Regulation*

The design, installation, and operation of all gas burning equipment in Ontario is under the Energy Branch of the Department of Labour. This was the largest, most complex, automated gas burning installation in Canada and required extra care.

(c) *Gas Supply*

Consumers' Gas constructed a 30-inch pipeline, from a pipeline station northeast of the city, southward down a transmission line right-of-way to the Consumers' Gas station 3 miles from the plant. From this station, a 24-inch line was constructed to the station where a metering and pressure reducing station was installed by Consumers' Gas. Gas was supplied at a nominal 250 psi and reduced to 60 psi to the generating station.

More details concerning the equipment and burner management system may be obtained from Reference 1.

Performance of Equipment in Converted Units

Satisfactory performance of all equipment was recorded with a normal number of commissioning bugs. It was reported that loss of generation due to control problems is less than that experienced with coal problems. Coal impellers on dual-fueled front-fired units were modified to take the gas gun through the centre. The impellers are still subject to burning or warping when coal conditions are poor.

Mixed firing of coal and gas, either together or alternatively, created problems with fly ash handling, as the ash now collected moisture produced from gas firing and plugged up hoppers and silos.

The conversion proceeded well and gave the station a new lease on life and led to flexible and improved station operation, especially general housekeeping and cleanliness.

Improvement Noted

The graph in Exhibit 18-2 shows the dramatic decline in ground level concentrations in downtown Toronto. This occurred for many reasons, two of which were the tall stack and conversion to burn natural gas. We did not know at the time that we received an extra bonus in the reduction of acid gas emissions which may be causing trouble farther afield.

References

1. W. K. Burton, W. Goodbrand, J. W. Lawson, "Design, Installation, Commissioning and Operation of R. L. Hearn Generating Station Converted to Natural Gas Firing, Thermal Power Section, CEA, Montreal, Quebec." Ontario Hydro Report, March 1974.

2. J. W. Lawson, P. Eng., "Recent Improvements in Air Pollution Abatement at the R. L. Hearn TGS," Ontario Hydro Internal Report, October 1972.

3. "Controlling Air Pollution in Metropolitan Toronto," Environment Conservation in Ontario, Department of Energy and Resources Management, 1971.

4. "Air Pollution Study of R. L. Hearn Generating Station for the Hydro-Electric Power Commission of Ontario," Stone & Webster, March 1968.

5. A. Phillips and J. R. Ryan, "R. L. Hearn TGS Air Quality Monitoring Program and 1979 Results, Report No. CTS-07144.3-2," November 1980.

Exhibit 18-2

Trends in Sulfur Dioxide Concentrations at 67 College Street

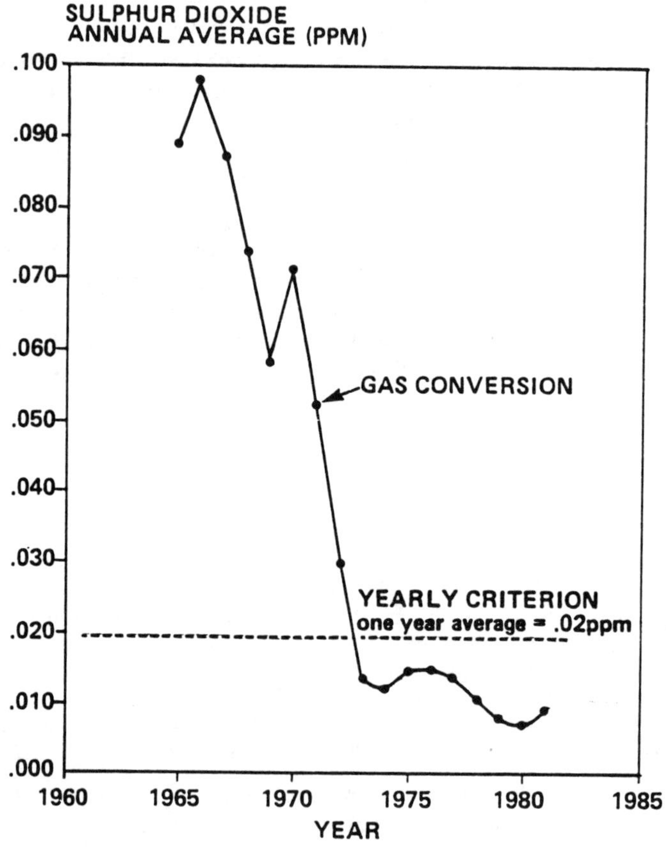

Section VI

Technological Developments and New Applications

Selected Authors

Chapter 19

New Gas Combustion Technologies for Emission Control

STEPHEN D. BAN,
EUGENE E. BERKAU,
and JOHN M. PRATAPAS
Gas Research Institute

Natural gas is a versatile fuel but its potential for control of emissions from all combustion processes has not yet been fully explored. For example, natural gas can in many cases be directly substituted economically for other less clean-burning fuels to reduce emissions and meet local, regional or federal regulations. This is referred to as select gas use (SGU) in Figure 19-1.

Conventional coal combustion results in three major types of emissions; nitrogen oxides, sulfur oxides, and particulates (ash and carbon). Substitution of gas for coal will result in a corresponding reduction in particulates and SO_X through simple dilution and will be directly related to the amount of coal displaced by gas. NO_X emissions are also likely to be less, since comparable gas combustion generally produces a lower NO_X level than coal combustion.

However, if gas is used so as to enhance the effectiveness of combustion, NO_X can be selectively reduced further, as indicated in Figure 19-1. Gas-enhanced combustion also involves substitution of gas for coal but requires that the gas be introduced under combustion conditions that better utilize its unique form value.

For the same amount of fuel used, gas-enhanced combustion will result in greater reductions of NO_X and similar reductions in SO_X and

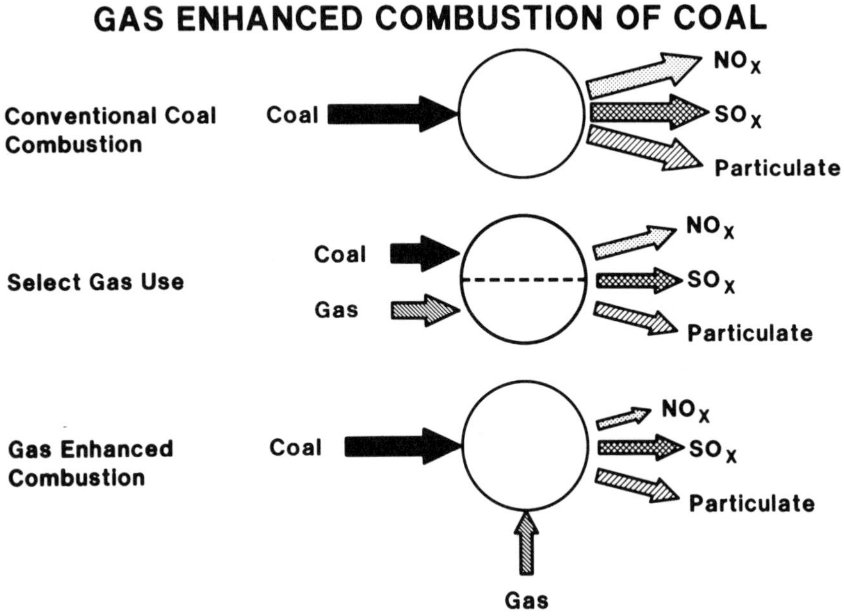

Figure 19-1

particulate when compared to select gas use. The development and field verification of gas-enhanced combustion technologies and their impact on boiler operations is critically important to the future use of natural gas in this area.

Looking beyond "select use," gas is a fuel that is uniquely suited to provide versatility and performance on an as needed basis to deal with the ever changing environmental regulations and the widely varying fuel, design, and operating conditions of existing power plant systems. Technologies for gas enhancement of coal and oil combustion are known, but significant technical issues remain to be explored and need to be proven to have practical solutions on an industrial scale. The technical and economic questions which exist in relation to emission reductions and boiler operations form the basis for GRI's ongoing research program.

This chapter will briefly describe GRI's current and future research program (1) to systematically examine the cost-effectiveness and benefits of gas technologies relative to the more commonly considered technologies for SO_X and NO_X control; and (2) to develop and field test promising gas technology options for control of NO_X and SO_X emissions from combustion sources.

The focus will be on present and future GRI research relative to the performance goals that have been set for gas-enhanced combustion technologies. As will be seen, there are many factors which must be considered in the selection of gas-enhanced combustion and other control options to assure that emissions meet the necessary levels in a cost-effective manner.

Finally, the schedule and an overview of GRI's plan to complete the principal research activities will be presented relative to gas-enhanced combustion technologies for coal fired utility and large industrial boilers over the next three years. This plan includes: (1) comparative benefits evaluation of gas-enhanced combustion technologies relative to alternative SO_X and NO_X control options; (2) field verification of gas use in a reburn application for NO_X control; and (3) pilot scale and field evaluations of the enhancement of a dry SO_2 control technology through use of gas combustion to improve cost-effectiveness.

Background

Table 19-1 provides a perspective on fossil energy consumption and associated emissions of acid rain precursors, sulfur and nitrogen oxides. Based upon 1980 fuel use data, the following observations can be made:

- Coal and oil combustion are the principal sources of SO_X and NO_X emissions in the U.S. from fossil fuel combustion, contributing 79% of the total man-made emissions.
- Coal combustion alone accounts for 80% of the SO_X, 31% of the NO_X and 58% of the total.
- Oil and gasoline and other fuels together result in 56% of the NO_X and 35% of the total SO_X and NO_X.
- By contrast, gas contributes 13% to the NO_X and 7% of the total NO_X and SO_X emissions while accounting for 30% of the total fossil fuel energy consumption.

These figures demonstrate the environmental benefits of burning natural gas and explain the historical focus on coal combustion as the

Table 19-1

ENERGY CONSUMPTION AND SO_x/NO_x EMISSIONS INVENTORY BY FUEL TYPE (1980)

	% of Total			
	SO_x	NO_x	SO_x and NO_x	Consumption
Natural Gas	Nil	13	7	30
Coal	80	31	58	21
Oil	18	25	21	19
Gasoline	1	28	13	19
Other[a]	1	3	1	11
10^6 Tons/Year	23	19	42	—

[a](LPG, Liquid/Solid Wastes, Refinery Gas)

primary target for SO_2 reductions in acid rain precursor control strategies (e.g. the 10 million ton per year reduction of SO_2 proposed in Senate bills).

At the same time, it is important to point out that total U.S. emissions of NO_X and SO_X are almost equal. This is noteworthy because recent scientific findings on the role of NO_X and other pollutants in acid precipitation has led to increased emphasis on the need for nitrogen oxide reductions. One form of legislation proposed for acid rain control provides that each ton of NO_X reduction is equivalent to a ton of SO_X reduction.

An alternative form of the draft legislation treated two tons of NO_X control the equivalent of one ton of SO_X control. Although the implications of proposed acid rain legislation are only speculative at this time, the potential for gas to be used for NO_X and SO_X control cannot be ignored.

In addition, environmental actions by Federal, State and local authorities continue to focus on solutions to other combustion-related pollution issues such as nonattainment, new source permitting requirements, air toxics, visibility degradation, etc. It is clear that natural gas could contribute significantly to addressing all of these pollution issues. Accordingly, planning is ongoing and current research is in progress to ensure that gas-enhanced combustion technologies are developed and proven to allow gas to be widely utilized in the most efficient way possible.

Select Gas Use

To provide some perspective on the potential for select gas use for emission control purposes, GRI and A.G.A. initiated a study (Figure 19-2) in October of 1982. The study, conducted by Environmental Research and Technology, Inc. (ERT), examined the environmental and other benefits to be derived by selectively burning natural gas and other fuels in electric power plants.

For the purpose of this study, select gas use (SGU) was defined as the combined use of gas and coal or oil in two ways: (1) at different times in boilers having dual fuel capability and, (2) in separate boilers fired in parallel (please note that co-firing of boilers with more than one fuel was not quantitatively analyzed by ERT but will give comparable results).

In their analysis, ERT compared life cycle costs of gas substitution versus other fuel switching and/or scrubber options for achieving

---

GRI/A.G.A. ENVIRONMENTAL BENEFITS OF SELECT GAS USE

SIX ELECTRIC POWERPLANT CASE STUDIES

- TWO FUEL PRICE PROJECTIONS

LIFE CYCLE COST COMPARISONS

- SELECT USE
- LOW SULFUR COAL
- FLUE GAS SCRUBBING

Figure 19-2

alternative environmental performance goals (e.g., a 50% reduction in SO_2 emissions) from six case studies of actual oil and coal-fired power plants. Followers of acid rain legislation debate will recognize that a 50% control level has been proposed for existing large sources (powerplants) without scrubbers in several Senate bills.

In the final report of the technical and economic evaluations, ERT stated, ". . . there are a number of cases where select gas use is likely to be a serious competitor with other fuels, largely for environmental reasons." ERT also pointed out that burning low-sulfur western coal is the major competitor to SGU if local or regional regulations allow its importation. Table 19-2 summarizes ERT's findings for one of the case studies examined.

In the base case, a utility was assumed to fire two 660 MW steam turbine generators with 3.4% sulfur coal. One option examined was to blend existing coal with low sulfur coal to lower the average sulfur

Table 19-2

RESULTS OF SELECT USE GAS STUDY FOR MIDWESTERN UTILITY COAL:

3.4% SULFUR/50% SO_x REDUCTION

Options	Incremental Annualized Cost (Millions 1983 $)	$/Ton SO_x
Low Sulfur Coal	45-50	428-476
Select Gas Use	52-75	495-714
Wet Scrubber	81-84	771-800

Coal: \$1.90 – \$3.20/10^6 Btu (1983)
Gas: \$4.10/10^6 Btu (1983)

content to 1.7%. The select use approach was to replace half of the heat input derived from coal with natural gas. In the scrubber option, one unit utilized a scrubber, while the second unit burned 3.1% sulfur coal. Because two fuel price projections were used in calculating the levelized costs, a range of annualized costs and \$/ton SO_X is shown. Using the lower fuel price projection for 3.4% sulfur coal (1983 price—\$1.90/10^6 Btu), both the SGU (1983 price—\$4.10/10^6 Btu) and low sulfur coal (1983 price—\$3.20/10^6 Btu) options are significantly less expensive than scrubbing. For this same assumed fuel price index, SGU is also very competitive on a cost effectiveness basis with low sulfur coal (\$498/ton SO_X versus \$428/ton, respectively).

It is important to note that these and other case studies were quite sensitive to future gas and coal prices and capital costs. Additional research was recommended to more fully quantify the complete set of environmental and other benefits of select gas use for additional site specific applications and regions of the country. In addition, ERT also recommended near-term research relating to burner design, instrumentation and controls, optimized firing configuration and potential operational problems (e.g. abrasion, slagging, and corrosion).

The implications (Figure 19-3) of this jointly sponsored study of the environmental and other benefits of select gas use can be summarized as follows:

(1) Through the limited case studies analyzed, SGU was shown to be a cost competitive option for achieving environmental objectives such as reducing emissions of SO_X and particulates;

(2) The particular advantages and applicability of SGU will be *highly dependent* upon site specific factors and needs;

(3) These factors and the obvious advantages of making gas an "integral" fuel rather than a substitute fuel for achieving emission control in coal combustion suggested that select gas use advantages can be expanded through gas-enhanced combustion technologies; and

(4) Selecting the most cost effective options—either for additional research or field applications—requires a systematic and comparative benefits analysis of control alternatives to identify R&D issues and to "sell" the use of gas as a viable option.

IMPLICATIONS OF GRI/A.G.A. SELECT GAS USE STUDY

- **Select Gas Use Can Be Cost-Effective**
- **Applicability and Cost-Effectiveness Highly Site Specific**
- **Gas Use Opportunities Can Be Expanded Using Gas-Enhanced Combustion Technologies**
- **Comparative Benefits Analysis Is Required to Evaluate Options**

Figure 19-3

GRI's Research Program to Develop and Apply Gas-Enhanced Combustion Technologies for Emissions Control

Although gas has long been recognized to be the cleanest burning fossil fuel, its full potential for use in reduction of emissions has not been systematically investigated or evaluated. This is due, at least in part, to the lack of information needed to identify specific opportunities and to quantify benefits that might be realized if gas were strategically and technologically utilized for pollution control purposes. While gas has been used to produce reducing atmospheres in heat treating furnaces, its use to reduce combustion-generated NO_X has not been considered until relatively recently.

In response to this need, GRI has implemented a broad-based research effort (Figure 19-4) to identify, develop and apply promising select use and gas-enhanced combustion technologies. The overall goals of this program are to establish the necessary economic information and technology bases whereby the most promising and beneficial uses of gas technologies to mitigate pollution can be identified and implemented for the benefit of the gas industry and its customers.

GRI RESEARCH PROGRAM TO DEVELOP AND APPLY
GAS-ENHANCED COMBUSTION TECHNOLOGIES
FOR EMISSIONS CONTROL FROM
COMBUSTION PROCESSES

Figure 19-4

The program has the following specific technical objectives shown in Figure 19-5:

(1) To identify potential opportunities for using gas and gas-enhanced combustion technologies in pollution control.

(2) To quantify the potential benefits and cost effectiveness of these technologies relative to other control options and establish research performance goals.

(3) To develop the most promising gas-enhanced technologies, and

(4) To verify their performance through testing in field applications.

There are three program elements (Figure 19-6) which correspond to these objectives. The *combustion source inventory* program will provide the basis for identifying candidate boilers and other specific opportunities for gas-enhanced combustion technologies. A *gas benefits analysis* program is attempting to quantify the specific gas markets for select use and gas-enhanced combustion technologies. The *gas applications research* program is directed to laboratory and field studies to verify the advantages of gas-enhanced combustion technologies and to resolve technical/economic issues.

Combustion Source Inventory

The Combustion Source Inventory (Figure 19-7) is a 50 state fuel consumption and SO_X, NO_X and hydrocarbon emissions inventory that was developed under contract for the Morgantown Energy Tech-

GRI RESEARCH OBJECTIVES

- Identify Opportunities for Use of Gas for Emissions Control
- Quantify the Benefits Relative to Other Options
- Develop Gas-Enhanced Combustion Technologies
- Verify Advantages of Gas Use in Emissions Control

Figure 19-5

PROGRAM ELEMENTS

- Combustion Source Inventory
- Gas Benefits Analysis
- Gas Technology Applications

Figure 19-6

COMBUSTION SOURCE INVENTORY

- 1980 Fuel Consumption
- SO_x, NO_x, Hydrocarbon Emissions
- Major Point Sources
- Area Sources
- County/State/Region Totals

Figure 19-7

nology Center (METC) of the Department of Energy, primarily for policy analyses of alternative acid rain legislative proposals. Canadian emissions are also included in the inventory.

The inventory can be categorized as a "top down" approach by virtue of the fact that fuel totals by state from the DOE State Energy Data Report are used to establish an energy balance for each consuming sector. The five consuming sectors covered are electric power plants, industrial, residential, commercial and transportation. For each sector, estimates of SO_X, NO_X and hydrocarbon emissions are accounted for at varying levels of detail; e.g. emissions of a specific source within a county to total emissions from all sources at a county, state, regional or national level.

It is beyond the scope of this chapter to fully describe the data base and the reporting and manipulation capabilities that currently exist or are under development by GRI. We have installed the DOE data files into an interactive computer system referred to as the Graphical Exposure Modeling System (GEMS). GEMS was developed under the auspices of the U.S. Environmental Protection Agency's Office of Toxic Substances and provides a simple interface to statistical analysis, graphic display capabilities (plus other functions) with data manipulation support.

GRI has funded further software development to enhance the mapping capabilities for subsequent use in identifying areas where gas could be used to mitigate emission problems. One example of the data base being used is depicted in Figure 19-8. This is a map of the U.S. showing the locations and corresponding annual emissions of the 50 largest coal-fired utility power plants.

This map can be overlayed with a map of gas pipelines (Figure 19-9) to provide some perspective on the current availability of gas to these locations for use in gas-enhanced combustion technologies and to develop a gas-based control strategy. County level maps for a particular region of interest can then be generated along with specific reports of fuel consumption and emissions data, leading to identification of specific target technologies of broad applicability.

This will be utilized to examine, among other things, the relationship between coal and oil consumption and emissions problems by identifying areas of high SO_X and NO_X emission densities, where select gas use and gas-enhanced combustion technologies could be viable mitigation approaches. The data in the combustion source inventory provides essential fuel and emissions data for comparative evaluations of the costs and environmental benefits of gas-enhanced combustion technologies in a given region, as well as to identify the broad applicability of specific technologies.

Gas Benefits Analysis

The objectives of the Gas Benefits Analysis program element are to develop and apply a systematic procedure to evaluate and quantify the relative costs and benefits of using SGU or gas-enhanced technologies as environmental controls. As depicted in Figure 19-10, specific emission source problems identified through the combustion source inventory or other means are to be analyzed in the context of current or pending regulatory concerns and requirements.

The technical, economic, and institutional feasibility of each gas technology alternative is evaluated on a comparable basis with other pollution control options to project (as quantitatively as possible) the benefits of the most likely candidates for successful commercial application.

Based on the evaluation results, data gaps and research performance goals are identified and prioritized in order of competitiveness with other options and benefits to the gas industry and gas users. Economic and environmental benefits of using gas for pollution control can then be calculated in terms of cost savings, pollution reduction and gas markets.

TOP FIFTY U.S. COAL-FIRED UTILITY POWER PLANTS

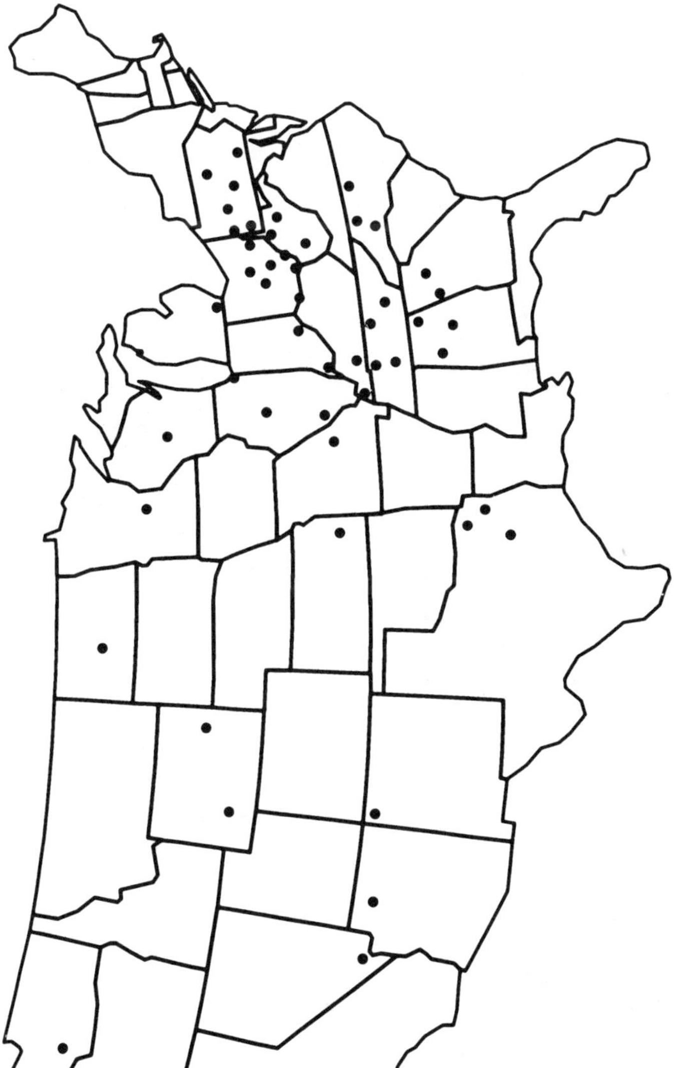

Figure 19-8

MAJOR NATURAL GAS TRANSMISSION PIPELINES AND TOP 50 COAL-FIRED POWER PLANTS

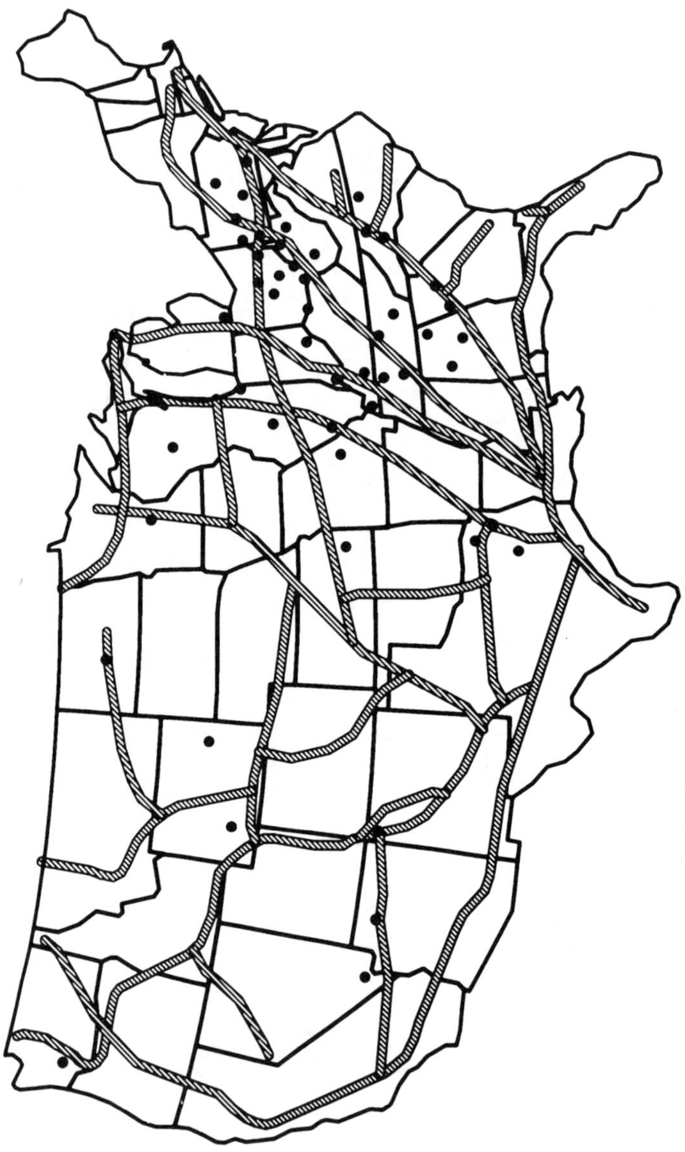

Figure 19-9

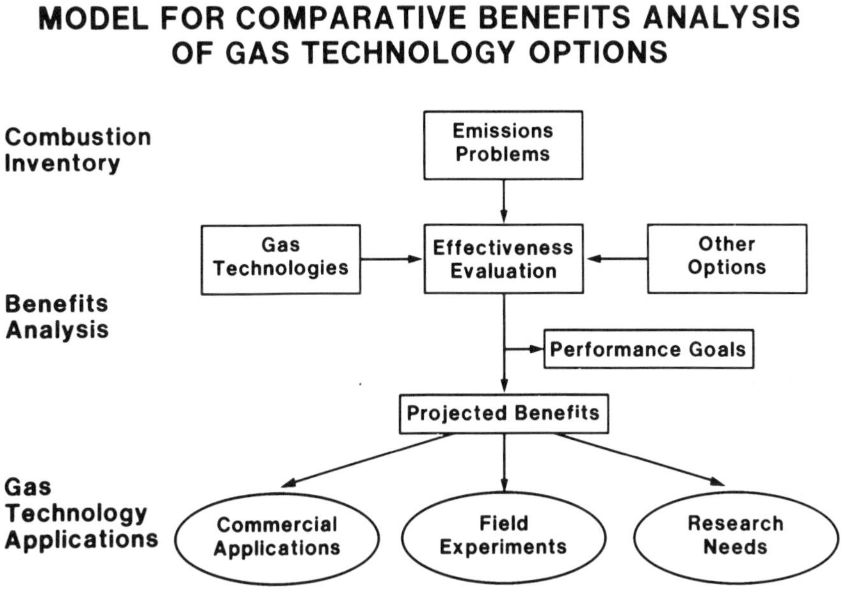

Figure 19-10

The results of this analysis can lead in three distinct directions. In some cases a gas technology or select use opportunity can find immediate commercial application. Alternatively, the benefits analysis might identify technical issues requiring additional research. In other cases, a gas technology might be ready for field evaluation to verify the advantages of gas under real operating conditions.

Current activity is in the initial stages of development of a complete list of opportunities to use gas enhancement to mitigate emission problems. This will include an assessment of current and proposed control strategies to determine whether gas-based technologies can provide a superior option. Current candidate technologies, applications and the emissions of concern are listed in Table 19-3. Differing stages of application of the benefits analysis model has been completed for each of these gas technology options.

Table 19-3

CANDIDATE TECHNOLOGIES FOR GAS-ENHANCED COMBUSTION

Technology	Application	Concern
Reburn/Afterburn	Boilers/Furnaces	NO_x
	Coal Water Slurry	NO_x
	Municipal Wastes	Particulate/Toxics
	Industrial Wastes	Particulate/Toxics
	Wood Stoves/Boilers	Particulate/Toxics
	Hazardous Wastes	Toxics
Dry Sorbent Injection (DSI)	Boilers/Furnaces	SO_x
Reburn/DSI	Boilers/Furnaces	NO_x/SO_x
Auxiliary Gas Firing	Utility Boilers	Derating, Availability

Gas Applications Research

Current research (Figure 19-11) is focusing on gas-enhanced combustion technology directed to in-furnace control of acid rain precursor emissions of SO_X and NO_X from large coal fired boilers. Currently, federal acid rain policy is emphasizing the need for more research into causes, effects and mitigation alternatives.

EPA, DOE, EPRI and others are investigating various mitigation control technologies applicable to large coal fired boilers. The two technologies most considered to be amenable to gas enhancement are "reburning" for NO_X control and Dry Sorbent Injection (DSI) for SO_X control. Gas may enhance the performance of both these technologies.

Reburning is depicted in Figure 19-12 relative to conventional combustion and involves diverting a portion of the fuel downstream into a second fuel-rich "reburn" zone to create conditions for NO_X reduction. Air is added in the third or final "burn-out" zone to complete the overall combustion process at temperatures that impede any further NO_X production.

Overall NO_X reductions of 50% or more are expected when compared to conventional combustion. For more than a decade it has been known that nitric oxide (NO) can be reduced in flames. Recently, the Japanese have applied this concept to in-furnace NO_X

GRI'S CURRENT GAS-ENHANCED COMBUSTION TECHNOLOGY RESEARCH EMPHASIS

- Acid Rain Precursors – SO_x and NO_x
- Large Coal Fired Boilers
- Gas Reburn Technology – NO_x
- Gas-Enhanced Dry Sorbent Injection – SO_x

Figure 19-11

SCHEMATIC OF REBURN TECHNOLOGY FOR NO_X CONTROL

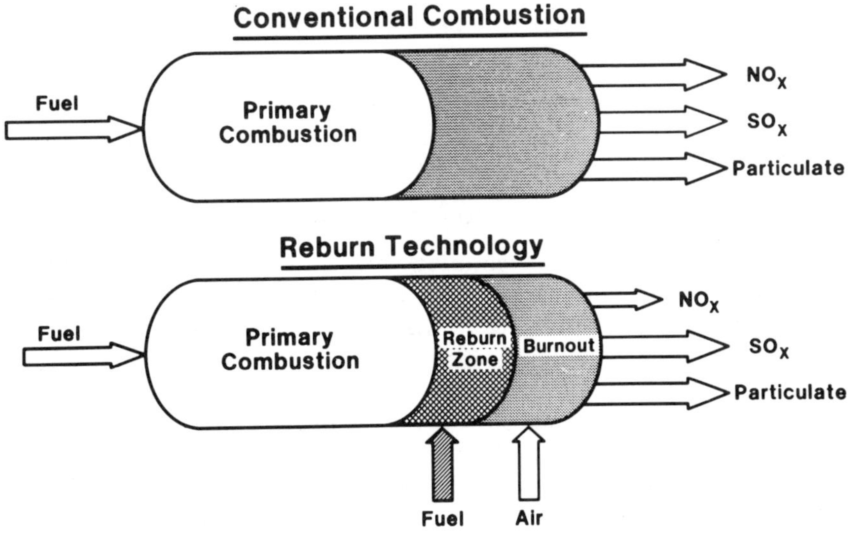

Figure 19-12

reduction through a system known as the Mitsubishi Advanced Combustion Technology (MACT). The Japanese have concluded bench-scale, pilot-scale and commercial tests on reburn in Japanese boilers and report that "generally speaking, in tangentially fired units NO_X levels of 30 ppm for gas, 80 ppm for oil and 150 ppm for coal have been realized without deteriorating quality of combustion and flue gas."

However, the use of coal or oil as a reburning fuel in a retrofit conversion is likely to introduce operational problems in certain boilers due to poor carbon burnout as well as slagging in the platens or primary superheater region resulting in loss of efficiency. As will be discussed later, gas is considered to be an ideal fuel for reburning to avoid these concerns while achieving lower exhaust NO_X levels. Gas use as a reburn fuel also offers a significant potential for operational improvements.

Figure 19-13 depicts dry sorbent injection technology for SO_X capture relative to conventional combustion. The process can be considered to involve three steps:

(1) Calcination of a dry sorbent, such as limestone, by injecting into the high temperature environment to form chemically active high-surface area particles (CaO);

(2) Mixing of the calcined sorbent with the products of the primary combustion while maintaining the reactive surface area; and

(3) Sulfation, in which SO_2 is captured by reaction with the calcined sorbent particle to form calcium sulfate.

A specific form of DSI technology has been under development by EPA for a number of years and has been referred to as Limestone Injection Multi-stage Burners, or LIMB. In a general sense, this approach does not require injection of the sorbent through the multi-stage burners, so the technology is more generically referred to as dry sorbent injection, with the specific EPA case being LIMB.

EPA and EPRI have sponsored considerable experimental testing to define the process parameters controlling sulfur capture. Based upon a critical review performed for GRI on available research results, there is scientific evidence that the combustion of natural gas could be used to enhance sorbent reactivity, resulting in reduced sorbent injection requirements with increased SO_2 capture.

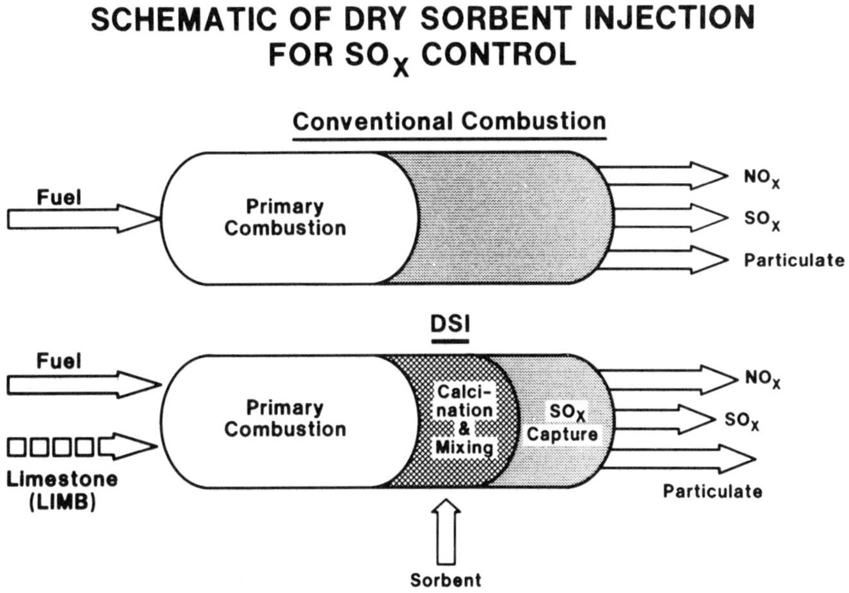

Figure 19-13

Several roles for gas are possible: (1) To improve the sorbent calcination; (2) to transport and distribute sorbent improvers; and (3) to control the thermal environment and carbon burnout in the upper regions of the boiler.

Table 19-4 is a partial listing of technologies, either commercially available or under development for control of SO_X or NO_X emissions from large coal fired boilers. These represent the principal alternative control methods competing with gas technologies.

Scrubbers for large coal fired power plants have been installed primarily as a result of new source performance standards. Although capable of achieving 80%-95% capture of SO_2, they require high capital expenditure (between \$140-\$250/kw in new installations and at least 10%-40% more in retrofit applications).

In a retrofit application, scrubbers are typically the least cost-effective control option in terms of cost per ton of SO_2 removed. Alternatively, coal cleaning and low sulfur coal are commercially available techniques for controlling or limiting SO_2 emissions but

Table 19-4

TECHNOLOGIES FOR CONTROL OF SO_X OR NO_X EMISSIONS FROM COAL-FIRED BOILERS

Technology	Pollutant	Status
Scrubbers	SO_X	Commercial
Coal Cleaning	SO_X	Commercial
Low Sulfur Coal	SO_X	Commercial
Direct Sorbent Injection (DSI)	SO_X	Developmental
Low NO_X Burners	NO_X	Commercial
Gas-Enhanced DSI	SO_X	Developmental
Gas Reburn	NO_X	Developmental

have limited applicability and may not be effective enough to achieve required reductions at a specific plant.

Direct dry sorbent injection in the form of LIMB is being developed by EPA as a cost effective approach for achieving about 50% SO_X capture. Capital costs of about $85/kw for retrofit applications are reported, however, the large sorbent (limestone) requirements (Ca/S ratios of 2 to 3) greatly diminish the operating cost advantage.

Low NO_X burners for coal fired boilers are commercially available and have primarily been used to meet new source performance standards and best available control technology requirements associated with prevention of significant deterioration determinations.

The retrofitability of low NO_X burners is influenced to a large degree by furnace geometry. A recent EPRI study estimates that "only about 38% of U.S. coal-fired capacity (units of 100 MW or larger) can foreseeably use the low-NO_X burners now commercially available or under development." Based upon pilot scale testing, the NO_X reduction capability of advanced low-NO_X burners is projected to be about 45%–60%.

Technologies for simultaneous control of SO_X and NO_X emissions from coal fired boilers (Table 19-5) are still under development at this time. The DSI technology can include low NO_X burners as an option for simultaneous controls.

Table 19-5

TECHNOLOGIES FOR SIMULTANEOUS CONTROL OF SO_x/NO_x EMISSIONS FROM COAL FIRED BOILERS

Technology	Pollutant	Status
DSI in Conjunction with Low NO_x Burners	NO_x/SO_x	Developmental
Gas Reburn/Enhanced DSI	NO_x/SO_x	Developmental

Originally, low NO_X burners were considered essential for the limestone injection. Recent research has indicated that improved SO_2 capture is possible if sorbent is injected in the upper regions of the furnace rather than through the low NO_X multi-stage burners. These research findings support GRI's current research focus on gas reburn with gas-enhanced dry sorbent injection in the upper regions for simultaneous NO_X/SO_X control discussed later.

Figure 19-14 illustrates the variability in cost effectiveness for several of the technologies applicable to large coal fired boilers for SO_X control. This variability is primarily a function of initial sulfur content in the coal but is also influenced by numerous other parameters such as fuel and transportation costs, raw material (e.g. sorbent) requirements and costs, etc.

The key parameters provide the basis for targeting appropriate performance goals in relation to cost/ton SO_X removed and a percentage reduction target for gas-enhanced combustion technologies. For illustration purposes a desirable range for cost effectiveness with gas-enhanced dry sorbent injection has been drawn. The figure illustrates one of the major problems for decision-making in selecting the most cost/effective control approach for a specific plant and the need for evaluating cost benefits of control technologies on a comparative basis.

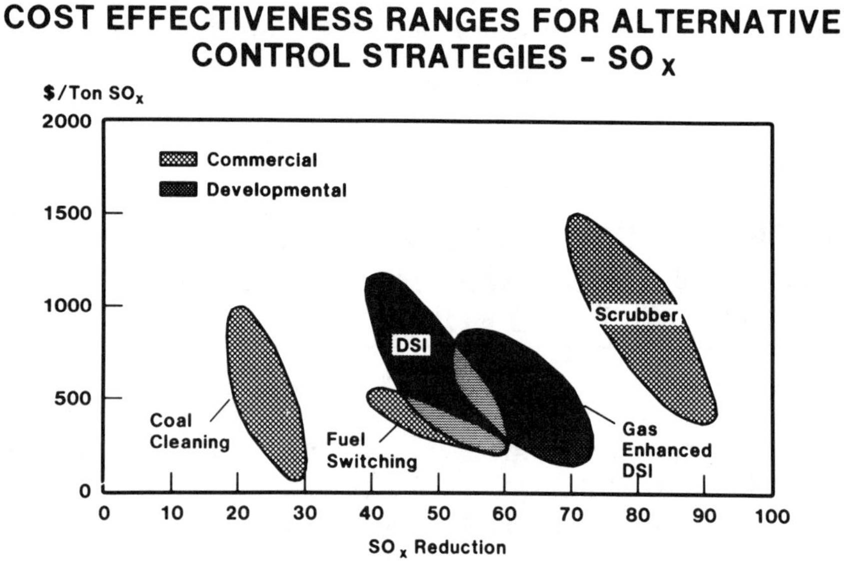

Figure 19-14

Based upon the results of our preliminary benefits analysis of state-of-the-art technologies for control of SO_X or NO_X emissions from large coal-fired boilers, tentative performance goals for gas-enhanced combustion technologies have been established (Table 19-6). For flue gas desulfurization with wet scrubbers, capital cost is estimated to be $200/kw for retrofit applications and 85% SO_2 reduction.

This will lead to a levelized cost effectiveness of about $1000/ton of SO_2 removed for medium sulfur coal. Published results indicate that LIMB, on the other hand, could require $85/kw in capital costs. However, because of higher sorbent requirements relative to scrubbers, levelized costs for 50% SO_2 removal are currently estimated to be $1000/ton for LIMB technology.

With respect to NO_X reduction, lwo NO_X burners are considered to be state-of-the-art technology with some advanced burners projected to provide 50% NO_X reduction at a levelized cost of $200/ton of NO_X removed. Capital costs are estimated to be $15/kw.

Table 19-6

PERFORMANCE GOALS FOR GAS ENHANCED COMBUSTION TECHNOLOGIES

(Basis: 500 MW; 2% Sulfur Coal; Pre-NSPS)

	State-of-the-Art			Gas Technologies	
Performance Parameters	FGD	DSI	Low NO_X Burner	Gas Reburn	Gas Reburn/DSI
Capital Costs ($/kW)*	200	85	15	10	75
SO_X Reduction (%)	85	50	0	10	70
CA/S (Molar)	1.5	2.5	–	–	1.5
NO_X Reduction (%)	0	0	50	60	60
Acid Precursor Reduction (%)	65	40	10	20	70
$/Ton Precursor	1000	1000	200	400	700

*Retrofit

The target performance goals for competitive use of gas reburn technology is to increase NO_X reduction at capital costs less than low NO_X burners. The specific commercial targets considered appropriate are 60% NO_X reduction at a levelized cost of $400/ton of combined NO_X and SO_X removal. [Remember—any displacement of coal with gas results in a proportional reduction in SO_2 emissions, i.e. 10% gas means 10% lower SO_2.]

Total reduction of NO_X and SO_X are targeted for 20% removal. Gas reburn/dry sorbent injection technology performance goals are targeted for 70% SO_X and 60% NO_X reduction with a capital investment of $75/kw. This will lead to a 70% reduction in total NO_X and SO_X emissions at a levelized annual cost of $700/ton.

Gas Reburn

The conceptualized application of gas reburn technology to a pulverized coal utility boiler is depicted in Figure 19-15. Within the fuel-rich region, hydrocarbon fragments from the reburning fuel will react with nitric oxide produced in the main flame zone forming other nitrogen species and molecular nitrogen. Secondary air is provided to establish an oxidizing regime which completes the combustion of the reburning fuel and the nitrogeneous species are converted to either N_2 or NO.

In a utility boiler sufficient residence time is the critical factor for conversion of nitric oxide to N_2 and to completely combust the reburn fuel. Reburn fuels which cannot be mixed and combusted quickly will most probably create carbon burnout, slagging and corrosion problems in most U.S. boilers.

Gas is an ideal reburning fuel for such boilers because of the (typically) short residence times between the primary combustion zone and the convective sections. Reburn is applicable to all large U.S. boilers in one of three categories: cyclone, tangential, or wall fired.

By comparison, low NO_X burner technology is potentially retrofitable only to some boilers, and it should be recognized that the controlled mixing process associated with low NO_X burner technology results in an increased flame length in the main firing zone. In many instances the distance between the front and rear boiler faces is simply too short to accommodate the increased flame length of a low NO_X burner.

In short, low NO_X burners are *not* universally applicable. On the other hand, at this time there are no obvious technical constraints limiting the general applicability of natural gas reburning.

Figure 19-16A presents data recently obtained in a bench-scale research furnace which illustrates the potential for gas reburn compared to an optimized primary heat release zone burning a low sulfur western coal. The data illustrate that for reburn fuel levels of interest (5% to 10%) natural gas reburn is 50% more effective in reducing NO_X emissions compared to coal as the reburn fuel and is always more effective than coal for all contemplated fuel utilization levels (Figure 19-16A). Figure 19-16B illustrates two other advantages of gas reburn over coal:

- sulfur oxide emissions are reduced because of a reduction in fuel sulfur input;

Gas Reburn – NO_x Control

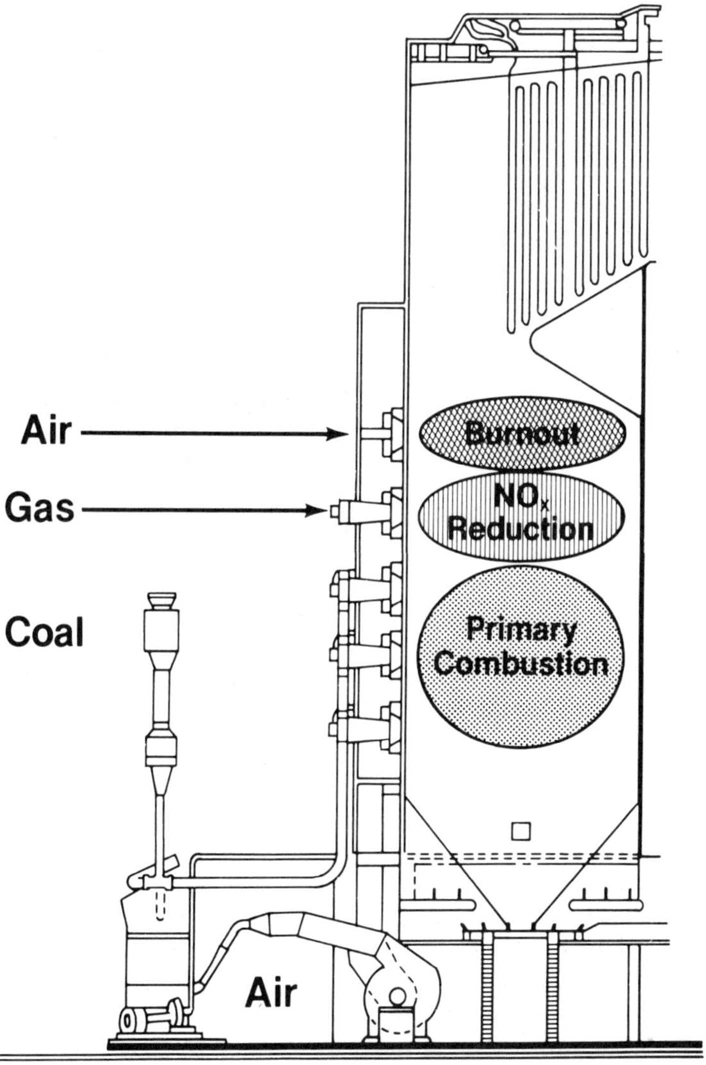

Figure 19-15

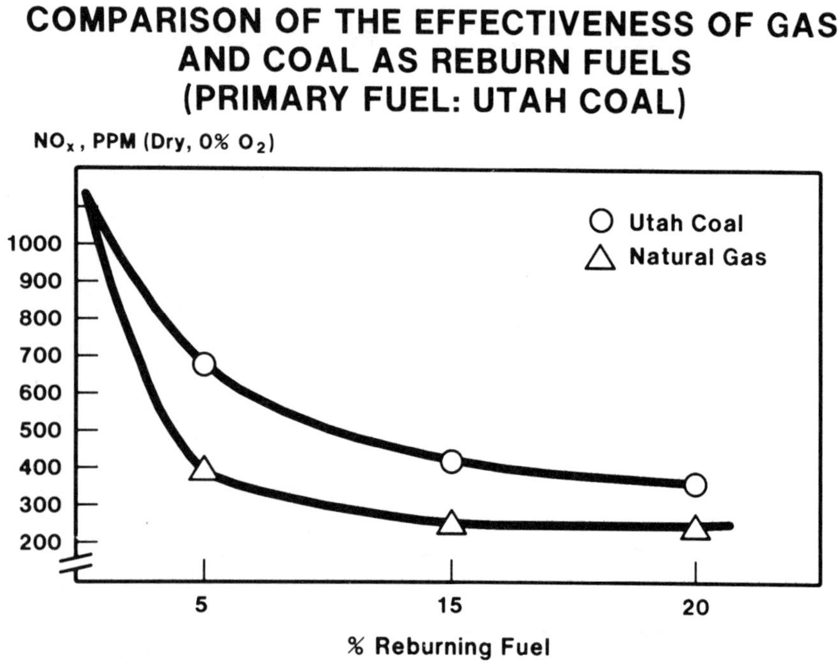

Figure 19-16A

- emissions of CO are reduced because gas as the reburn fuel is less difficult to burnout in the residence times available in boilers.

Gas reburn can have other advantages to the boiler operator. Particulate loadings will be reduced (Figure 19-17) and this could improve availability and reduce maintenance costs, a very significant financial benefit.

Reburn with gas has considerable potential as a cost effective NO_X control technology for existing coal-fired utility boilers. Recent research efforts have been directed towards optimization of the effectiveness of the gas reburn fuel in order to reduce both the pollutant emissions and overall excess air levels.

This requires efficient fuel and air injection systems which ensure that the reburn fuel and the overfire air are well-mixed in the appro-

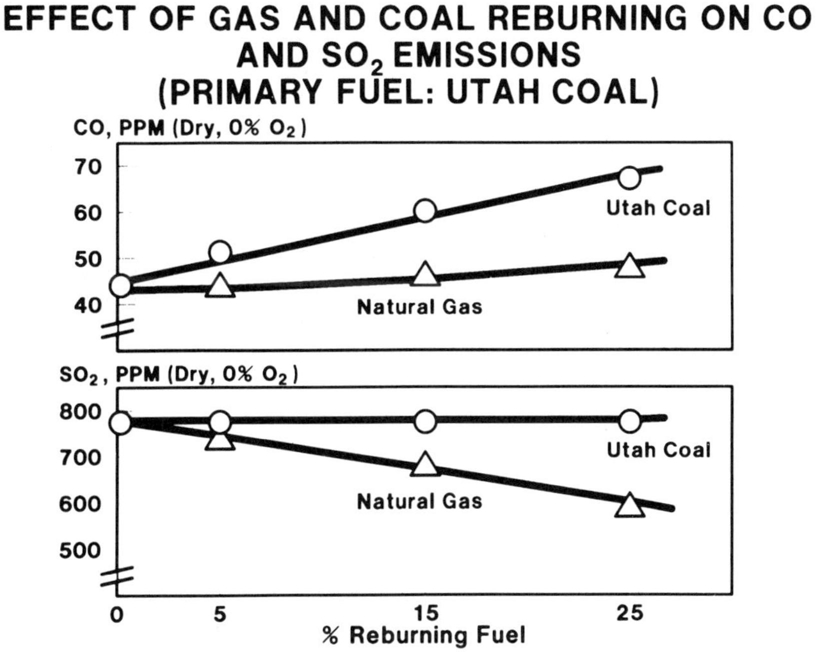

Figure 19-16B

priate temperature regimes. These areas and the development of engineering designs and costs for retrofit are emphasized in the current research program for gas reburn (Figure 19-18).

Gas-Enhanced DSI

GRI sponsored research of gas-enhanced dry sorbent (limestone) injection has only recently been initiated on a bench scale. The SO_2 capture process in a utility boiler is depicted in Figure 19-19 and involves the three critical steps mentioned earlier:

(1) Calcining of the limestone;

(2) Mixing and protection of the reactive surface from sintering or fusing with the coal ash;

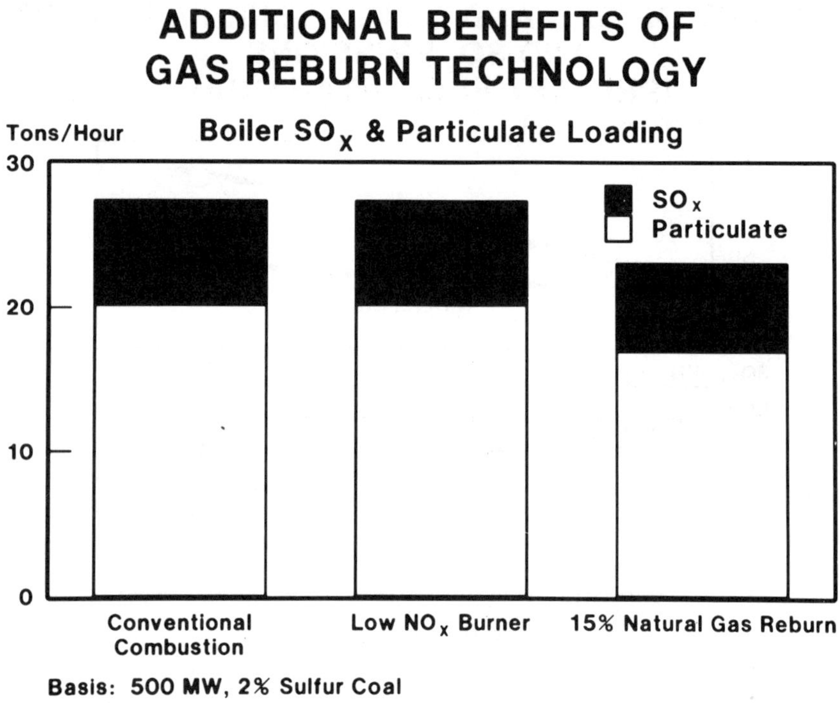

Figure 19-17

(3) Sulfation of the CaO to maximize SO_2 capture in the allowable temperature window (1650°-2250°F) which is located in the upper furnace regions of U.S. boilers.

The key issue in utility boilers is mixing and residence time within the temperature window necessary for SO_2 capture. The thermochemical characteristics of gas and controllability of gas combustion make it an ideal fuel for optimizing the calcination process, for enhancing the mixing of the particles, and for protecting the reactive sorbent surface from sintering or fusion by temperature control.

Calcination refers to the loss of CO_2 from a carbonate or water from a hydrate and it produces calcium oxide. The ability of calcium oxide to capture SO_2 is strongly dependent upon sorbent morphology in total surface area, pore size distribution, etc.

GAS REBURN FOR NO_x CONTROL CURRENT RESEARCH EMPHASIS

- Mixing of Gas Reburn Fuel, Overfire Air, and Coal Combustion Products
- Optimum Temperature Regimes
- Amount of Reburn Gas
- Engineering Design/Costs for Retrofit

Figure 19-18

Surface area is a strong function of the peak temperature attained by the particle provided calcination is complete. Thus one potential application of natural gas is the controlled and enhanced calcination of the sorbent to produce a high surface area by creation of the optimal thermochemical environment.

Figure 19-20 is a picture of an unreacted calcined limestone particle magnified 15,000 times. The high surface area is provided by the multitude of crevices and interstices which have been generated during the calcination process.

Figure 19-21 shows a fly ash/sulfated calcine mixture. Some fly ash particles appear to be coated with smaller sorbent or ash fragments. The high surface area of the sorbent has been reduced, limiting the availability of surface to react with SO_2. The net result is the need to use a large excess of sorbent (molar ratios of calcium to sulfur of 2–3) to obtain 40%–50% SO_X capture.

Recent experiments have shown that certain compounds may be required to enhance the ability of low-cost sorbents to capture sulfur, thus providing a further opportunity for gas. In this instance a gas-fired controlled calciner would be used to deposit the promoter on the sorbent particles using vaporization condensation techniques. Figure 19-22A shows some preliminary data that indicates improved SO_2 capture and Ca/S ratios for promoted sorbents processed

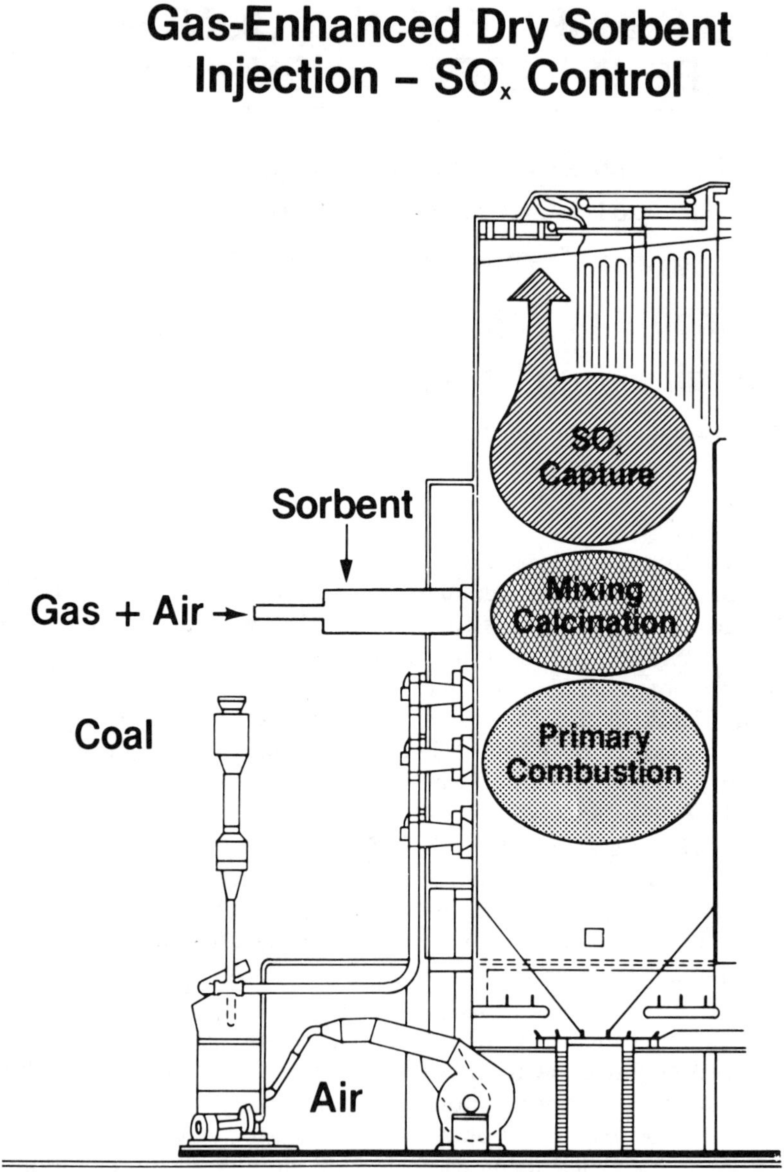
Gas-Enhanced Dry Sorbent Injection – SO_x Control
SO_x Capture
Mixing Calcination
Primary Combustion
Sorbent
Gas + Air
Coal
Air

Figure 19-19

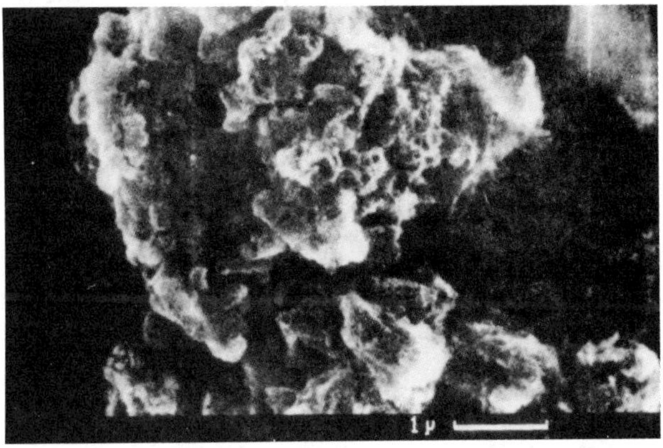

15,000X

Figure 19-20
SEM Photomicrograph of Calcine Particles Produced from Hydrated Lime

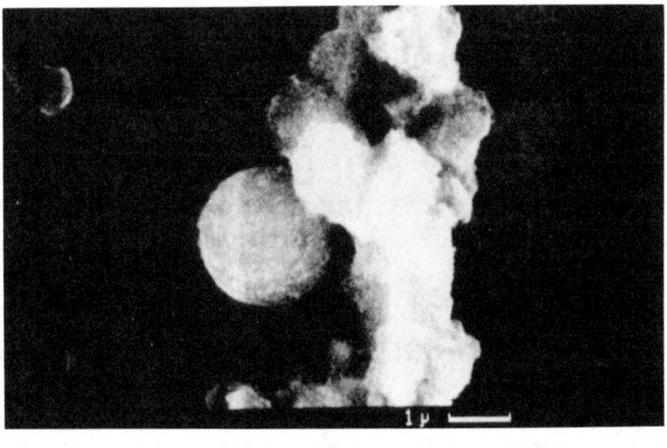

10,000X

Figure 19-21
SEM Photomicrograph of Sulfated Calcine Particles in Combination with Fly Ash from a Medium-Sulfur Coal

POTENTIAL OF NATURAL GAS/SORBENT INJECTION SYSTEM TO INCREASE SO_2 CAPTURE [1]

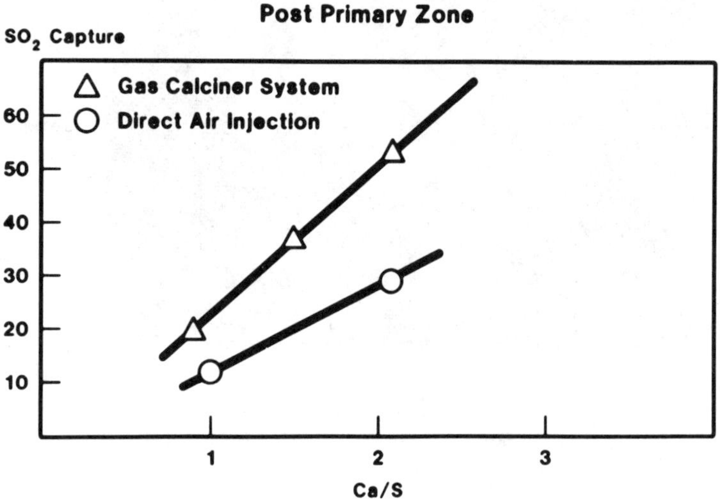

(1) Source: EERC

Figure 19-22A

GAS-ENHANCED DRY SORBENT INJECTION CURRENT RESEARCH EMPHASIS

- Enhancement of Sorbent Surface Area and Reactivity
- Protection of Sorbent Particle from Ash and Sintering
- Efficient Mixing of Sorbent with Combustion Products
- Gas Requirements for Enhanced SO_x Capture

Figure 19-22B

through an external gas calciner system versus direct injection into the post primary combustion zone.

Mixing the active sorbent with the combustion products, prevention of surface area loss due to thermal sintering or interaction with molten ash particles and the provision of the maximum residence time in the temperature window necessary for SO_2 capture are key issues in the application of DSI to utility boilers. The controllability of gas combustion makes it an ideal fuel for the production and utilization of an active sorbent.

GRI's research goals are to reduce the Ca/S ratio to less than *2* for *70%–75%* SO_2 reduction. The current research emphasis (Figure 19-22B) is to perform initial studies that will focus on increasing the sorbent surface area during calcination for high reactivity or the use of promoters.

The use of gas combustion alone to create an optimal controlled atmosphere to promote SO_2 capture will also be explored. The final phase of this research is to develop means for using gas to rapidly mix the enhanced reactive calcined sorbent with the flue gases under conditions simulating full scale boiler operation to evaluate gas fuel requirements for enhanced SO_2 capture.

Combined NO_X/SO_X Control

In-furnace NO_X reduction occurs at approximately the same temperatures for which SO_X capture by injected limestone is effective. Figure 19-23 presents a schematic of a combined NO_X/SO_X control applied to a coal-fired utility or large industrial boiler. NO_X control is provided by gas reburning.

One of the burners provides fuel rich conditions for in-furnace NO_X reduction. The gas calcined sorbent is then to be injected together with the overfire air for burnout and SO_X capture. This combined gas-enhanced system has several potential advantages compared to selective SO_2 control by DSI. The sorbent is calcined and injected remote from the main heat release zone to avoid contact with molten ash particles and deactivation.

As previously indicated, sorbent injection at lower furnace temperatures favors higher surface areas and therefore more active sorbent particles. Use of overfire air to transport the sorbent avoids the cost and complexity of other means (such as flue gas injection) and maintains overall excess air control.

Gas Reburn/Dry Sorbent Injection SO_x/NO_x Control

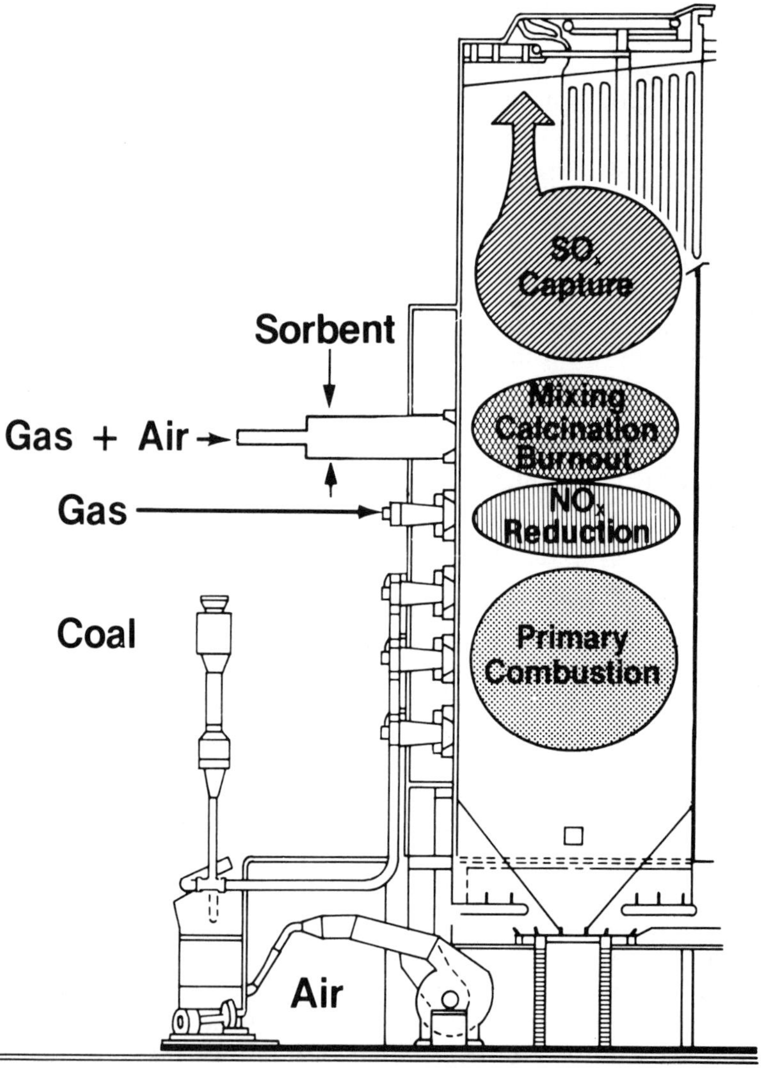

Figure 19-23

By achieving simultaneous control of NO_X/SO_X, this approach provides not only for enhanced environmental benefits but has the potential of being more cost effective than the individually applied SO_X and NO_X control technologies.

As one example of the potential advantages of gas-reburn/enhanced DSI, a primary concern of boiler operators with DSI technology is the significantly increased mass loading of particulate matter in the boiler and the extent of the modifications necessary to existing particulate control systems. Figure 19-24 graphically demonstrates that 15% natural gas reburning with DSI reduces the required sorbent injection rate for SO_2 capture and incremental particulate loading (each by 15%) relative to conventional DSI (LIMB).

BENEFITS OF GAS REBURNING WITH DRY SORBENT INJECTION 50% SO_2 CAPTURE

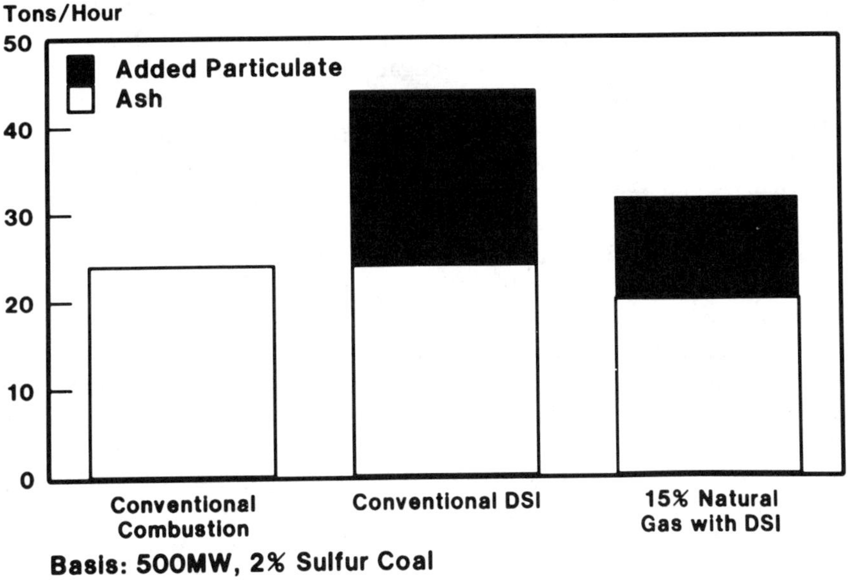

Figure 19-24

This is because the SO_2 and particulate levels in the furnace become 85% of that for conventional coal combustion as a result of 15% gas use. The net effect is an overall 30% reduction in particulate loading for 50% SO_2 control.

Further, if GRI's gas-enhanced DSI research does lead to increased sorbent utilization, the particulate loading from gas-enhanced DSI could approach that of a conventional coal firing. This would minimize the costs of modifications to install particulate control devices and mitigate related boiler operational problems.

The key issues to be resolved are the technical feasibility and flexibility of cost-effective integration to meet the conditions necessary for both gas-enhanced NO_X reduction and SO_2 capture. The specific emphasis in GRI's research plan for simultaneous SO_X/NO_X control will be defined based upon the results of the gas-enhanced DSI bench scale research currently underway.

Schedule for Future Research

The schedule for planned research is shown in Figure 19-25 for the 1984–1988 time period. There are four major categories of research. The Benefits Analysis began with the initial GRI/A.G.A. study of select gas use and continues with current activities to evaluate the comparative benefits, e.g. costs savings, pollution reduction, gas-fuel requirements, of gas-enhanced combustion technologies relative to the most likely SO_X and NO_X control options for coal-fired boilers.

This research was completed in 1985 and establishes the priorities and emphasis for GRI's field testing of gas-enhanced combustion technologies.

The results are also being used to develop a users computer model for comparative evaluation of the various SO_X and NO_X control options for specific applications to coal-fired boilers. This model is scheduled to be completed and verified by early 1987. Current plans are to continue to expand the user models for other applications of gas-enhanced combustion technologies in the future.

Research on gas reburn technology relative to coal reburn and low NO_X burners for NO_X control in a bench and pilot scale coal-fired furnace has been underway since early 1984. The current work on the 3 megawatt pilot scale furnace is designed to optimize the gas reburn technology under conditions simulating those of actual boiler operations and is scheduled for completion late this year.

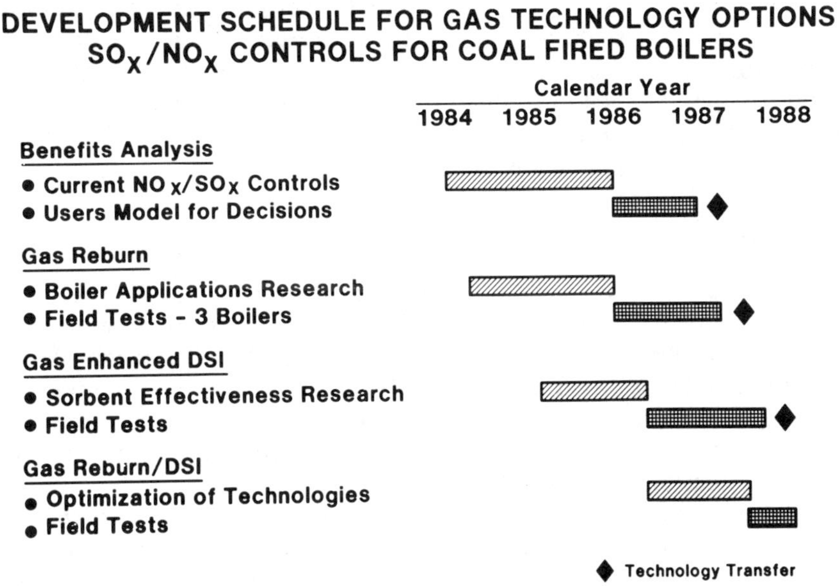

Figure 19-25

The results will be used to aid in planning field tests for three coal fired boiler types to be initiated in 1986. The pilot scale unit will continue to be used to support the field testing. Currently being considered are tests in cyclone, wall and tangential-fired boilers.

Completion of the field tests in the first boiler is planned for early 1987 and active discussion of options is ongoing with potential cosponsors, e.g. EPA, EPRI, DOE, utilities, etc.

GRI research on gas-enhanced DSI was initiated in 1985 on the same bench scale coal-fired furnace. The initial results from the studies of improved calcination and enhanced sorbent reactivity are to be obtained by mid-1986. The bench and pilot scale data along with the benefits analysis will provide the basis for planning a field test on a large coal-fired boiler.

Planning for the field tests will begin in early 1986. The tests will be scheduled for completion in late 1987. The field test results of gas reburn for NO_X control and gas-enhanced DSI for SO_X control will establish the baseline data for integrating these technologies for simultaneous NO_X and SO_X control.

Pilot scale studies to integrate and to optimize the gas reburn/sorbent injection technologies are to be initiated in the middle of 1986. Field testing the simultaneous NO_X/SO_X control of a coal-fired boiler by gas-enhanced combustion technology is planned to begin in early 1987, and should be completed in late 1988 if previous field testing sites are available and suitable.

Future Outlook

A primary goal of the program is an objective evaluation of the benefits and applicability of gas-enhanced combustion technologies relative to other options for control of SO_X and NO_X emissions from coal-fired boilers. Based on the results of research to date, it appears that gas-enhanced combustion technology can be used to improve the cost-effectiveness, versatility and applicability of reburn and dry sorbent injection controls.

If ongoing research proves to be successful and confirms that gas-enhanced combustion technologies can provide a cost-effective option for control of SO_X and NO_X emissions from coal-fired boilers, further development in a field environment will be pursued vigorously.

To this end, GRI is currently planning several alternatives. The joint funding (with EPA) of a field evaluation of reburn technology is planned to begin in 1986 on an industrial boiler with options for sorbent injection. The specific site for this evaluation will be chosen on the basis of proposals submitted in response to a request for proposals which EPA released late in 1985.

GRI is also a potential partner in an industrial consortium that is planning a field evaluation of both reburn and sorbent injection technologies on a utility boiler. Similarly, GRI and EPRI staff are discussing the possibility of cooperative funding of field research. Discussions have also been initiated with DOE as a part of the DOE Clean Coal Technology Initiative.

Conclusion

Gas-enhanced combustion technologies may offer owners of coal-fired boilers flexibility to tailor emission controls to specific NO_X and SO_X reduction requirements at lower capital costs and improved cost effectiveness. In addition, to aid the gas industry in providing the optimum controls for site specific situations, the development of a cost/benefit methodology to comparatively evaluate commercially available emission control options is necessary.

While GRI research of gas-enhanced combustion technologies is currently at the bench and pilot scale, the potential benefits of gas to provide flexibility and low-cost options for emissions controls and operational improvements are being documented. GRI field testing plans will establish how broadly gas-enhanced combustion technologies can be applied and will improve acceptance in the market.

A primary goal of the program is to take the necessary research steps to assure successful field application. Consequently, active pursuit of co-sponsorship from industrial partners in field testing activities is underway. GRI solicits the input of interested parties in providing a test site, co-sponsoring a field test or otherwise participating with GRI in the development and field testing of these promising gas-enhanced combustion technologies.

Bibliography

Proxmire, W., and Humphrey, G., Acid Rain Control Bill, S.503, introduced in Senate, February, 1985.

Mason, J., "The Current Status of Research on Acidification of Surface Waters," the Royal Society, London, June, 1984.

Gas Research Institute/American Gas Association, "Evaluation of the Environmental and Other Benefits of Select Use of Natural Gas." GRI Report 82-0091. Prepared by Environmental Research and Technology, Inc., November, 1983.

Gas Research Institute, "Evaluation of Recent SO_X and NO_X Emissions Inventories," GRI Report 84/0120, Prepared by Argonne National Laboratory, May, 1984.

U.S. Department of Energy, "Analysis of Potential Combustion Source Impacts on Acid Deposition Using An Independently Derived Inventory," DOE/MC/19285-1505 Vol I (DE84012005), Prepared by PEDCO Environmental, Inc. and P.W. Spaite Co., December, 1983.

Spaite, P.W., "The Role of Local Sources and Oil Burning in Acid Precipitation," Presented to Committee on Energy and Natural Resources, August 19, 1982.

Crocker, T. D. and Regens, J. L., "Acid Deposition Control—A Benefit-Cost Analysis: It's Prospects and Limits," *Environmental Science and Technology*, November 2, 1985.

Gas Research Institute, "Comparative Application Studies and Economic Evaluations of Gas Reburn/Dry Sorbent Injection Technologies," GRI Draft Report, prepared by Energy and Environmental Research Corp., June, 1985.

Takahashi, Y. *et al.*, "Development of MACT In-Furnace NO_X Removal Process for Utility Steam Generators," American Power Conference, Chicago, April, 1982.

Gas Research Institute, "A Selective Review of Papers and Technology Presented at the First Joint Symposium on Dry SO_2 and Simultaneous SO_X/NO_X Control Technologies," GRI Draft Report, prepared by Energy Systems Associates, February, 1985.

Greene, S. B., *et al.*, "Bench Scale Process Evaluation of Reburning for In Furnace NO_X Reduction," presented at the ASME Joint Power Conference, September, 1984, Toronto, Canada.

Electric Power Research Institute (1983), "Developing The Options For Emissions Control," *EPRI Journal*, November, 1983.

Lachapelle, D., "EPA's LIMB Technology Development Program," *Chemical Engineering Progress*, May, 1985.

Dahlin, D. S. and Kirchgessner, D. A., "Surface Characterizations Microanalysis of Sorbents and Ash/Sorbent Mixtures," EPRI/EPA Symposium, San Diego, CA, November, 1984.

Breen, B. P. and Gabrielson, J. E., "A Review of In Furnace NO_X Reduction and SO_X Reduction by Sorbent Injection," presented at the EPRI/EPA 1982 Joint Symposium on Stationary Combustion NO_X Control, November 2, 1982, Dallas, Texas.

U.S. Department of Energy, "Select Use of Coal and Natural Gas In Utility Boilers, A Manual For Evaluating Environmental, Technical, and Legal Aspects." Unpublished Report prepared by PEDCO Environmental, Inc., March, 1984.

Chapter 20

Select Use of Natural Gas for Waste Management

PROFESSOR O. K. MAWARDI
Case Western Reserve University

The increase in regulations affecting landfills and the difficulty of siting new ones has led many municipalities and large institutional waste producers, e.g. hospitals, schools, housing complexes, to seek alternative procedures of solid waste disposal. At the same time, the recent rise in the cost of all fuels has caused public utilities to seek alternative sources of fuels. One solution which meets both criteria of waste disposal and moderately priced fuels is the waste-to-energy process that converts waste into a number of useful energy forms.

As a result, it is expected that the number of waste disposal plants will rise sharply once it is established that these plants are economically viable. A concern that stems out of the proliferation of these waste disposal plants is the deleterious effect on the environment of the combustion products from these plants. The preliminary encouraging finding of the environmental benefits of the select use of natural gas in coal and oil burning plants[1,2] prompted research to examine the potentials of applying this combustion procedure to waste-disposal plants.

This chapter reviews preliminary assessment of the select use of natural gas for waste disposal. It discusses a survey of waste disposal policies in and outside the U.S., the economic and other benefits of select use applied to a waste disposal plant, and gives recommendations for this method of waste disposal. The emphasis is on the co-firing combustion mode of the waste with natural gas. The Btu contribution of the natural gas is therefore an important fraction of the total thermal output from the waste energy-recovery plant. This mode of operation is to be distinguished from that using natural gas for start-up. It is also different from the practice followed by the smaller waste-disposal plants, viz., those with disposal capacity of 50 tons per day or under, to use a supplementary fuel beyond the start-up phase in order to maintain the heat rate constant.

When assessing a waste-disposal system one has to consider several broad issues of technical, economic and political character.

(a) *Technical Considerations*

The advent of landfill gas technology[3] has considerably reduced the nuisance value of landfills for municipal solid-waste disposal. Through this technology 11 billion cubic feet of gas were extracted in 1984 from landfills.[4] But the increasing shortage of new landfill sites in the vicinity of our larger cities urgently calls for the construction of solid-waste disposal plants that can process the waste *rapidly and economically*. Incineration meets the two above mentioned criteria.

With existing technologies, solid-waste disposal systems are capable of producing a wide variety of energy and fuel products. The most marketable of these products are steam and/or electricity and natural gas. Next in diminishing importance are refuse-derived fuels (RDF) and materials recovered from waste such as metals, paper and glass. Steam is the predominant recovered energy product (see Table 20-1). The spectrum of uses extends from district heating and cooling to process heat for industrial applications. Table 20-1 is compiled from data gathered in the U.S.[5] It does not include landfill gas which is not a combustion-derived fuel.

The modern incinerator is not merely a refuse burner, but a sophisticated utility type boiler which uses solid waste as a fuel. Furnace design based on knowledge of combustion and steam generation principles is incorporated in the incinerator and while the furnaces may vary in design, cost and efficiency, the common goal is the reliable conversion of waste into steam.

For the most part, the equipment available falls into one of three categories[6] which are:

(1) Control air type which consists of two refractory lined chambers. Solid waste is introduced in the primary chamber and is

Table 20-1. Waste Derived Energy Products Waste Product User

Waste-Product	*Industrial*	*Public Utility*	*Commercial*	*Institutional*
Electricity	0	25%	0	0
Steam	77.7%	16.6%	87.5%	100
RDF	22.2%	41.6%	12.5%	0

supplied with sufficient air to gasify the waste. The gases are directed to a secondary chamber where additional air is added and combustion is completed. These types of incinerators are fitted with fossil fuel fired burners in the afterburning section to insure a clean stack.

(2) Refractory furnace/dutch-oven type which consists of a multi-chamber box with moving air cooled stoker grates. The final hot gases are directed to a heat recovery boiler and then to some type of air pollution control device. Excess air is generally high; in the order of 150 to 300 percent.

(3) Water- or air- cooled furnace type which are the most modern state-of-the-art units. These units are similar to the utility boiler plants that utilize a water- or air-cooled refractory wall furnace integral with a boiler convection section.

The refuse-derived fuels (RDF) when densified can be burned in a regular boiler. For instance, facilities of municipal electrical utilities in Iowa and Wisconsin[7] have burned densified RDF in their boilers to produce high pressure-high temperature steam which can be applied then in a cogenerating capacity for steam heating and/or generation of electricity.

On the other hand, the combustion of unprocessed RDF in dedicated boilers produces medium-temperature and pressure steam which is best suited for industrial application and heating but not for electrical generation. Typical of the latter type of installations are the ones in Akron, Ohio and in Hempstead, N.Y. The unprocessed RDF can also be used with advantage in old inefficient (electric generating) plants that would otherwise have been phased out because of their inefficient mode of operation.

(b) *Pollution Controls*

Waste substances vary greatly in their chemical composition, state of aggregation, moisture and heat content. Thus incineration processes in a municipal waste disposal plant yield a great deal of variability in the emission. Three basic types of air pollution control equipment, however, have been used so far:[8] electrostatic precipitators, bag filters and high-energy scrubbers.

Of the three control techniques, electrostatic precipitators have given the most successful and consistent results with respect to particulate emission. The performance of electrostatic precipitators is very sensitive to the amount of moisture in the flue gases. Whenever the flue gases are cooled by the boilers and the only moisture in the gases originates from the wastes and the combustion process, the precipitator performs very well. Should the furnace gases be cooled by water sprays prior to reaching the precipitator, the emission control performance worsens appreciably.

Experience to date with fabric filters is very limited. Performance of the baghouse filters, however, appears to be marginal.

Several tray-type and venturi-type scrubbers have been installed on municipal-sized incinerators since the mid to late 1960's. Neither type of these scrubbers were able to meet the Federal emission standards (for particulates and gas emissions). It was believed at first that the scrubbers' performance depended on the chemical characterization of the effluent flue gases.

Subsequent tests[9] in which the scrubber water was recycled and corrected for pH and fresh water make-up provided did not improve the performance of the scrubbers. Increasing the pressure drop did not necessarily increase the collection efficiency. The scrubber failures have raised questions as to their ability to meet the emission requirements unless systematic investigations clarify the underlying phenomena controlling the performance of the scrubbers as gas and/or particulate removers.

(c) *Corrosion and Select Use*

Waste energy-recovery installations are troubled with corrosion accompanied by slagging and tube deposits in the same general areas in which these problems occur in some utility boilers. The similarity between corrosion patterns observed in waste-fired systems and those in conventional generators have led investigators to conclude that the causes might be similar.[10,11] Liquid phase corrosion in superheaters, and reducing zone corrosion in waterwall areas and dewpoint are generally accepted as explanations for the cause of wastage in these areas.

Preliminary attempts in the combined combustion of coal with natural gas in the select use mode[12] appeared to indicate a reduction of slag formation in the boilers. It is also speculated that the potential for cold-end corrosion, the acid dewpoint is closely related to NO_X emissions from the waste combustion. The amount of sulfur present in municipal waste is usually quite low, less than 0.1%, consequently sulfur emission is of no concern.

The mechanisms for NO_X generation are thermal formation and fuel nitrogen formation.[13] There are well understood methods of NO_X reduction and there are experimental evidences to indicate that select use does indeed reduce the NO_X emission.[14]

Several investigators have reported that accelerated corrosion often approaches catastrophic proportions when alkali metal chlorides are present with low melting sulfur-bearing compounds.[15] Explanations of the cause of corrosion in a waste-fired steam generator can be described by the role of chloride, sulfur and heavy metal compounds. Chloride and chlorine are contributed mainly by the polyvinvyl chloride (PVC) plastic and salted food residues.

Although the chemistry of the process is quite complex, it appears that the presence of chloride in a deposit on the tube surface can play an active corrosive role if the environment cycles between oxidizing and reducing conditions. It is believed that the preponderance of reducing agents in the select use mode of combustion may retard the corrosion and is well worth investigating.

(d) *Economic Considerations*

The total cost of waste disposal is not a simple case of the costs of transportation and landfilling. Another factor that has to be taken into account includes the scarcity of available sites for disposal. This is especially true in light of the Resource Conservation and Recovery Act[16] (RCRA), passed by Congress in 1976 (amended in 1978), of the 1979 Toxic Substance Act, of the 1970 Clean Air Act and of citizen opposition to having any kind of landfill anywhere in their neighborhood.

In the past, low cost land, minimal regulation and lax enforcement permitted low-cost landfilling for most solids for about $5 per ton of waste. The combined effect of RCRA together with strict enforcement of landfill operation has increased significantly the landfill costs. Figure 20-1[17] shows the increase of the landfill cost with time. The "knee' in the curve reflects the initial jump in costs due to RCRA, while escalation due to inflation and other factors shows a steady increase in the costs at the rate of 8% per year.[18]

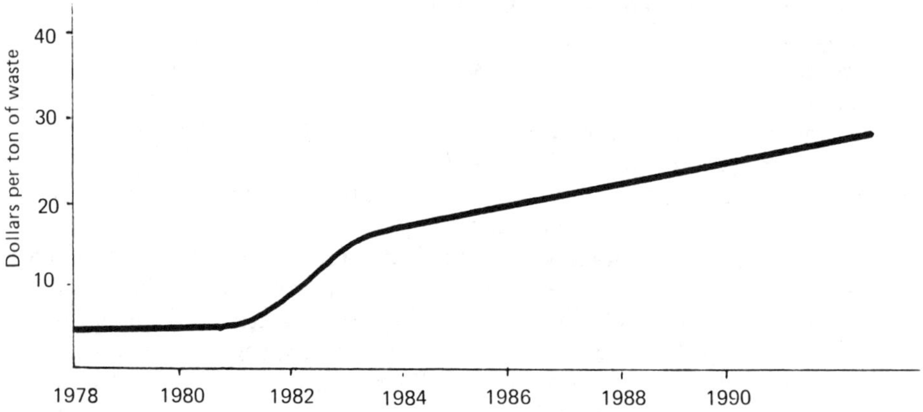

Figure 20-1. Anticipated Trend in Landfill Costs

During the initial period of construction, start-up and shakedown of the waste-to-energy facility there will be little if any revenue. The break-even point where the revenues will offset the annual costs occurs after several years. This will be shortened to the extent that the cost of energy alternatives continues to rise. Additionally, the environmental benefits of the select use procedure may be achieved at a lower cost than via alternative control methods.

(e) *Regulatory Impediments*

Although the remarks mentioned in this section do not pertain to the select-use practice directly, they are of crucial importance to the waste energy recovery. It is essential, therefore, to understand the implications of these issues because that facet of select-use practice considered here is very closely tied to waste disposal.

There are a number of factors which account for the relative scarcity of waste energy-district heating systems. One of them is the high cost of installing new steam distribution pipelines in addition to the high capital investment for the waste-disposal plant. As a result, the growth of such systems is almost limited to those locations where steam lines are already in existence.

But existing steam lines are found mainly in densely populated, downtown regions where there are often strict air pollution standards and where the traffic of refuse-filled trucks travelling in congested city streets to a downtown site does not seem practical. A compromise must be reached, therefore, in which the site is close to but not in the center of town. There is no solid evidence to indicate that the combustion of solid waste or of RDF is worse for the environment than conventional fossil fuels.

At any rate, federal regulators have thus far allowed the exemption of waste disposal-energy recovery plants from meeting the emission NO_X requirements in critical areas. The Fuel Use Act probably technically allows select use with RDF, but the complexity and uncertainty of complying with the Act is a significant impediment from the point of view of many electric utilities. Repeal of the Act would be an important step for select use with RDF.

Although the present guidelines definitely encourage urban resource recovery sites, it is not clear whether these rules would still apply should the number of such plants continue to increase. Clearly, this is a situation most favorable to the select fuel concept.

In cities where steam lines are in place and operated by investor-owned utilities, there are certain problems. State regulatory commissions generally require that fuel savings from the use of waste-disposal plants be passed on to customers. This practice destroys much of the incentive of utilities to develop such plants.

Another inequity often imposed by State Utilities Commissions relates to the ability of the investor-owned utility to limit its service to district heating regions which are economically efficient. Since steam cannot be transported beyond a few miles, reasonable rates can be maintained only if limited to shorter distances. Consequently, if customers in outlying regions demand services at the same rate, the system cannot be kept economical. This situation is in vast contrast to electric service where the limiting zone aspect is nearly absent. It is also believed that this might be the reason ascribed to district heating systems being unprofitable.

The federal government has attempted to address some of these issues through the Public Utility Regulatory Policies Act of 1978. This act is supposed to encourage conservation of electricity and fossil fuels. Provisions in the Act permit the Federal Energy Regulatory Commission to review state utility commissions' rulings and to intervene whenever there are opportunities to further energy conservation practices.

Along this line of thinking, cogeneration is emerging as a very powerful procedure to intensify energy recovery. At the turn of the century, the standard procedure was to have steam from boilers drive turbo-generators to produce electricity, while the exhaust steam was channelled to a local district heating system; otherwise, the exhaust energy would have been wasted.

It was customary, therefore, for a public utility to sell both electricity and steam. With the low cost of energy, district heating could not compete with the other energy sources. Furthermore, with the exodus to suburbia, district heating just about disappeared since it was not economical to distribute steam to a widely scattered load, since many suburbs were located farther than 2-3 miles from the power plant, i.e. beyond the limit of operation of a district heating system.

With the rapid increase in the cost of fuels in the last decade, there has been a renewal of interest in cogeneration. There are approximately 1090 cogeneration facilities[19] either in operation or in the design/construction phase in the United States. The estimated combined output electrical capacity of these facilities is 28,000 MW. Of this capacity 37.6 percent is believed to be gas fueled.

Aggressive state regulatory actions in California and elsewhere have initiated innovative ways to overcome constraints in utility regulations. For instance, since 1980 California investor-owned utilities have offered avoided-cost prices for electric energy from waste disposal. The avoided cost is defined by the Federal Energy Regulatory Commission (FERC) as "The incremental costs to an electric utility of electric energy or capacity, or both, which but for the purchase from the qualifying facilities, such utility would generate itself, or purchase from another source."

The avoided cost principle is presently practiced in the entire country as required by PURPA. The FERC has also ruled (1981) that certain qualifying cogenerators are exempt from state and federal regulations. In particular, cogeneration facilities are exempt from certain provisions of the Federal Power Act and from the Public Utility Holding Company Act of 1935.

Because of these regulations, it is projected that utilities will become important partners in the resource recovery development business. As a result, one can identify a few concrete benefits to the public:

(1) Revenues from power sales will help reduce the waste disposal costs or "tipping fees."

(2) It takes on the average ten years for a utility to develop a new conventional electric generating facility. This long lead time makes it difficult for a utility to respond to an increased energy demand from a region. Smaller waste-to-energy plants, however, can be completed in the shorter period of three to five years.

(3) Finally, cogeneration reduces the dependence of the nation on imported oil.

The above discussion has enumerated some of the salient advantages (and disadvantages) of waste-to-energy conversion plants. In particular, the preferential treatment given to waste disposal plants was pointed out. It is safe to assume, however, that RDF do pollute the environment. Therefore, it will not be long before anti-pollution regulations and practices will be enforced on these plants once their number escalates to the point of being a pollution hazard. The study of pollution control methods hence becomes very timely.

Survey of Known Practices in Waste-To-Energy Conversion

Current priorities in the United States with respect to energy and materials from municipal solid waste include (1) the use of refuse as a supplementary fuel, (2) methane from landfills, (3) pyrolysis, and (4) resource recovery. An extensive inventory of resource recovery facilities was compiled by the National Center for Recovery, Inc. (1982). Although the inventory includes 34 relatively large facilities, the concept of energy recovery from waste has not been as fully accepted in the U.S. as it has been overseas. Landfill gas (LFG) projects are being initiated in numerous communities in the U.S. As of the end of 1983, at least 27 facilities were producing LFG with several more in the design/construction stage. The estimated production in 1984 was 2.8 Bcf of High Btu landfill gas and 8.2 Bcf of Medium Btu gas.

A comprehensive survey of practices in European countries conducted a few years ago reveals the advanced state of technical development in this field. The key findings in this survey are summarized below.

Landfill generally remains the preferred method of disposal in Western Europe where suitable sites are available, but these are becoming increasingly difficult to find, particularly near large cities.

Incinerators are in common use, especially in urban areas, but costs have increased so rapidly that several governments are discouraging the construction of new plants. Increasingly stringent air pollution regulations necessitating the use of gas scrubbers (e.g. in West Germany) can cause the costs to climb even higher.

Provision for heat recovery is made in many incinerators, especially larger plants. The technical difficulties (notably corrosion) are now reasonably well understood. The problem of matching supply with demand remains a significant barrier to heat recovery, especially for electric generation. The recovery of heat for district heating and/or industrial use is preferred, particularly when an incinerator can provide the base load throughout the year. Rising fuel costs are making the economics of heat recovery more attractive.

Most countries are interested in the possibilities of exploiting new technologies such as materials and fuel separation and pyrolysis. On the whole they are waiting for data from the U.S., but in some places development is proceeding at a rapid pace.

Countries in which front-end materials and fuel separation processes are being developed include England, Italy, Spain, Sweden and West Germany. Of particular interest is the burning of refuse-derived fuel in cement kilns. Incidentally, this application of the select-use concept is being investigated here in the U.S. by the cement industry.

Countries in which pyrolysis processes are being developed include Denmark, England, Netherlands, Sweden and West Germany.

Some work has been done on materials recovery from past incinerators residues especially in England and France but incinerator operators have been reluctant to install large-scale systems. In several countries, household sorting and separate collection systems are being looked into quite seriously.

Economics of Select Use for Waste-to-Energy Recovery

At present, there are no known waste energy recovery installations in the U.S. that fully employ the select-use concept. Most installations use natural gas (or oil) for start-up. The consumption, as expected, is a small percentage (5%) of the energy generating capacity of the incinerator.

On the other hand, there is a substantial amount of information on the economics of waste energy recovery plants of different types (starved air, fluid bed, combustor, waterwall and standard). The select-use concept (SU) should improve the economics of a waste/energy plant for a number of well-defined reasons:

(1) SU will reduce the emission of SO_X and NO_X compounds. In the absence of precise quantitative information, it will be assumed that the reduction is a linear function of the natural gas used. It will further be assumed that only electrostatic precipitators will be installed. The cost of these control devices are well-known.

(2) SU will reduce the amount of ash produced. The assumption made is that the ash reduction is proportional to the amount of natural gas used.

(3) SU can improve the overall efficiency of the boilers especially when the waste-to-energy plant operates in a cogeneration mode and with a fluctuating load. SU can therefore transform the output in a virtual load levelling fashion which is known to improve the efficiency of the system.

Assumptions

The cost analysis requires assumptions about equipment costs and labor. Because there exists a substantial amount of experience for incinerators of modest capacity (100 tons/day), this size has been used as a reference. The following assumptions have been made:

(1) All equipment has been amortized at 10% over a useful life of 20 years.

(2) The installation operates 240 days a year.

(3) The cost of capital for the plant is fixed over the life of the plant and is not affected by inflation.

(4) The operating costs are not fixed but rise (or fall) with inflation. However, the capital cost represents almost 70% of the first year's total annual cost. Therefore, the effect of inflation is minimized because only the remaining 30% of the cost rises at the inflation rate.

(5) The inflation rate is taken as 4.2%.

(6) The average estimated from a variety of waste yields approximately a 10% contraction for ash production. This means that a 100 TPD installation which processes 26,000 tons per year yields 867 tons of ash per year.

(7) The cost of hauling the ash is on the basis of $38/ton for the first year.

(8) The effect of SU on the boiler is to introduce a 2% improvement in efficiency.

(9) Enough natural gas is used to lower the emission of NO_X by 20%. This reduction is reflected in a lowering of the cost[22] of the air pollution and ash removal installations by 20%. It also lowers the cost of hauling the ash by 38 $\times$ (867 $\times$ 20% = $6,590 per year.

(10) The amount of natural gas used is estimated on the simplified premise that the reduction in pollution, viz. 20% = $\frac{\text{natural gas Btu}}{\text{natural gas Btu} + \text{Waste Btu}}$. But the gross fuel input per year for a 100 TPD of waste with an equivalent Btu value of 7100 per lb is reckoned to be 332,270 MMBtu/year. Therefore the natural gas contribution per year is 6645H MMBtu. This is equivalent to an expenditure of $284,420/year for the first year on the basis of 1984 prices[23] of $4.28 per MMBtu.

(11) The price of electricity[24] is $18.34 per MMBtu and that of steam[25] is $7.15 per 1,000 lbs.

(12) The combustion efficiency[26] for the boiler is taken to be 75% and that for the electric conversion efficiency[27] 30%.

Cash Flow

The cash flow for the first year and for two cases, a conventional waste-to-energy installation and the corresponding SU option is indicated below. The basic installation is a 100 TPD plant consisting of two waterwall incinerators each of 50 TPD capacity.

CAPITAL INVESTMENT (in $1000)

		Conventional[28]	*SU Installation*+
Main Installation (Excluding Pollution Controls)		7,600	7,600
Air Pollution Control		500	400
Air Removal Installation		250	200
	Total	8,350	8,200

DEBT SERVICE (in $1000)

		Conventional	*SU Installation*
Capital Amortization (20 years)		417.5	410
Interest (10%)		835.0	820
	Total	1,252.5	1,230

OPERATING EXPENSES (in $1000)

		Conventional[29]	*SU Installation*
Power (electric)		101.60	101.60
Natural gas		27.5	284.42
Water and Sewer		7	7
Chemicals		20	20
Labor		240.00	240.00
Ash Disposal		33	26
Maintenance		45	45
G & A		89	89
	Total	563.1	813.02

+The assumption is made that essentially the same installation is used and through SU, the emission and ash production is reduced by 20%.

REVENUES

		Conventional	*SU Installation*
Steam Sales++		1,771,52	2,167.20
Tipping Fees*		78.00	78.00
	Total	1,849.52	2,245.20

CASH FLOW (in $1000)

		Conventional	*SU Installation*
Debt Service		1,252.50	1,230.00
O & M		563.10	813.02
	Total	1,815.60	2,043.02
Revenues		1,849.52	2,245.20
	Balance	33.92	202.18
Equivalent profit $/ton of waste		1.30	7.78

The trend of revenue projection vs the expenses are shown in Figure 20-2. The estimated expenses are for constant 1984 dollars and have been pro-rated to expenses per ton of waste. Inflation factors have been calculated in accordance with Reference 24.

Energy Model for Waste-to-Energy Conversion System

In order to accommodate a large number of inputs, we designed a computer model. The program was planned to be user-friendly so as to permit the inclusion of information on:

(1) The plant characteristics (efficiency, base level, amount of gas use, etc.)

(2) The waste characteristics (tipping fees)

++The conventional plant operates at 75% efficiency with an output of 381.176 Mlbs of steam. The SU installation operates at a higher efficiency: 77% and a corresponding output 20% higher, i.e. 466.559 Mlbs. The capacity utilization for both cases is assumed to be 0.65.

*Taken as $3.00 per ton of waste.

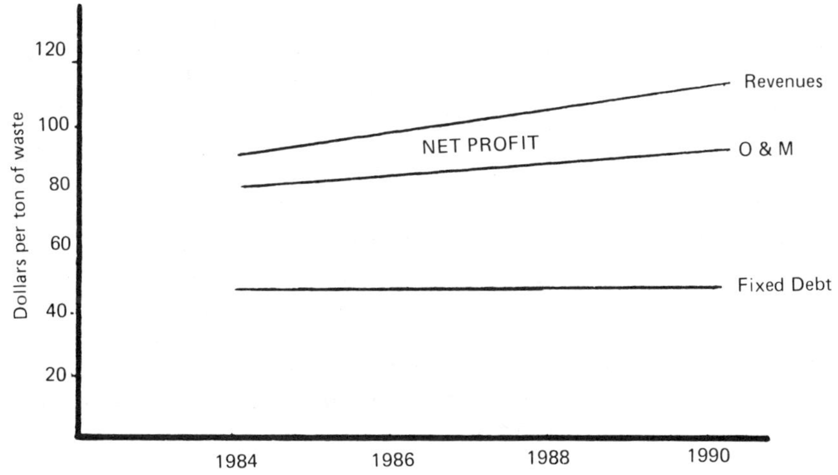

Figure 20-2. Revenue Projection for the Plant. All Quantities Shown Are in Nominal Dollars.

(3) The equipment used (incinerators, shredders)

(4) Emission and particulate equipment control

(5) Method of financing (interest, life of loan)

The output from the program permits the determination of:

(1) Trade-off point; i.e. the percentage of Btu (contributed by natural gas) at which point the waste-to-energy system becomes profitable.

(2) The dependence of emission control cost on the use of natural gas.

Conclusions and Recommendations

The analysis in this study indicates that the revenues of a waste energy-recovery plant are increased when SU is introduced. The net profit for a 100 TPD plant is $1.30 per ton of waste processed. Introducing SU this profit becomes $7.78 per ton, i.e., nearly $6.50 per ton. This projected rise in the profit would result from the burning of natural gas with an equivalent Btu content of 20% of the waste Btu.

The beneficial advantage of SU is more strikingly demonstrated if there were a strict enforcement of the NO_X mission control as is the case in Japan or Germany. This study assumes that the air pollution control equipment deals mainly with particulate. Had one included scrubbers for NO_X emission control the cost for a conventional plant becomes much higher than a SU installation.

Additional benefits of SU which have not been explicitly included in the analysis refer to the expected reduction in slagging and corrosion in the plant. The improved performance of the installation means lower maintenance and longer life of the equipment. The impact of SU on the combustion process is still not too well understood. There is a need for a serious effort to conduct a systematic research on this process. The results of such work are bound to lead to a wider acceptance of select use of natural gas.

References

1. See for instance the Proceedings of the "International Conference on Combined Combustion of Coal and Gas," Cleveland, Ohio 1982 (Case Western Reserve University and American Gas Association).
2. American Gas Association, "Fact Sheet August 17, 1984," cites 20 facilities which employ or have employed select use.
3. Energy from Solid Waste, F. A. Domino, Editor, (Noyes Data Corporation, New Jersey) 1979.
4. "Methane from Landfills" (1984), Report by American Gas Association Policy Evaluation and Analysis Group and Renewable Sources of Methane Subcommittee.
5. T. H. Barnett *et al.*, "Incentives for Using Municipal Waste as an Energy for District Heating," Tech. Rept., Gershman, Brickner & Bratton, 1981.
6. G. Swinehart, "State-of-the-Art of Combustors for Municipal Waste Disposal," Tech. Report, O'Connor Envirotech Corp., 1980.
7. Ref. 5.
8. Present Status and Research Need in Energy Recovery from Wastes, American Society of Mechanical Engineers, Proceedings of Conference (1976), p. 27, *ff.*
9. P. J. Waznys and C. J. Desio, in Proceedings of Conference of Research Needs in Energy Recovery from Wastes, (1976), p. 48, *ff.*

10. W. Neson, *et al.*, "Corrosion of Superheaters and Reheaters of Pulverized-Coal-Fired Boilers," Trans. ASME, Series A82, 194, (1960).

11. A. L. Plumley, *et al.*, "A Review Study of Fireside Corrosion in Utility and Industrial Boilers," Canad. Chem Proces. 51, 52 and 70 (1967).

12. R. J. Drummond, "Combination Fuel Burning at James River Plant," Paper No. 9, International Conference on Combined Combustion of Coal and Gas, 1982.

13. W. R. Niesen, Combustion and Incineration Process (Marcel Dekker, New York) 1978.

14. Emission Control Technology for Industrial Boilers, A. E. Martin, Editor (Noyes Data Corporation, New Jersey) 1981.

15. A. J. B. Cutler *et al.*, "The Role of Chloride in the Corrosion Caused by Flue Gases and Their Deposits," Trans. ASME Jour. Eng. for Power A 93, 307 (1971).

16. RCRA—Public Law 94-580, October 21, 1976, Amended 1978.

17. W. A. Sanders *et al.*, Public Works 74, May 1981.

18. Environmental Impact Statement Criteria for Classification of Solid Waste Disposal Facilities and Practices—U.S. Environmental Protection Agency SW-821, December 1979.

19. American Gas Association Fact Sheet, May 25, 1984.

20. European Technology for Obtaining Energy from Solid Waste, D. J. DeRenzo, Editor (Noyes Data Corporation, New Jersey) 1978.

21. Ref. 4.

22. A. M. Squires in "Power Generation and Environmental Change," Ed. D. A. Berkowitz & A. M. Squires, (MIT Press, Cambridge, Mass.) 1969.

23. American Gas Association, August 1984 Fact Sheet.

24. American Gas Association, TERA Analysis, November 16, 1983.

25. Represented price of steam, such as charged by the Sumner County, Tenn. waste energy-recovery system.

26. Estimate provided by O'Connor Envirotech Corp.

27. Edison Electric Institute, Cogeneration and Small Power Production, (Washington, D.C., February 1980).

28. Data provided by STV Corp.

29. Ref. 28.

Chapter 21

Natural Gas Use to Facilitate Coal and Waste Combustion

A. E. S. GREEN, *University of Florida*
E. J. MIZE, *Peoples Gas System Inc.*
D. S. De, *Indian Institute of Technology*
B. A. GREEN, *Furnace Consultant, Micanopy, FL*
P. WEST, *Sunland Training Center*
M. SANDIFER, *Indian Institute of Technology*
C. NIVEN, *University of Florida*

Nstural gas has provided about 20% to 30% of the national energy used for several decades and during the 70's production was up to 24 trillion cubic feet per year. Concerns as to the possible exhaustion of our natural gas reserves as embodied in the Fuel Use Act of 1978, conservation measures, construction of coal fired boilers and the recent oil glut have led to a decrease in the use of natural gas to the 18 Tcf/yr level. Despite many projections of large price increases, natural gas prices (per thermal unit) have tended to remain below the price of residual oil, as might have been expected.

In 1984 the sources of energy used in the United States were 42% petroleum, 25% natural gas, 23% coal and 10% all other (nuclear, hydroelectric, solar, wood, wastes, etc.). In this work we consider using a small portion of our current production of natural gas to facilitate further reductions in our use of oil, which we continue to import in large quantities. Thus we propose to take advantage of natural gas's unique physical and chemical properties by using it as an assist fuel for the combustion of lower quality hydrocarbons.

We were originally led to consider this blended fuel concept in 1980 in connection with a study of increased coal use and oil back-out in Florida[1] and a small business cost problem.[2] As a result of experimental, theoretical and economic studies since then,[3-11] we are convinced that there are advantageous synergisms ("a united action producing a greater effect than the sum of the parts") in the simultaneous combustion of natural gas with lower quality fuels such as pulverized coal, coal liquid slurries, residual oil, urban refuse, sludge, wood wastes and a variety of industrial wastes. We focus primarily upon advantageous physical interactions leading to gains beyond the simple additivity of the heats of combustion. Thus we seek combustion arrangements which lead say to $1+1=3$ or $0+1=2$.

Recently the possibility of replacing oil by coal water mixtures (CWM) has received strong attention[11-14] and is nearing the large boiler demonstration phase. Coal water mixtures pump like oil, maintain the infrastructure of utility and industrial oil use and afford environmental benefits through the physical and chemical coal cleaning during slurry preparation. Unfortunately CWM use in oil designed boilers can lead to 30% to 50% deratings,[15-17] reduced carbon conversion efficiencies, slagging, fouling, flame instability and other problems. In a previous study[11] we examined various methods for assisting CWM combustion to overcome or mitigate these problems.[18-20] These assist options have been examined along with the option of (1) staying with oil (SWO) or (2) going to coal water slurry without assist (WOA). This comparison included assist by (3) oil (CWO), (4) methanol (CWM), (5) oxygen (OXY), (6) fine pulverization (FIP), (7) improved combustion arrangements (ICA), (8) natural gas (NGA) and (9) two-stage combustion with natural gas assist (2SG).

The results of an assessment of the relative merits of these options are summarized in Table 21-1. Here grades are given for the various benefits associated with each option including financial or economic (F), national security (S), environmental (E) and political (P). Two additional options, (10) pulverized coal alone (PCA) and (11) pulverized coal with gas assist (PCG) have been added in consideration of the current problem of the small oil-CWM price differential. These eleven options are given net grades G_1 corresponding to a linear composite score and G_2 corresponds to a root mean square composite score. The corresponding rankings indicate that greater attention should be given to gas assisted coal combustion.

Table 21-1. Coal Assist Options

Options	*F(0.4)**	*I(0.3)**	*E(0.2)**	*P(0.1)**	*G1*	*G2*	*R1*	*R2*
1 SWO	B	D	B	C	2.3	2.47	10	9
2 WOA	C	B	C	B	2.4	2.45	8	10
3 CWO	C+	C	C+	C+	2.35	2.36	9	11
4 CMW	C	B	B	C	2.5	2.55	4	4
5 OXY	C	B	C+	C+	2.45	2.48	7	7
6 FIP	C	B	B	C	2.5	2.55	4	4
7 ICA	D	B	B+	C+	2.25	2.48	11	7
8 NGA	B	B	B+	C	3.0	3.02	1	1
9 2SG	B	B	B+	C	3.0	3.02	1	1
10 PGA+	C+	B	C	C	2.5	2.53	4	6
11 PCG+	B	B	C+	C	2.8	2.82	3	3

*Numbers in parenthesis denote weights.[11]
+Assumes compliance PC is delivered to boiler.

Gas Assisted Combustion

The R&D attention which has been given to oil backout alternatives since 1973 has focused considerable effort on coal pyrolysis, coal combustion as well as upon the combustion of gas, liquid and solid hydrocarbon fuels in general. An important property of a hydrocarbon fuel is its H/C ratio as illustrated in Figure 21-1. Methane (CH_4) the main constituent of natural gas has the highest H/C ratio. Other hydrocarbon fuels, including not only the conventional ones but also wood, solid waste, sewage sludge, etc. are also shown. The science of combustion points to the major role of free radicals such as H, OH, C in accelerating combustion kinetics. Second only to molecular hydrogen, an expensive secondary by-product of primary fuels, methane is the most energetically favorable hydrogen rich fuel. Hydrogen is used in synthetic fuel production for the pyrolysis of coal and other fuels (i.e. hydropyrolysis) to increase the H/C ratio of product fuels. So also methane can assist in the pyrolysis of other HC fuels (methanolysis) to increase their H/C yields.

Many experimental and pilot scale results of coal pyrolysis have

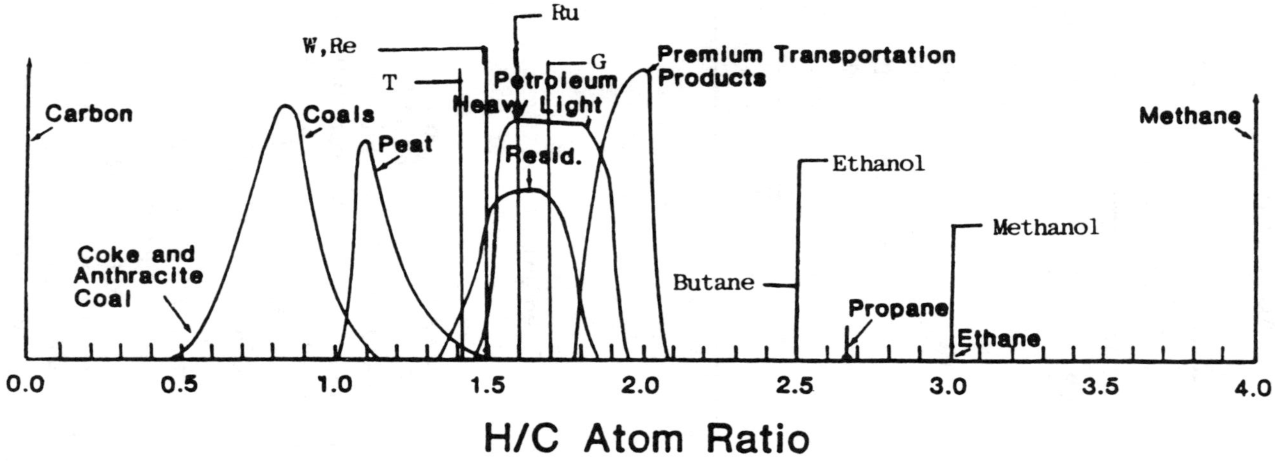

Figure 21-1. Hydrogen/Carbon Atom Ratios for Various Hydrocarbon Fuels

The positions of Methanol, Ethanol, G—Garbage, Ru—Rubbish, W—Wood and T—Trash are also indicated with water and bound oxygen discounted.

been organized in terms of highly simplified mathematical models. The detailed physical and chemical mechanisms which are actually involved in coal pyrolysis and combustion and coal water slurry evaporation, pyrolysis and combustion are illustrated in Figure 21-2. This realistic conceptual model is adapted from a schematic model of coal pyrolysis by Steinberg, Fallon, and Sundaram,[21,22] a Brookhaven National Laboratory synthetic fuels group and a SWM combustion model by Sakai et al., a Mitsubishi combustion group.[23]

An attempt to mathematically model the kinetics illustrated in Figure 21-2 has been made by Green et al.,[10] using a representative molecular model of coal which is compatible with proximate and ultimate analyses,[24] x-ray diffraction and other analyses of coal structures,[25] devolatilization data taken at MIT,[26] and Brookhaven pyrolysis data.[20-21] The last data sets give yields of the various gases, methane, ethane, ethylene, the light liquids benzene, toluene and xylene, and of heavy liquids or tars when coal pyrolysis is carried out in H_2, CH_4, CO_2, Ar, He and N_2. The molecular model which identifies coal by the formula $C_{138}H_{104}O_{14}A_2$, where A_2 is a representative ash, together with reasonable kinetics gives encouraging results in relation to these data. Essentially this coal molecule is partitioned into subgroups whose bonds are broken at higher temperatures, so that in pyrolysis the coal decays into constituents according to a large set of coupled reaction equations.

Figure 21-3 illustrates the total mass loss of coal particles as functions of time for various pyrolysis temperatures in relation to the MIT devolatilization data. Note that the mass loss into volatiles increases strongly with the rate of temperature increase and the magnitude of the final temperature. In Figure 21-2 it is seen that following atomization and evaporation of the water, CWM droplet combustion becomes the problem of PC particle combustion. The steps in coal particle combustion include (1) pyrolysis into volatiles and tars, (2) thermal breakdown of coal tars into combustible gases and oils and a residual char-ash, (3) fast combustion of the volatiles and (4) slow burning at the surface and fissures of the char-ash via surface reactions with the inwardly diffusing air oxygen.

An informative discussion and macroscopic model of CWM combustion has been given by Sakai et al.[23] (see Figure 21-4) who compare CWM with PC combustion. They conclude that: (1) The mean diameter of atomized CWM droplets is generally larger than typical PC particles. A CWM droplet thus contains many PC particles

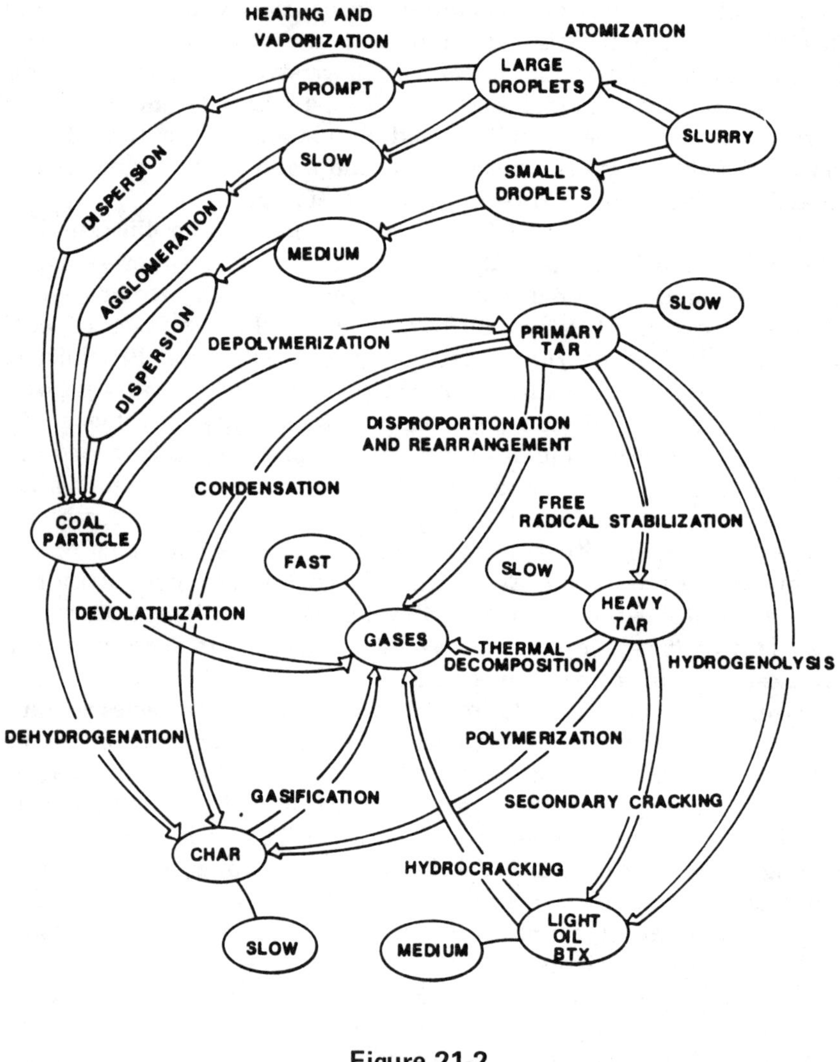

Figure 21-2

Schematic diagram of coal-water slurry and coal particle pyrolysis and combustion adapted from Steinberg et al.[21,22] and Sakai et al.[23]

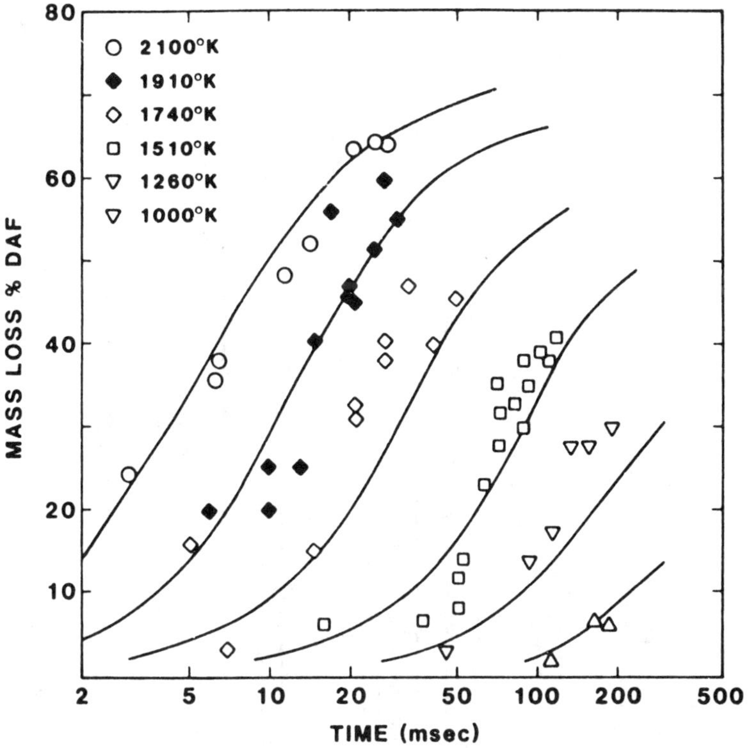

Figure 21-3

Pyrolysis yield vs. time for bituminous coal particles at various temperatures,[26] and fits with molecular pyrolysis model.[10]

which may agglomerate during the drying process. (2) Water evaporation is required hence the ignition point of CWM is further from the burner than the ignition point of PC. Special techniques are needed to achieve stable CWM flames. (3) The mixing of CWM droplets with combustion air is very strong, leading to a short flame length and altered fuel nitrogen conversion to NO. (4) The hydrogasification reaction C+OH CO+H reduces the emissivity of a CWM flame in the volatile matter combustion zone. This also extends the carbon burnout time. The lower flame temperature lowers NO_X formation. (5) By appropriate burner design the flame stability of a CWM flame

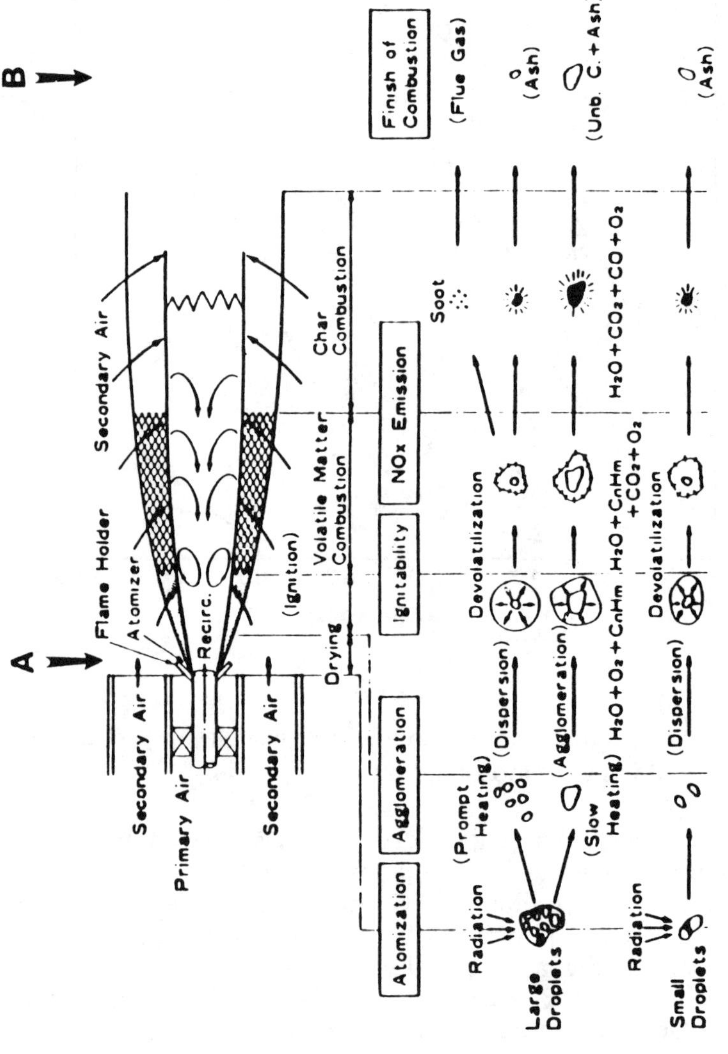

Figure 21-4

CWM combustion model (adapted from Sakai et al.[23] The arrows A and B show where "foreburning" with natural gas can lead to physical synergisms.

can be maintained at the same level as that of a PC flame. (6) Unburned carbon loss for CWM is higher than for PC for the same coal particle size distribution. Sakai et al. identify: (1) atomizing characteristics; (2) agglomerative properties; (3) instability and flammability; (4) NO_X formation; and (5) combustibility as ruling factors in CWM combustion.

We are a long way from achieving precise mathematical models of the schematic physical model illustrated in Figure 21-2 or the macroscopic model in Figure 21-4 proposed by Sakai et al. Until we reach such a stage it will be necessary to rely on experimental tests to reach firm conclusions as to the results of CWM combustion. Since there is a consensus[15-17] that combustion of CWM in oil boilers will be associated with 30%–50% deratings and combustion problems, the next section is devoted to a program underway in Florida to mitigate these problems by gas assist.

Gas Assisted Combustion in a Water Tube Boiler

Potential synergisms in the combustion of pulverized coal (PC) or coal water slurries with natural gas have been discussed in a number of our prior works.[2-11] The opportunity to demonstrate such synergisms for realistic combustion arrangements came in 1985 with the support of the Peoples Gas System Inc., the Florida Gas Transmission Co. and the Florida Natural Gas Association. Following notification of the award, negotiations were initiated to obtain a suitable boiler and a site with suitable facilities. Problems arose in both connections but fortunately a functioning watertube boiler at the Sunland Training Center (STC) approximately 4 miles from campus, was made available to us through the kind cooperation of Max Jackson, Superintendent, and other officials at STC.

Boiler and Facility

The Struthers Wells-Titusville WTP-20 535 hp two drum boiler installed in 1954 (National Board No. 6644) has a steam capacity of 20,000 lb/hr at 250 psi, a heating surface of 2678 square feet and a furnace volume of 568 cubic feet. Figure 21-5 shows the dimensions and principal components of the boiler. Originally designed for firing oil, the boiler was retrofitted in 1965 to also burn natural gas.

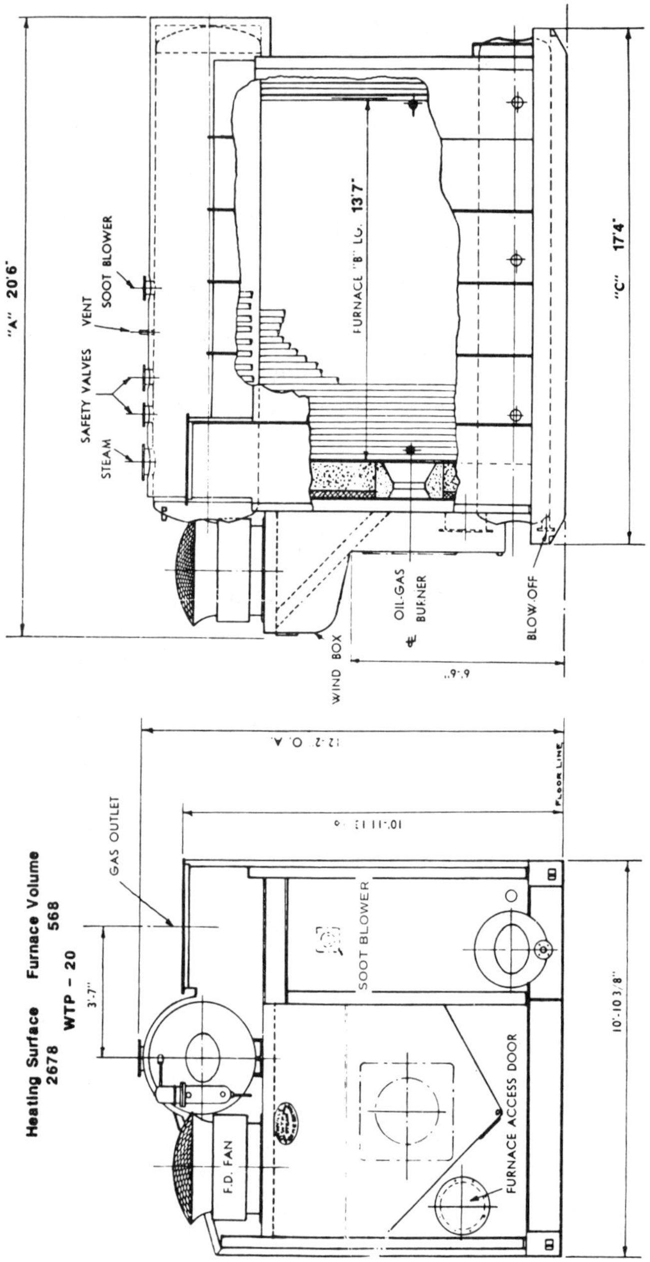

Figure 21-5
Sunland Training Center WTP-20 Watertube Boiler (No. 3) manufactured by Struthers-Wells in 1954

It has a Type COM blower made by L. F. Wing with a 7.5 hp Star Kimble motor, a Fireye Electronic Corp safety valve, a Copes Vulcan 15A air control system and a Vulcan soot blower. Because of reduced steam requirements and a decision to go to small satellite boilers for heating classes and residences STC no longer has a requirement for this boiler.

The STC boiler is much bigger than we originally envisaged for our program and since funds were limited we adopted a "minimum modification philosophy" in planning our initial proof of concept tests. Our plan is now essentially to instrument the facility to serve as a quantitative calorimeter. Thus the boiler will be used to determine the energy recovery when various fuels are burned with various levels of gas assist. We will examine particularly what might be called "foreburning" and "afterburning." The foreburning is to rapidly evaporate, pyrolyze and ignite near the atomizing zone the primary fuel which will be of lower quality than natural gas. The intent is to assist the primary fuel to burnout more rapidly and completely in its passage through the oil boiler. Afterburning gas assist is intended to help burnout the gaseous and particulate products of the primary fuel. Towards these ends the donut shaped gas burner installed in 1965 was modified to provide concentrated gas heat to the atomizing primary fuel and to supply gas for afterburning.

Modified Gas Burner

Lessons learned in several exploratory experiments were incorporated into the modification of the gas burner. The design objective was to establish three gas streams: (1) from ten 1/8" pipes directed axially toward 10 openings drilled in the diffuser plate, (2) from the inner ring of holes of the gas burner deflected tangentially forward by blocking plates and, (3) from the outer ring of holes directed radially outward. To control the gas feed rate in each of these streams the following measures are incorporated in the design. Stream 1 can be controlled by using various sizes caps on the 1/8" nozzles. Stream 2 can be controlled by the use of blocking plates with different milled sections over the forward ring of holes. Stream 3 is controlled by using blocking plates with holes which can be translated laterally to either match the back ring holes or block the back ring holes.

Three air streams are also provided for: (1) the air through the holes in the diffuser plate to feed gas stream 1, (2) the air through the swirl openings in the diffuser plate which help feed gas stream 2 and, (3) the air which goes around the burner ring to feed gas stream 3 for afterburning. We are planning to install an adjustable cylindrical baffle to control the air proportions in streams 2 and 3.

Combustion Trial

The gas feeding arrangement at Sunland Boiler No. 3 was changed to facilitate quick disassembly and access and the modified gas burner was installed. The newly installed pipe joints were checked and leaks were eliminated. The start up and operation of the boiler which had not been fired for several months proceeded routinely.

When the swirl vanes were in the open position the flame was central mostly around the axis of the burner but it was yellowish. With gradual shifting of the vanes from open to the central position the flame at first remained centered around the axis and became bluish. With further shifting of the vanes towards the shut position the flame shifted towards the periphery and became slightly yellowish again. With the vanes in the shut position the flame was in the form of a peripheral ring and there was no flame at the axis of the burner. After about half an hour the burner end refractory wall of the furnace became red hot. The yellow colour of the flame gradually became blue and the flame stabilized in the form of a diverging cone. After about an hour of operation the steam pressure was around 60 psi and at that stage some steam had to be exhausted to release pressure.

The combustion trial at Sunland appeared to indicate that the modified gas burner could generate an intense stable flame mainly concentrated near the burner, the atomizing zone for oil, COM, or CWM firings.

Data Acquisition and Automated Control

A Zenith 151 desk top microcomputer is used for data acquisition, monitoring, and limited control functions in conjunction with other instrumentation (together to be referred to as the Data Acquisition System (DAS)). Remote sensors for temperatures, fluid flows, steam output and flue gas composition are intended to provide comparison data on boiler's performance for different fuels and combustion arrangements. Later the data will be used as feedback for automatic control of the flue, air damper, and of the rate of natural gas and CWS entering the system. Each of the remote sensors provides a voltage output and is wired directly to the Doric front end module (FEM). This FEM which provides signal conditioning for 20 inputs is mounted on the boiler and connected to the Doric Dataport (DD236) via a four conductor cable. The DD236 polls these inputs on a regular basis. When requested, it transmits the data in digital form to the PC.

The thermocouples and the flue gas analyzers (O_2, CO, T_f) provide millivolt outputs and are wired directly to the FEM. The DD236 takes each thermocouple output and converts it to degrees Fahrenheit. The flow meters provide a pulse output proportional to the flow volume. These signals pass to the PC which converts the signal to an equivalent flow rate.

It should be noted that two types of software are being developed for this project. The first type is for interfacing and recording functions which can easily be tested before the equipment is installed in the field. The second type of software will be used for control.

Circuits for interfacing the PC to the positioning motors are under design and construction. Table 21-2a lists the instruments available for our measurements with the STC watertube boiler and Table 21-2b lists the measurement to be made with these instruments.

Experimental Firing

For this firing four sets of thermocouples protected with sheaths were installed inside the furnace at distances of 1.5 ft, 3.0 ft, 4.5 ft and 6.0 ft from the burner mouth along its axis. Each set consisted of three thermocouples, one for measuring the gas temperature on the axis of the flame, one 8 inches below and one 16 inches below the axis of the flame. The thermocouple sheaths were supported by brick to make them rigid. The insulated extension cables of the thermocouples were covered with tiles to prevent damage by the flame. The cables were taken out of the furnace through a hole made in the wall of the furnace manhole and were connected to a temperature recorder through a junction box. The ignition switch of the furnace was put on with 35% capacity and air vanes at position 2 from open. Within 3-4 minutes the pilot flame was visible and main flame was ignited. About half an hour after ignition the temperature from all the twelve thermocouples was recorded. The measurements were

Table 21-2a. List of Instruments

Manufacturer	*Model*	*Description (Uses)*
DORIC	DATAPORT 236	Analogue to Engineering Unit (Digital Form) Conversion Multiplexing (Flame Temps)
SYBRON	SERVOMEX 777	Flue Gas Sensing (0, CO, Tf)
FOXBORO	E83W03SSSP	Vortex Flowmeter (Steam)
WILLGOOD	F-300	Flowmeter (Slurry)

Table 21-2b. Quantitative Measurements

Flame Temp (16 pts)	F or C Thermocouples to Data port 236 to PC
Steam Volume 10 cfs	Foxboro Model E83W Vortex Flowmeter to PC
Slurry Volume 20 cfh	Willgood Model F-300 Flowmeter to PC
Flue Gas O Content	% Servomex 777 Combustion Analyzer
Flue Gas CO Content	ppm Servomex 777 Combustion Analyzer
Flue Gas Net Temp	F or C Servomex 777 Combustion Analyzer

then repeated by placing the vanes at positions 1 to 6 from open. With the shifting of the vanes the flame spread gradually away from the axis of the burner. The highest temperature reading of 2073°F was recorded by the thermocouple placed on the axis of the burner at a distance of 2 ft from the burner nozzle with the vanes at position 1. The lowest temperature was recorded by the thermocouple nearest to the burner and 16 inches from its axis. The relatively low temperature at this position may be attributed to the presence of stagnant cold gas in the lower part of the furnace near the burner end. With shifting of the vane position from open to close the maximum temperature was indicated by the intermediate thermocouple nearest to the burner.

The experiment was intended to standardize the temperature measuring devices and to optimize the vane positions for maximum intensity of heat near the burner. With vane position at 1 the flame appeared to be highly concentrated near its axis and the thermocouple sheath on the axis of the flame nearest to the burner became red-hot and cracked exposing the thermocouple tip. The temperature at various locations have been initially organized with the help of an assumed Gaussian Plume Model (GPM)

$$T = To \exp -[\alpha^2 r^2 + \beta^2 (z - z_O)^2]$$

Using a nonlinear least square data fitting routine we have determined the parameters which best fit our initial observations. Figure 21-6(A) and (B) illustrate two representative results on a computer generated contour diagram. Table 21-3(A) gives the GPM parameters for the set of gas flame observations.

Firing With Residual Oil

The next firing of the Sunland boiler was intended to initiate a study of possible improvements in the combustion performance of residual oil by co-firing with natural gas. The combustion improvement would be ascertained by monitoring the flame temperature, flue gas composition, steam output, photography of the flame and by observing the opacity and colour of the flue gas coming out of the chimney. Since the boiler had not been fired on oil for about 4 years and had been out of adjustment the following steps were taken:

(1) Damaged thermocouples were replaced.

(2) Oil preheating and atomizing steam lines were put in order.

(3) The oil supply line to the burner was checked for uniform flow.

(4) The oil burner nozzle was checked for uniform flow.

(5) The main and pilot gas valves were opened.

(6) The pilot flame was visible and the atomizing pump was switched on. For about two minutes there was no flame, then suddenly an oil flame spread through the entire fire box. The flame, however, went out in a few seconds. Dense black smoke came out of the chimney.

(7) The starting sequence was initiated again and the second time the flame continued to be partially stable with the momentary appearance of a red flame.

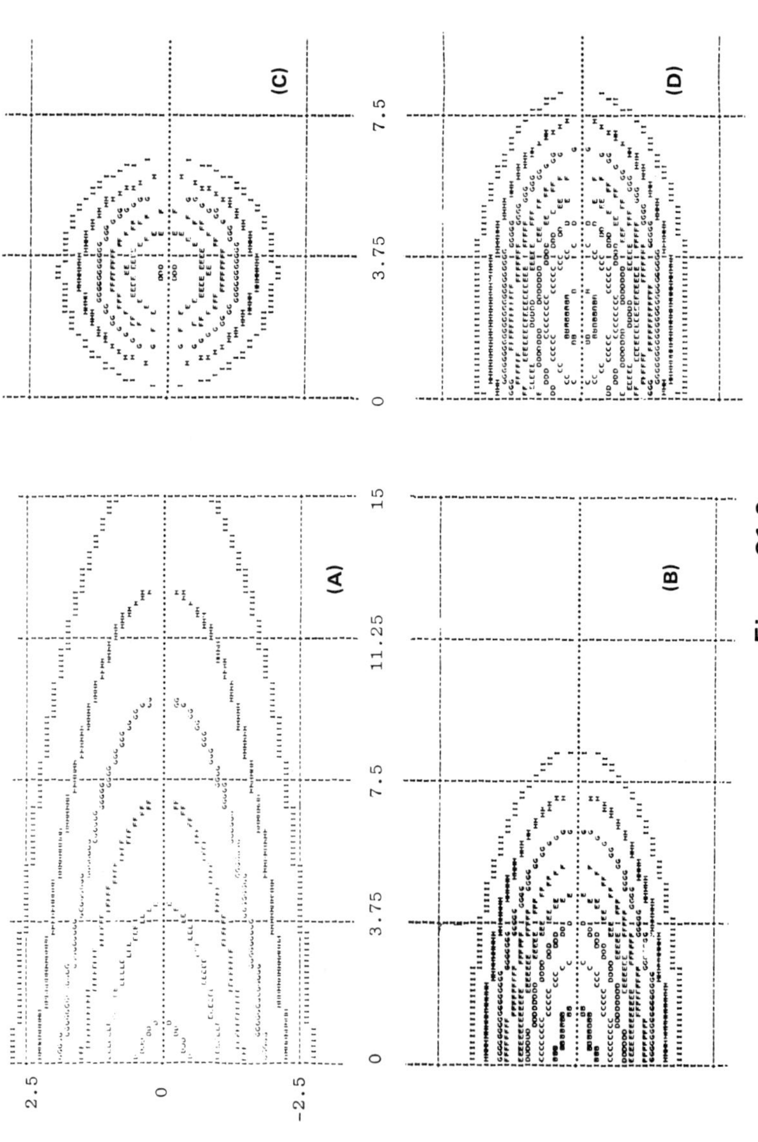

Figure 21-6

Isotemperature contours: (A) Gas Vane at 2; (B) Gas Vane at 3; (C) Oil Vane at 2; (D) Oil Vane at 1. A = 2000°F, B = 1600°F, C = 1400°F, D = 1200°F, F = 1000°F, G = 800°F, I = 400°F

Table 21-3. Parameters Gaussian Plume Model

	T_o	α	β	z	χ	V. Pos.	Cap.
(A) Gas Flame							
00	1.89	.657	.152	.141	.32	2	35%
01	2.19	.455	.045	−7.64	.471	5	50%
02	1.38	.434	.088	.822	.241	3	50%
03	1.69	.536	.13	1.25	.271	3	50%
04	2.38	.49	.049	−7.98	.40	1	50%
05	2.17	.466	.044	−8.05	.57	1	35%
06	2.19	.459	.045	−7.72	.58	6	35%
07	2.19	.455	.044	−8.2	.61	5	35%
08	2.21	.446	.044	−8.67	.53	4	35%
09	1.61	.418	.054	−5.78	.25	3	35%
10	1.898	.601	.144	.489	.31	1	35%
11	1.83	.58	.156	1.2	.31	2	35%
(B) Oil Flame							
12	1.85	.642	.206	2.19	.38	1	
13	1.41	.578	.367	3.32	.20	2	
14	1.32	.499	.294	3.11	.25	3	
15	1.45	.569	.252	2.3	.21	4	
16	1.74	.556	.295	2.92	.18	5	
17	1.80	.472	.265	3.19	.15	5	

NOTE: T_o in units of 1000°F, z in ft, α, β, in ft^{-1}

(8) As the refractory lining of the fire box gradually became heated the quality of the flame improved. After about ten minutes the flame became highly luminous, emitting a dense white smoke.

(9) The temperature recorder was then connected to the thermocouples for measuring the flame temperature.

(10) The suction probe of the CO/O_2 analyzer was inserted into the stack through a 3/8-in. diameter hole.

The temperatures were measured, flue gas analyzed and photographs of flame were taken simultaneously with the vane position at open (position 1). The measurements were repeated with vane positions 2 to 6 from open. Figure 21-6(C) and (D) are examples of temperature contours for the oil flames. Table 21-3(B) gives the GPM parameters.

Analyses of the flue gas showed more than 14% O_2 and more than 400 parts per million CO in all cases. While burning a fuel CO should not form as long as the percentage of oxygen is more than 4 percent. Simultaneous presence of high oxygen and high CO in the flue gas could be explained by the fact that atomization of the oil was extremely poor resulting in very slow rate of combustion and subsequent excess oxygen in the flue gas. The presence of a highly oxygen deficient zone in the flame where steam and oil were in intimate contact lead mutually to form CO and H_2. The white colour of the flue gas coming out of the stack was an indication of the presence of excess steam and the yellowish colour was possibly indicative of sulfur. When the rate of atomizing steam supply was reduced the percentage of oxygen in the flue gas lowered slightly and the colour of the flue gas also improved. The boiler was opened for inspection and thermocouple adjustment. The interior was covered with black deposits as might have been expected from lack of proper air-fuel adjustment and sustained combustion.

In recent years petroleum refineries are extracting more gasoline and light heating oils, leaving higher sulfur, asphaltine, and metal content in the residual oil fraction.[29] Thus current residual oil might also be helped by gas assist. With this possibility in mind we have modified the STC No. 3 boiler controls so that we can fire gas and oil simultaneously with independent controls. In effect in our initial experimental program we will use residual oil as a surrogate for coal liquid slurry. We hope to develop accurate measuring techniques which will enable us to determine whether we have achieved a cleaner and more efficient burning of residual oil.

Coal Liquid Slurry Preparations

Our experience with residual oil is primarily designed to prepare us for the firing of coal liquid mixtures, in particular coal oil mixtures (COM) and coal water mixtures (CWM). While we originally planned to concentrate on CWM, because of the low spot market price of oil the COM option should also be considered. Coal liquid fuels have many of the properties of residual oil except (1) viscosities are usually higher (~1000 centipoise) and (2) ash contents are much higher (~5% instead of 0.1%). We have under negotiation several commercial coal liquid sources in Florida and hope to have high fusion temperature-low sulfur slurry fuels available when we are finally ready for our combustion trials.

Before we can fire with coal liquids we must modify the oil boiler system to accommodate the ash generated during coal liquid slurry combustion and construct a fuel handling and atomizing system to accommodate the abrasiveness and viscosity of coal liquid slurries. Our final proposed solution to the coal ash problem is described later in the chapter. Our interim solution employs several mitigating measures which are intended to make it possible to carry out quantitative observations in the very near future. These measures are:

(1) To accommodate the fly ash we are installing a simple spray system in the flue section before the stack. We have designed a cyclone dust separator for this same purpose but because of construction time and cost we have put this aside. We have also considered the use of a baghouse or an electrostatic precipitator but these are beyond our current means.

(2) To deal with the accumulated bottom ash we will utilize a vacuum cleaner cyclone system to be operated through the existing manhole on the left bottom side of the boiler and an additional hole to be made on the bottom right side of the boiler.[30]

(3) To clear the tubes and drums in the convective pass section we will use the soot blower with steam provided by the adjacent firetube boiler which is on the steam line with the test boiler.

(4) To clear the vertical tubes in the water tube section we will follow a two-phase strategy: (a) after the bottom ash is cleared by measure (2) and the furnace has cooled we will manually sweep down the tubes exposed in the furnace and remove the accumulated ash by (2) above; (b) when the time between firings becomes shorter than the cooling time we will establish channels in the front furnace face at the upper right and the middle left to introduce a manually manipulat-

ed removable soot blower to be operated with compressed air or steam. Again the vacuum cleaning-cyclone arrangement of (2) will be used to minimize the introduction of fly ash into the inaccessible convective section.

The construction of the interim ash handling system and the slurry piping and handling system are in the process of assembly. The slurry system will provide CWS fuel to the boiler and to the future vertical vortex combustor along with implementation of appropriate safety measures. A schematic of the slurry system is shown in Figure 21-7(A). The objective of this system is to effectively deliver CWS from three 55-gallon barrels for test burns of one hour. The system is broken into two separate parts. The containment-to-surge-tank portion is designed to take the interruption of flow when changing containment barrels. The surge-tank-to-atomizer part is designed for continuous flow.

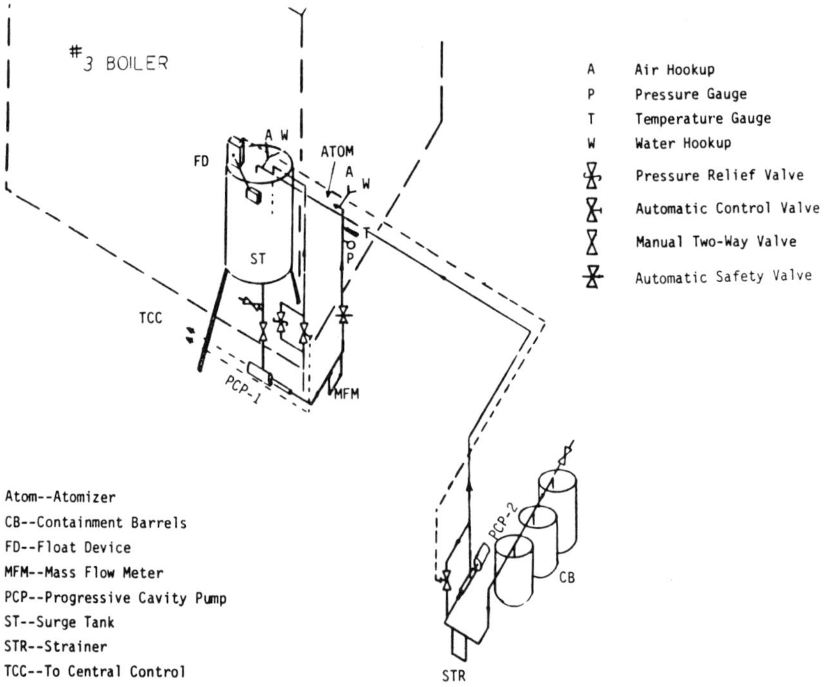

Figure 21-7(A). Slurry System

For close access to the piping system, the barrels of CWS are placed outside the physical plant. A progressive cavity pump (PCP-1) is used to pump the CWS up the elevated surge tank. The location of the PCP-1 is between the containment barrels and the surge tank. A signal to activate the by-pass control valve is sent automatically from a float device on the surge tank. The PCP-1 runs at a constant speed to supply the maximum CWS and the by-pass control valve regulates the flow rate to the surge tank.

The CWS piping system must be free of oversized particles and sludge masses. A strainer is placed early within the piping system to insure no clogging of sensitive components. Basket strainers with 3/16-in. holes are used. Periodic cleaning of the strainer will be necessary.

A surge tank is installed above PCP-2 to maintain a continuous CWS supply to the pump. The positive pressure gravity feed eliminates the chance of cavitation. A float device on the surge tank activates the by-pass control valve to maintain a constant level of CWS in the surge tank when changing supply barrels.

A second progressive cavity pump (PCP-2) supplies the CWS to the atomization loop. A by-pass control valve to the surge tank balances the pressure and flow rate. A high pressure release valve is installed in the atomization loop. A Wilgood 300 bearingless liquid flow meter is installed before the atomizer. Temperature and pressure gauges are included after the flowmeter and bypasses PCP-2 to give temperature and pressure readings before atomization. A Delavan SL 707 nozzle with a stainless steel pintel and a Parker Hannifin 684-0614 nozzle will be used in initial CWS firings.

Schedule 40-¾-in. steel pipe and flexible hydraulic hose are used in the piping system. The entire piping system is fitted with air and water connections placed in appropriate locations so as to permit complete washout of the CWS system. The water used to flush the piping system is collected at drain points and contained until appropriate disposal.

Steam Measurement and Disposal

For lack of need, the steam generated by the test boiler at STC will be vented to the atmosphere. The steam pipe between the main manifold and the head valve will be modified as shown in Figure 21-7(B). The pipe will be disconnected from the main manifold and the

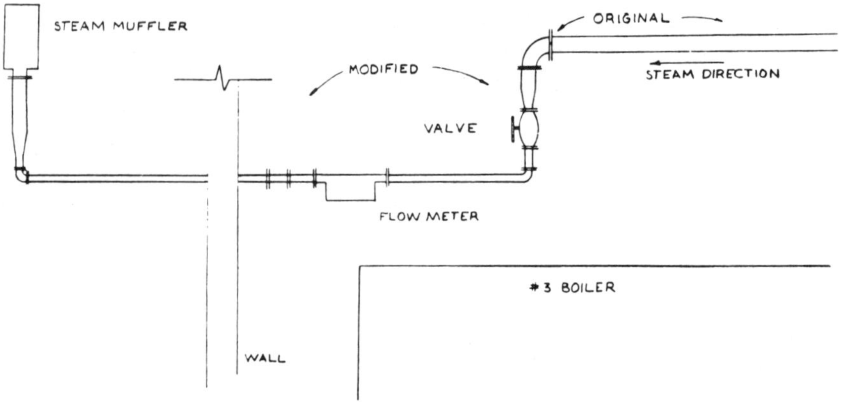

Figure 21-7(B). Steam System

manifold capped. The steam pipe will be reduced from the original 6-in. to 3-in. so that a Foxboro 3-in. vortex flowmeter can be installed. A five turn valve will also be placed in the modified section of pipe before leaving the building. On the outside of the building an Allied Witan S60 steam muffler will be attached. The steam muffler and piping system is designed to handle up to 15,000 lbs/hr of steam at a pressure of 100 psi. This disposal and measurement system should accommodate all the tests planned at STC during the proof of concept phase of our work.

In view of our initial steps at quantifying gas-residual oil firings we are tentatively planning to begin with gas-coal mixture firings. This will minimize the initial introduction of ash into our oil boiler system and give us the opportunity to assess our ash mitigation measures. It will also enable us again to test our heat balance and efficiency assigning procedures and give us experience in measuring carbon burnout. Following this work we plan to burn coal water mixtures with gas assist.

Boiler Efficiency and Carbon Burnout

Our general strategy to achieve accuracy for heat balance measurements and calculations is to bring the boiler to a steady state of steam generation on natural gas at various standard outputs (e.g. 50%, 75%, 100%). Then we will decrease the gas feed rate and increase the

CWM with the drum water level held constant, adjusting the air levels as needed. The variables needed to establish the heat balances will be recorded and the boiler efficiency calculated. These include the gas, slurry and air flow rates, steam generation rate, flame temperature, flue gas temperature, CO_2, CO and O_2 content of the flue gas and the inlet water and air temperatures. The loop will be closed in that each sequence will be recycled back to the all gas condition to check reproducibility. The bottom ash and the fly ash will be sampled and tested for carbon burnout. Our objective is to obtain sufficient quantitative data which can serve as a basis for phenomenological modeling of the effects of gas assist. A more detailed measurement protocol is now being formulated.

Gas Assist in Waste Disposal with Energy Recovery

Hospital Waste Disposal

In 1981 Florida Hospital in Orlando purchased two used batch type incinerators (C-760 Consumat Systems Inc) to expand its prior incinerator capability. The two systems were refurbished and installed in a preengineered metal building on a reinforced concrete slab with necessary equipment piers and footings. To recover the energy, a 150-hp York Shipley firetube boiler was installed between the two incinerators. Refractory lined sections are used in the stacks of each incinerator and refractory lined ducts feed the hot flue gases to the heat recovery boiler which has all the necessary boiler controls. An induced draft fan draws the hot incinerator flue gases through the boiler and the cooled flue gases are finally vented through an additional stack.

A standard starved air C-760 which is without heat recovery or natural gas assist in the secondary chamber is illustrated in Figure 21-8. Heating values of various waste are listed in Table 21-4. In practice the predominant wastes produced at the Florida Hospital (at a rate of about 10 lb per occupied bed per day) include trash, rubbish garbage and pathological wastes. Typically the mixed waste has a heating value of 7,000 to 8,000 Btu/lb and a moisture content from 10%-25% by weight. The addition of segregated pathological waste from the Altamont Spring branch of the Florida Hospital enhanced the amount of high moisture Type 4 waste in the mixed waste stream.

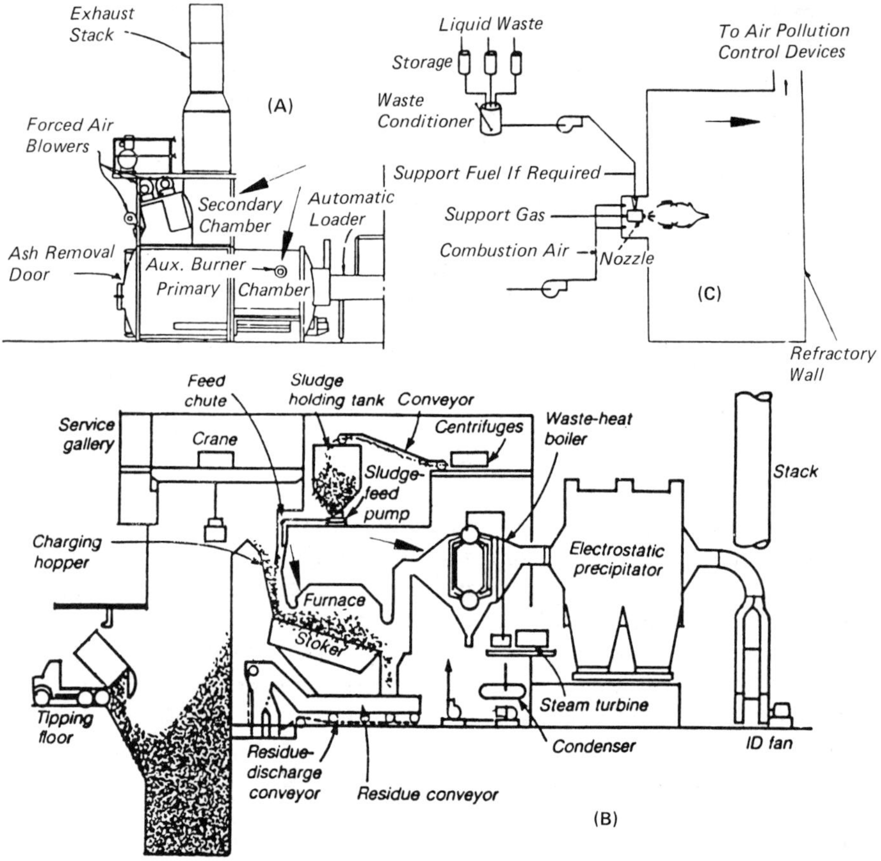

Figure 21-8

Incinerators: (A) Type used in Orlando Florida Hospital (from Post, Buckley, Schuh and Jernigan[31]). Arrows show where gas assist is helpful. (B) One channel of the Glen Cove refuse/sludge system. (C) Liquid injection incinerator for hazardous wastes (from Proctor[44]).

The wide range of waste heating values and the varying demands of the hospital for steam and heat favored the use of gas assist in several forms. In the Florida Hospital natural gas burners are installed in (1) the primary incinerator chamber, (2) the secondary afterburning chamber, and (3) the heat recovery boiler. The burner in the primary chamber is mainly used for start up and ignition of waste. The gas burner in the secondary chamber from the stacks. The gas burner in the firetube boilers is fired when needed to enhance steam output and also for flue gas clearing.

This twin incineration boiler system has been operational since 1984 and has provided the major steam and heat needs for the Florida Hospital ever since. Two 600-hp Johnson firetube boilers previously provided heat during harsh cold spells which occasionally hit Florida.

The extent to which gas-fuel synergisms discussed in previous sections have been achieved in this simple and versatile incinerator/heat recovery system has not yet been quantified. The wastes are reduced to less than 10% of the original volume, greatly reducing the landfill requirements. The simplicity and effectiveness of this waste disposal/energy recovery solution, in an age which glamorizes "high technology," has been an obstacle to further progress.

Community Refuse Incineration

The Florida Hospital Incinerator/Energy Recovery System using gas assist was assembled using batch incinerators developed in the early 70's. More recently, incinerator modules have been developed which permit continuous processing of waste. In such systems a skid steer tractor pushes the waste into an automatic loader which injects the waste into the primary combustion-gas production chamber where transfer rams move the material slowly through the system. The high temperature environment in the gas production chamber is provided with a controlled quantity of air so that gases from the process are not burned in this chamber but fed to the upper or pollution control chamber. Here the gases are mixed with air and controlled to maintain a proper air fuel ratio and temperature for entrance into the heat exchanger where steam is produced. Consumat continuous incinerators pass the inert material from the primary combustion process in the form of ash into a wet sump. It is then conveyed into a closed bottom container and hauled to a landfill.

Table 21-4. Waste Classification

Type		*Moisture Content*	*Incombustible Solids*	*Btu Value Lb*	*Lbs per Cu Ft*
0	TRASH, a mixture of highly combustible waste: paper, cardboard cartons, wood boxes, and combustible floor sweepings, from commercial and industrial activities. Contains up to 10% by weight of plastic bags, coated paper, laminated paper, treated corrugated cardboard, oily rags and plastic or rubber scraps.	10%	5%	8500	8-10
1	RUBBISH, a mixture of combustible waste: paper, cardboard cartons, wood scrap, foliage and combustible floor sweeping, from domestic, commercial and industrial activities. Contains up to 20% by weight of restaurant or cafeteria waste, but little or no treated papers, plastic or rubber wastes.	25%	10%	6500	8-10
2	REFUSE, consisting of an approximately even mixture of rubbish and garbage by weight, common to apartment and residential occupancy.	50%	7%	4300	15-20

3	GARBAGE, consisting of animal and vegetable wastes from restaurants, cafeterias, hotels, hospitals, markets, and like installations.	70%	5%	2500	30-35
4	HUMAN AND ANIMAL REMAINS, consisting of carcasses, organs and solid organic wastes from hospitals, laboratories, abbatoirs, animal pounds, etc.	85%	5%	1000	45-55
5	BY-PRODUCT WASTE/GASEOUS, LIQUID OR SEMI-LIQUID, such as tar, paints, sludge, fumes, etc., from industrial operations.	Varies	Survey Required		
6	SOLID BY-PRODUCT WASTE, such as rubber, plastics, wood waste, etc., from industrial operations.	0	0	Varies by Material	Varies

Continuously operating incinerators run constantly and have better operating systems. Standard modules from Consumat Systems[32] have waste processing capabilities from 2100 lb/hr to 8400 lb/hr and in simple combination can handle 100–400 tons per day. Gas assist is usually not utilized in Consumat package incinerators. However the Florida Hospital experience plus the considerations given in earlier sections of this report point to environmental and operational benefits of gas assist.

A refuse separation and recycling program can be valuable in increasing the efficiency of refuse incineration and in energy recovery. However, it would take a very conservation minded community to make refuse separations and recycling work.

Hanson et al.[33] describe a number of technological innovations including co-combustion of densified refuse-derived fuel (dRDF) with coal. Here solid waste is loaded on a conveyor which feeds a primary 5-inch trommel screen. Material not passed is processed by a flail mill reducing large objects to less than 5 inches. By a lofting process an air classifier separates the lighter sub 5-inch components which are principally paper and plastic. This fuel material is processed into cylindrical plugs one-half inch in diameter and one inch long which have approximately one-half the fuel value of coal. This dRDF can be coburned in a stoker-fired coal system in an integrated utility system. Alternatively it can be co-fired with sludge and natural gas in a heat recovery incinerator.

The heavy residue rejected by the air classifier in the production of dRDF is passed over a drum magnet for recovery of magnetic scrap. The residue may be further processed or used as landfill.

Sludge/Refuse-Gas (SRG) Co-Combustion

There have been many studies of waste incineration[34–39] including feasibility studies of sludge/refuse co-combustion. However, Glen Cove, N.Y. has a fully commercial sludge/refuse facility that cogenerates 2.5 MW.[40] The plant has two combustion channels each rated at 125 tons/day with 10:1 feed rates of refuse to sludge. The HHV of the mix is about 4000 Btu/lb. Centrifuging concentrates the sewage solids from 5% to 20% solids. Positive displacement progressive cavity pumps deliver sludge to the refuse feed hopper of each furnace. The sludge is fed to the back side of each charging hopper and uniformly deposited into the refuse. Figure 21-8(B) illustrates one channel of the Glen Cove system.

At Glen Cove bulky combustible wastes first pass through a shear that reduces the material to the 6- to 18-inch range before deposition into the storage pit. The fuel mixture from the storage pit and sludge holding tank are fed to two stoker-fired refractory furnaces. These are fired at 1700°F without auxiliary fuel generating over 34,000 lb/hr of 600-psig saturated steam in a dual-drum vertical tube convection boiler. Waterwall loops located on each furnace along the stoker line add 10% to the steaming capacity of the boiler. The steam feeds a turbine generator producing about 2.5 MW. Half of the electric output is used by the plant.

Unlike the twin Florida Hospital incinerator the twin Glen Cove incinerator does not incorporate gas augmented afterburning chambers. These would provide added heat to the waste boiler when needed and help cleanse the final combustion gases. Such provisions would undoubtedly provide extra flexibility and better emissions control.

In sludge/refuse incinerators the higher heating value of the refuse is used to assist in the evaporation of sludge water which greatly increases the heating value of the sludge. High performance sludge dewatering devices such as belt presses, filter presses and advanced centrifuges can alternatively be used.[41–43] In this case the incinerator can be of a more compact design—such as a fluidized bed combustor or the first stage vortex combustion chamber described earlier. In the latter case gas assist can be extremely helpful in foreburning for flash evaporation of the residual water and rapid pyrolysis and ignition and afterburning for air pollution and odor control.

Niessen[43] discusses direct flame afterburner systems for incinerators most of which use natural gas. He calls attention to the need for more attention to afterburner design not only for improved pollution reduction but also for improved heat recovery.

Wood Wastes

Since 1973 a great deal of attention[44–48] has been given to the potential use of wood as an alternative to oil. These efforts considered both intentionally grown biomass and wood waste. Much of the R&D effort associated with the former program has declined with the dismantlement of the synfuel program and the recent oil glut. However, wood waste disposal remains a continuing environmental problem along with urban refuse.

Mill residues ranging from large slabs, edgings and end trim, to stringy hog-wooded bark and fine sawdust are traditional sources of wood waste. Pulp and paper mills leave mountains of unused bark. With recently developed whole tree chippers a great variety of forest residues could serve not only as a source of energy but to foster timber stand improvement. It is estimated[48] that thinning 100,000 acres per year would yield the energy equivalent of 5 million barrels of oil per year.

Hardwoods typically have an average oven dry heating value of 8500 Btu/lb and their dry bark 8000 Btu/lb. Oven dry softwoods and their barks average about 8900 Btu/lb. At 50% moisture content 4250 Btu/lb may be taken as representative. Whole tree chips delivered at $20.00 a green ton at a combustion efficiency of 65% is equivalent to $3.60 per million Btu, a reasonable thermal price considering that waste disposal is accomplished at the same time.

Water content is a major variable in wood waste utilization and here again gas assist can serve as a useful co-fuel to accommodate the particular characteristic of any load in a heat recovery incinerator. The technology of hogged wood waste burning has advanced considerably in recent years. Gas assist for drying, pyrolysis and ignition and gas afterburning for emission control purposes adds the extra degree of control needed to accommodate a large diversity of wood waste fuels. Rather than waste wood and consume landfill space, the time has come to use wood waste for our energy needs.

Nationally, waste wood from the forest industry such as tops and stumps, cull trees and nonmerchantable species could supply about 10% of current energy use. In Florida with its large forest industry, waste wood could supply about 20% of current energy use. Gas assist in the initial and afterburning zones can help make this otherwise wasted resource into a valuable and environmentally acceptable fuel.

Hazardous Wastes

Proctor at the University of Florida[49] has made a technical assessment of alternative technologies for hazardous waste management. He divides hazardous waste physically into gaseous, liquid, solid, slurry, or sludge forms. From a hazards standpoint Proctor identifies seven classes: (1) reactive, (2) toxic, (3) oxidizer, (4) ignitable, (5) corrosive (acid), (6) corrosive (basic), (7) miscellaneous. Each class has unique characteristics, and often a specific hazardous waste stream may reside in more than one class. For example, a material that is corrosive may also be ignitable, or a material that is reactive may also be toxic. The miscellaneous class includes known carcinogic and mutagenic materials.

The technologies for disposing of hazardous waste include (1) thermal, (2) biological, and (3) chemical technologies. The first may be subdivided into (a) incineration, (b) molten salt, (c) plasma arc, (d) wet oxidation and (e) supercritical water. Our concern here will be within the (1a) thermal-incineration category.

The intent of thermal treatment is to destroy the molecular structure of a hazardous material and rearrange the atoms into compounds that are nonhazardous. Alternatively the thermal process should reduce the volume of hazardous waste for further treatment, for resource recovery or to be landfilled. Thermal processes offer the potential for heat recovery as outlined in the previous subsections.

The following types of incineration have been considered for hazardous wastes: (1) liquid injection incineration, (2) boilers, (3) rotary kilns, (4) cement kilns, (5) multiple hearth, (6) fluidized bed incinerators and (7) pyrolysis. Figure 21-8(C) illustrates a liquid injection incinerator. Many hazardous wastes are good candidates for incineration because of their high heating value. Thus the energy necessary to maintain the steady-state operation of a hazardous waste incinerator can often be supplied by the waste itself. For wastes of lower energy than that necessary for stable combustion co-burning with a higher energy waste or fuel to maintain a stead-state operation is essential. It is here that natural gas can serve as a vital assist fuel. The combustion gases must also be cleaned to meet air pollution standards and here too gas assist is helpful.

The performance of a hazardous waste incinerator is measured by the Destruction Removal Efficiency (DRE) of Principal Organic Hazardous Constituents (POHCs), where

$$\mathrm{DRE} = 100\ [1 - (W_{out}/W_{in})]$$

Here W_{out} is the POHC output (mass-emission rate released to the atmosphere). W_{in} is the POHC input mass feed rate of the POHC in the waste stream.

The amount of a POHC in a hazardous waste can be, and most often is, only a trace constituent in a waste stream (e.g., metals in a nonhazardous solvent used to strip the paint off an aircraft). Sometimes, however, the entire waste stream may be hazardous (e.g., benzene contaminated in an industrial process). DRE performance standards have been established by the U.S. EPA. For organic hazardous constituents incinerators the DRE must meet or exceed 99.99%.

Boilers which by design are excellent heat recovery units can sometimes use hazardous wastes as fuels. Incinerators are specifically designed to achieve high destruction removal efficiencies of hazardous wastes. While incinerators have refractory walls which are quite hot during operation, boilers usually have metallic tubular walls which are cooled by water passing through them. Despite the differences between boilers and incinerators, a boiler when operated correctly can serve as an excellent hazardous waste treatment unit.

Liquid hazardous wastes coincinerated with oil are the predominant waste type disposed of in boilers. The results can be a significant reduction in fuel costs, and a significant reduction in liquid waste disposal costs. Care must be exercised in the operation of the boiler to avoid severe corrosion of boiler components which usually occur in cooler regions.

The use of boilers in hazardous waste management is very attractive because of the large number of boilers in operation. In recognition of this fact, the U.S. EPA has outlined the use of industrial boilers for the disposal of PCBs. Co-firing with natural gas and afterburning with natural gas obviously afford extra insurance in hazardous waste management while broadening the options of materials which can be burned.

Two-Stage Firing in a Vertical Vortex Combustor

The major problem in directly retrofitting an oil boiler for coal is the problem of ash build-up on and the erosion of the heat transfer surfaces. The use of a first-stage combustor which simultaneously serves as an ash separator and producer of hot gases, rich in CO and possibly H_2 to be afterburned in the second stage—the oil boiler—has long been our solution of choice.[2-11] To achieve high ash separation a coal-gas fed vertical vortex combustor (VVC) leading to a dry ash was considered to be the most practical solution although a slagging fuel must also be considered. Thus our VVC should be differentiated from the horizontal cyclone combustor which has been in the coal combustion literature for a long time,[50] but has come into disfavor because of high NO_X generation. To feed an oil boiler with a standard cyclone combustor would require major changes in the oil boiler itself to accommodate the large amount of slagging ash which would be injected into the boiler.

The University of Florida VVC Program

The first VVC in the Florida program was a 1 million Btu/hr combustor constructed in 1981. The schematic design of the system is shown in Figure 21-9(A). It was fired on many occasions from 1981-1984 with PC and CWM fuel and with its afterburning flame ignited as a diffusion flame in open air. The ash collected in the bottom hopper had various degrees of carbon burnout ranging up to 85%. Unfortunately, for lack of funds, this VVC was not instrumented for quantitative work. In the present instance the instrumental build up of our direct firing CWG program offers the opportunity to use the same instrumentation for two-stage studies and we have initiated such a program.

One obstacle to the design of a VVC is the absence of a technical literature on such an ash-separator-gasifier. There are good literatures on the standard cyclone combustor[50] and on cyclone dust separators,[51-53] including hot gas cyclone dust separators.[54] However, in the absence of prior studies of VVC-ash separators we must either develop the theory for such a system or lean on approximations based upon cyclone dust separator analyses.

Design Considerations

Ash separation in a cyclone is accomplished by settling due to the difference in density of the particle and the gas. In spite of the seeming simplicity of the equipment and governing principles, characterization of their particulate-removal performance has been the subject of many efforts, the performance being less than 100% and function of particle size geometry and operating conditions.

The estimation of the volume of a CGVVC combustor requires information on the rate of heat generation in such combustors for burning a particular fuel. It has been reported in literature that a cyclone combustor burning bituminous coal can achieve a heat release rate of 500,000 to 900,000 $Btu/ft^3/hr$. No such data are available for combustion of coal-water slurry. It has, however, been reported by Mchale et al.[55] and Sakai et al.[23] that the combustion efficiency of coal water slurry was close to that of pulverized coal when burnt in a pulverized coal furnace.

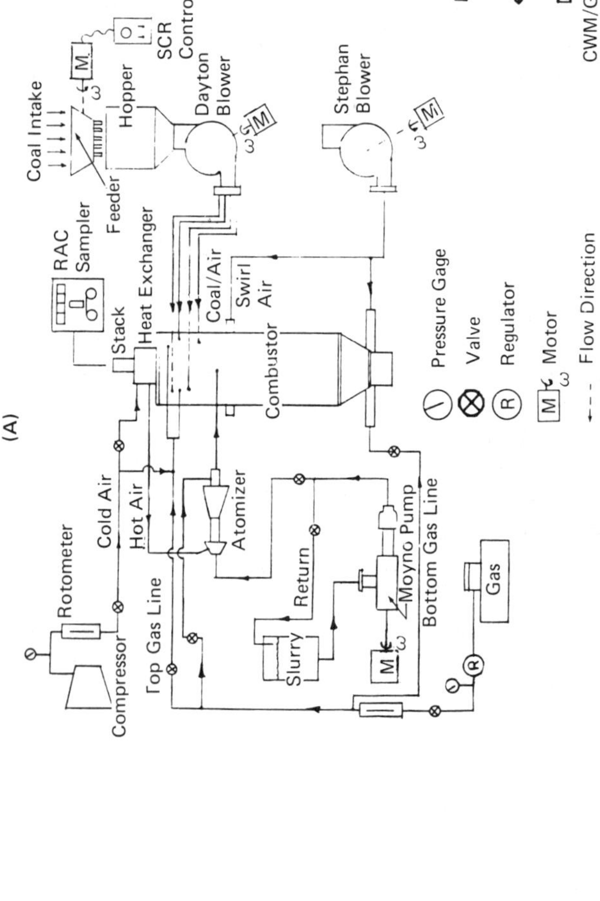

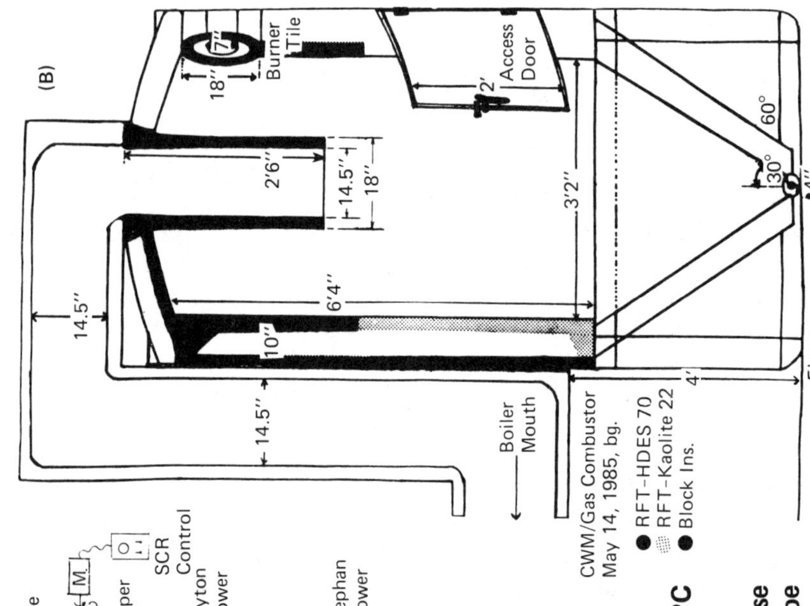

Figure 21-9. Vertical Vortex Combustors

(A) **1MM Btu/hr facility constructed in 1981 for PC and CWM firing with gas assist.** *(Above)*

(B) **10MM Btu/hr facility under construction for use with Sunland Training Center Water Tube boiler.** *(To the right)*

To design a CGVVC for a specific duty requires specification of its overall size, characterized by the principal diameter D and its shape and configuration which is determined by the ratios of inlet height, inlet width, outlet diameter, outlet height, cylinder height, total height and ash outlet diameter to the diameter of the cyclone. Selection of these eight parameters will in turn depend upon the duty to be performed, namely: (1) heat to be generated; (2) rate of flow, temperature, pressure of gas stream; (3) composition of the gas; (4) coal loading; (5) size distribution of coal; (6) degree of collection.

In order to develop an optimal design, it is necessary to have a model for the collection process which gives a grade efficiency function and method for calculating the pressure drop through the device. A model for a cyclone dust separator developed by Leith and Licht[56] shows good results in the prediction of grade-efficiency relationships. Their approach employs the concept of backmixing and average residence time, as well as the geometry of the cyclone and the natural length phenomena exhibited by vortical flows, to account for the turbulent and flow related redistribution of the uncollected particles in the gas. The results of their theoretical approach is an equation which accounts for the geometric design, cyclone diameter, particle size, and physical properties of both gas and solid; operating conditions, and the vortex law which characterizes the tangential velocity distribution. In the equation the grade efficiency was expressed in terms of two dimensionless groups; (1) a cyclone inertial parameter reflecting operating conditions of the cyclone and (2) a cyclone design number reflecting the configuration of the cyclone.

Table 21-5a summarizes the Leith-Licht design equations for a cyclone dust separator and Table 21-5b defines the terms. For CGVVC the total average residence time is taken as the average time required for gas to descend from the average level of entrance to the level of the bottom of exit pipe, plus the average residence time. The average minimum residence time can be calculated by assuming all the gas enters the cyclone at the midpoint of the entrance duct. Thus we obtain Eq (9) of Table 21-5b. The cut diameter as defined by Lapple[57] for a cyclone dust separator as the size of particles collected with 50% efficiency is given by Eq (10). Typically high efficiency cyclone dust separators have a cut diameter of 5 to 10 microns.

The pressure drop in a cyclone dust separator according to Shepherd and Lapple[58] is given by Eq (11).

In the design of a CGVVC the expansion of the volume of gas due

Table 21-5a. Nomenclature for Vortex Combustor

a	Height of cyclone inlet
b	Width of cyclone inlet
B	Diameter of cyclone bottom outlet
dp	Diameter of ash particles
$(dp)_{cut}$	Cut-diameter of ash particles
D	Diameter of cyclone combustor
De	Diameter of cyclone flue pipe
h	Height of cylindrical part of cyclone
H	Total height of cyclone
k	Empirical constant
K_a	a/D
K_b	b/D
N	Effective number of turns
Q	Volumetric flow rate of gas
S	Height of cyclone flue gas outlet
T	Gas temperature in cyclone
u	Inlet velocity of gas
ΔP	Pressure drop in cyclone
ρg	Density of gas
ρm	Density of manometric liquid for gas flow rate measurement
ρ_p	Density of ash particles
μ	Viscosity of gas
η_i	Efficiency of ash separation

Table 21-5b. Vortex Combustor Equations

$$\eta_i = 1 - \exp\text{-}2\,(\psi_{ci}K)^{\frac{1}{2n+2}} \qquad (1)$$

$$\psi_{ci} = P_p d_p^2\, U(n+1)/18MD^3/ab \qquad (2)$$

$$n = 1 - [1 - 0.67(D)^{0.14}]\left(\frac{T}{283}\right)0.3 \qquad (3)$$

where T is in °K and D is in meter

$$K = 8\,K_c/K_a^2 \cdot K_b^2 \qquad (4)$$

$$K_c = (V_s + V_{n1/2})/D^3 \qquad (5)$$

$$V_S = \frac{\pi}{4}\left(S - \frac{a}{2}\right)\left(D^2 - De^2\right) \qquad (6)$$

$$V_{nL} = \left(h - S\right)\frac{\pi}{4}D^2 + \left(1 + \frac{d}{D} + \frac{d^2}{D^2}\right)\left(1 + S - h\right)\frac{\pi D^2}{12} - \pi De^2\, 1/4 \qquad (7)$$

$$d = D - (D - B)\,(S + 1 - h)\,(H - h) \qquad (8)$$

$$l = 2{\cdot}3{\cdot}De{\cdot}\,(D^2/a{\cdot}b)^{1/3} \qquad (9)$$

$$t_{min} = \frac{(S - a/2)\,(D^2 - De^2)}{4Q} = \frac{V_S}{Q} \qquad (10)$$

$$(dp)_{cut} = \sqrt{\frac{9\,u{\cdot}b}{2N\mu\,(\rho_p - \rho g)}} \qquad (11)$$

$$\Delta P = \frac{u{\cdot}\rho g{\cdot}k}{2g\,Pm}\cdot\frac{a{\cdot}b}{D^2} \qquad (12)$$

Table 21-5c. Results of Sample Calculation

Data

Ultimate analysis of coal: Carbon 84.5%, Hydrogen 5.1%, Nitrogen 0.8%, Ash 4.5%, Oxygen 3.8%, Sulfur 0.8%

Slurry to methane ratio	3:1 on fuel value
Excess air factor	1.15
Diameter of vortex combustor	38″
Calorific value of CWS	10,000 Btu/lb
Calorific value of methane	22,100 Btu/lb
Heat generated in combustor	8.31 MBtu/hr
Coal water slurry burnt	620 lb/hr
Air supplied	1560 cfm at NTP
Air supplied to 1st stage	1014 cfm at NTP

DIMENSIONS OF COMBUSTOR

Diameter = 3.17 ft
Cylindrical height = 6.34 ft

Total height = 9.5 ft
Height of inlet = 1.0 ft

Width of inlet = 6 inches
Diameter of outlet flue = 1.98 ft
Height of outlet = 1.98 ft

EFFICIENCY OF ASH SEPARATION

Ka = 0.315		K_L = 0.157	
Vs = 10.09		V_N = 308.5	
ℓ = 6.69		K = 5235	
d = 20.2		K = 402	
n = 0.451		ψ_{ci} = 2.4×10^{-3}	
n_i = 0.57	for particles	5μ	
0.77		10μ	
0.91		20μ	

to rise in temperature must be taken into consideration while calculating the velocity of the jet in the vortex. Increase in velocity due to the expansion in volume may add to the pressure drop and may also increase the separation efficiency to some extent. For an accurate calculation of radial velocity of the particles in the vortex combustor it is essential to know the rate of decrease of particle size of the coal. Since such information is not readily available a mean value of the size of coal and that of ash has been used for finding the efficiency of ash separation.

Table 21-5c gives out tentative design calculations. Figure 21-9(B) is a diagram of our tentative first-stage combustor design for feeding the STC boiler. We must emphasize that no established theoretical model of a dry ash vertical vortex combustor is now available and hence simple modeling, judgment, resource limitations, need for versatility and other practical considerations have influenced our design.

The length of the ceramic connecting pipe between the VVC in Figure 21-9(B) and the STC boiler is somewhat of a problem. We had encountered a similar problem in connection with our exploratory work with the VVC in Figure 21-9(A) and have experimented with smaller VVC's in which the hot gas output comes out downward. We have also investigated the possibility of replacing the connecting pipe with a hot gas cyclone particle separator with a downward output. Such an intermediate hot chamber could be useful in SO_X control. Our exploratory investigations along these lines have shown promise.

Discussion of Natural Gas Assist Techniques

Our underlying thesis is that natural gas co-fired with a variety of lower grade combustible fuels can upgrade the combustion to permit improved levels of energy recovery and reduced levels of emissions. The gas must be used judiciously in relationship to the primary fuel to maximize the assist purposes natural gas can serve. We have already discussed "foreburning" as the use of gas to establish an intense high temperature zone in the CWM atomization region. This zone can provide flame stability over a large load range, flash evaporate the slurry droplets, rapidly pyrolyze and ignite the coal particles, generate CO and provide hydrogenic radicals which speed the combustion kinetics. Foreburning can also be helpful in liquid waste, sawdust and sludge burning for the same purposes.

We have also discussed "afterburning" assist with natural gas to help process combustible exhaust gases which would otherwise be below the flammable range. Large odor reduction efficiencies on hydrocarbons and organic vapors can be achieved by afterburning techniques [43] and particulate concentrations can be greatly reduced before exiting the furnace.

In our two-stage system we essentially carry out foreburning and afterburning in separate combustion chambers and accomplish pre-boiler ash separation at the same time. The possible use of such a system for hazardous waste disposal should be obvious.

"Preburning" might be used to categorize a number of other natural gas assist measures. For example the use of high pressure natural gas for primary CWM atomization by Essenhigh et al.[59] essentially makes use of compressor costs already included in the gas price. At the same time the gas helps in the foreburning. Another example which we have explored with the VVC in Figure 21-9(A) is the use of natural gas for secondary atomization by dissolving natural gas under pressure in the CWM before injection into the atomizer. The hope was that the sudden release of pressure after atomization would give a secondary boost to slurry droplet break up.[60] While our crude experiments were inconclusive Beers et al.[61] had partial success with a similar type of secondary atomization in which CO_2 is dissolved at high pressures in the CWM slurry. The possibility that CH_4, will give improved secondary atomization and help in foreburning should be investigated further. A third preburning gas assist would be to inject pulverized coal into an intense vortex gas flame possibly assisted by oxygen for a short period insufficient for burnout but sufficient to explosively rupture the coal particles. One might speculate then on the possibility of precombustion ash separation by a cyclone arrangement and the injection of the fragmented coal particles into the oil boiler. Some of the unusual bottom ash collections when we fired with the VVC of Figure 21-9(A) with PC were suggestive of this possibility.

Another type of gas assist which might be identified as "postburning" is the possibility of clearing deposits on the water tubes by gas firings. We are investigating this possibility in connection with deposits built up by our residual oil firings and our planned COM and CWM firings.

Another type of gas assist might be referred to as "coburning." For example if we wish to take a CWG fired oil boiler to its full rating to meet peak loads we might simply increase the proportion of gas to slurry. While some physical synergism might be accomplished in the foreburning and afterburning regions, basically we would simply be using the multifuel capability of the boiler and the greater capabilities of natural gas, to meet short duration needs. A similar coburning strategy might be used for seasonal emission reduction scenarios[62] again largely without physical synergisms. The burning of gas and coal in separate boilers for the purpose of plant emission compliance[63] can also be considered coburning although this, of course, is without physical synergisms.

Coburning assist techniques can largely be quantified with the help of economic analyses. However, at this time, foreburning, afterburning, preburning and postburning gas assist techniques can only be quantified by studies of gas combustion with other hydrocarbon fuels within a quantified energy recovery and emission measurement system.

Summary and Conclusions

There is a serious lack of quantified information on gas assist techniques which limits gas utilization for the purposes of facilitating coal and waste combustion. The primary initial thrust of our efforts using the Sunland Training Center's watertube boiler is to quantify foreburning synergisms by energy release measurements. Here the potential payoff is very large since by foreburning with gas assist we can bring many fuels which would be unusable in a preexisting boiler into useful flammability and energy recovery ranges. High water content fuels such as CWM, green sawdust, sludge are examples of such possibilities. Here the synergism might be illustrated as a case of $0+1=2$ or more scientifically $0+1>1$.

We also hope to measure afterburning benefits from the gas stream emanating from the aperture controllable rear ring of our gas burner or an alternative burner. This gas should be entrained in the outer air envelope from our windbox and should merge together with the burning coal particles near the rear of the oil boiler. If necessary we will install a gas burner in the rear wall of the STC boiler for enhancing this afterburning. Such gas staging should help in reducing flue hydrocarbons, particulates and also in lowering NO_X yields. In this latter regard we might speak of stoichiometric staging with the help of gas to accomplish the types of the low NO_X emissions which can now be achieved by air staging.[64] Here the synergism is related to the possibility that without afterburning the combustion products would be environmentally unacceptable, another example of $0+1=2$ (or $0+1>1$).

Our two-stage investigations permit us to study the possibility of additional benefits of CWG combustion. Here we have the flexibility of controlling the gas distributions in both stages to accomplish optimum preboiler ash separation, overall energy recovery and emission reduction. By using refractory lined piping we will get most of the coal generated heat from the first stage, while most of the coal ash will be confined to the first-stage ash pit. The possibility that we will need an additional hot gas cyclone separator between the two stages to remove residual ash is under consideration and we have made preliminary cyclone designs should this be necessary. Such a hot gas cyclone would also be a good chamber for limestone injection for SO_2 reduction and would allow us to achieve favorable temperatures for $CaSO_4$ formation.

While we are not currently investigating co-combustion of gas with waste we will soon be in a position to pursue quantitative investigations of this nature. Certainly it will be possible to co-burn with liquid hydrocarbon waste. It should also be possible to co-burn gas with finely powdered sawdust in our first-stage combustor and possibly concentrated sludge. We could also burn refuse; however, the construction or acquisition of a used batch-type incinerator would be wiser. We could then quantify energy recovery and emission reduction in cofiring refuse with gas. All of these investigations are possible once we have perfected our use of the STC boiler as a quantitative energy recovery and emission measurement system.

While we have not stressed trace flue gas measurements in our initial experimentation we also by flue gas sampling and access to sensitive gas chromatographs can measure environmental benefits of natural gas assist. Quantitative energy recovery and trace gas reduction data are essential for the future development of gas assisted coal and waste combustion.

The combustion literature has many allusions to natural gas assistance. However there is a great need to quantify a science and engineering of natural gas assistance to coal and waste combustion. Despite the current OPEC oil glut our national energy and environmental interests are clearly at stake.

References

1. A.E.S. Green (Ed.), *Coal Burning Issues*, University Presses of Florida, Gainesville, Florida (1980) with 32 coauthors.
2. B.A. Green, Discussion with A.E.S. Green on use of coal for firing pottery kilns to reduce firing costs (1980) and development of gas-coal combustor (1981).
3. A.E.S. Green (Ed.), *An Alternative to Oil; Burning Coal with Gas*, University Presses of Florida, with Contributions by J. R. Jones, M. J. Ellerbrock, J. M. Schwartz, S. J. Kuntz and B. Zeiler, (1981).
4. D. B. Vaidya, J. J. Horvath and A.E.S. Green, "Remote Temperature Measurements in Gas and Gas-Coal Flames Using the OH (O,O) Middle-UV Band," Appl. Optics, 21, 3357, (1982).
5. A.E.S. Green and K. M. Pamidimukkala, "Kinetic Simulation of the Combustion of Gas/Coal and Coal/Water Mixtures," Proceedings of First Conference on Combined Combustion of Coal and Gas, Cleveland, Ohio (1982).
6. A.E.S. Green, B. A. Green, B. Zeiler, J. Samuels, K. M. Pamidimukkala, K. Zawoy, and C. Winefordner, Development and Firing of Vortex Combustors for Burning Coal/Gas and CWM Mixtures, Unpublished (1981-1983).
7. J. J. Horvath, K. M. Pamidimukkala, W. B. Person and A.E.S. Green, "Spectroscopic Observations of Methane-Pulverized Coal Flames," Journal of Quantitative Spectroscopy and Radiative Transfer, 31, 189-201, (1984).
8. A.E.S. Green, "Oil Backout and Acid Rain," 130-152, "Acid Deposition Causes and Effects," A. Green and W. Smith Eds., Government Institutes, Rockville, MD 1983.
9. A.E.S. Green and K. M. Pamidimukkala, "Synergistic Combustion of Coal and Natural Gas," Energy, the International Journal.
10. A.E.S. Green, R. R. Meyreddy and K. M. Pamidimukkala, "A Molecular Model of Coal Pyrolysis," Int Journal of Quantum Chemistry, 18, 589-499 (1984).
11. A.E.S. Green, "Improved Combustion of Coal Water Mixtures," Proceedings of Seventh International Symposium on Coal Slurry Fuels Preparation and Utilization May 21-24, 1985, New Orleans, LA 1985, S. W. Chun Symposium Chairman, U.S. Dept. of Energy, Pittsburgh Energy Technology Center, Pittsburgh, PA.
12. C. A. Thomas, Symposium Chairman, Proceedings Sixth International Symposium on Coal Slurry Combustion and Technology June 25-27, 1984, Orlando, Florida, U.S. Dept. of Energy, Pittsburgh Energy Technology Center, Pittsburgh, PA.
13. C. A. Thomas Symposium Chairman, Proceedings Fifth International Symposium on Coal Slurry Combustion and Technology, Tampa, Florida,

April 25-27, 1983. U.S. Dept. of Energy, Pittsburgh Energy Technology Center, Pittsburgh, PA.

14. S. W. Chun, Symposium Chairman, Proceedings Seventh International Symposium on Coal Slurry Fuels Preparations and Utilization, *Ibid.* 11.
15. R. W. Borio and M. J. Hargrove, "The Effects of Coal Slurry Fuels on Steam Generators," Fourth International Symposium on Coal Slurry Combustion, Orlando, FL (1982).
16. O. K. Foo and E. M. Jamgochian, "Assessment of Coal Liquids Mixtures in Cooperating IEA Countries," June 1983. The Mitre Corp. Sponsored by International Energy Agency and the U.S. Dept. of Energy.
17. F. A. Horney and F. S. Nolte, "Boiler Derating for Coal-Water Mixtures," COAL TECHNOLOGY '83, 6th International Coal and Lignite Utilization Conference, Houston, TX (1983).
18. D. R. Taschler, Air Products and Chemicals, Inc.; F. S. Nolte and T. E. Stringfellow, Stearns-Roger World Corporation. The Impact of Oxygen-Enriched Combustion of Coal-Water Slurry Combustion, *Ibid.* 13.
19. Y. S. Pan, G. T. Bellas, R. B. Snedden and J. I. Joubert, Pittsburgh Energy Technology Center. "Coal Methanol Mixture Combustion Tests with Coals of Different Ranks," *Ibid.* 13.
20. J. Reason, "How Fuel Volatility Impacts Boiler Design, Operation." *Power*, Dec. 1983, pg. 51-53.
21. M. Steinberg, P. T. Fallon and M. S. Sundaram, "Flash Pyrolysis of Coal with Reactive and Non-Reactive Gases," Progress Report, Process Sciences Division, Dept. of Energy and Environment, Oct. 1, 1981-Sept. 30, 1982.
22. M. S. Sundaram, M. Steinberg and P. T. Fallon, Flash Pyrolysis of Coal.
23. M. Sakai, K. Tokuda, S. Kaneko, H. Takatsuda and T. Imamoto Mitsubishi, Heavy Industry "Research and Development of CWM Combustion," Proceedings Sixth International Symposium on Coal Slurry Combustion, *Ibid.* 12.
24. J. B. Howard, "Fundamentals of Coal Pyrolysis and Hydropyrolysis," *Chemistry of Coal Utilization* (Second Supplement Volume) John Wiley & Sons, New York 1981, p. 665.
25. P. H. Given (Ed.), "Coal Science," American Conference on Coal Sciences, Advances in Chemistry Series 55, American Chemical Society, Washington, DC (1964).
26. K. Kobayashi, J. B. Howard and A. F. Sarofim, "Coal Devolatilization at High Temperatures," Sixteenth Symposium (Intnl) on Combustion, p. 411 (1976).
27. M. Sandifer, L. Thompson and R. Conrath, Senior Design Project, Mechanical Engineering, University of Florida, April 1985.
28. V. Kahn, M. J. Hopmeier and J. A. Lara, Senior Design Project, Mechanical Engineering, University of Florida, April 1985.
29. E. M. Shelton and C. L. Dickson, "Heating Oils, 1984," National Institute for Petroleum and Energy Research, Bartlesville, OK (1984).

30. G. Docktor, T. Lawrence, and T. Pugh, Senior Design Project, Mechanical Engineering, University of Florida, July 1985.
31. Post, Buckley, Schuh and Jernigan, Inc., "Hospital Waste Management Study," October 1981, Orlando, FL.
32. Consumat Systems, "Energy from Waste, Modular Incinerator Systems, Municipal and Industrial," 1984.
33. H. P. Hanson, "Technological Innovations," Chap. 7 in *Coal Burning Issues, Ibid.* 1.
34. J. Skitt, "Waste Disposal, Management and Practice," County Waste Disposal Engineer, Staffordshire County Council. Halsted Press, 1979.
35. J. R. Holmes, *Practical Waste Management.* John Wiley & Sons, 1983.
36. A. Porteous, "Municipal Waste Derived Fuels." Production Combustion and Environmental Aspects. Proceedings of the European Congress on Energy Economy and Management in Industries. Algarve, Port. 1984.
37. L. Catalano (Compiler), "Managing Risks Means Go-Ahead For Solid-Waste Steam Plants." *Power,* Sept. 1984, pg. 67-70.
38. F. M. Lewis, R. Forbess, F. W. Greb, "Tracking Boiler Performance When Firing Solid Waste Fuels." *Power,* May 1985, pg. 59-61.
39. M. Hanforf, R. Hurdle, "Burning Solid Waste Will Bring County Maximum Benefits." Public Works, July 1984, pg. 46-47.
40. J. Makansi, "Power From Sludge." *Power,* Feb. 1984, pg. 16-31.
41. J. M. McFarland and D. L. Brink, Comprehensive Studies of Solid Wastes Management. Sanitary Engineering Research Laboratory College of Engineering and School of Public Health, University of California, Berkeley, May 1972.
42. J. A. Borchardt, J. Redman, E. Jones and R. T. Sprague, *Sludge and Its Ultimate Disposal.* Ann Arbor Science Publisher, Inc. 1981.
43. W. R. Niessen, *Combustion and Incineration Process, Applications in Environmental Engineering.* Marcel Dekker, Inc., New York, 1978.
44. T. H. Ellis, "Should Wood Be A Source of Commercial Power?" Forest Products Journal, Vol. 25, No. 10.
45. A. J. Baker, H. Clarke, "Wood Residue As An Energy Source—Potential and Problems." Reprint from "Trees—The Renewable Resource," Proceedings, Rocky Mountain Forest Industries Conference, March 13-14, 1975, Tucson, AZ.
46. N. P. Cheremisinoff, *Wood for Energy Production.* Ann Arbor Science Publisher, Inc. 1980.
47. J. I. Zerbe, A. Baker, "Forest Resources For Producing Energy," pg. 278-285, *Bioresources For Development*, Pergamon Press, 1980.
48. M. L. Hiser, "Wood Energy." Proceedings of Governor William G. Milliken's Conference, Nov. 1977. Ann Arbor Science Publishers Inc., Ann Arbor, Michigan.
49. C. L. Proctor, II, "Technical Assessment of Alternative Technologies For Hazardous Waste Management in Florida." The Hazardous Waste Manage-

ment Program Center for Biomedical and Toxicological Research, Institute of Science and Public Affairs, Florida State University, June 1984.

50. Babcock and Wilcox, *Steam*, New York, 1978.
51. P. N. Cheremisinoff, *Air Pollution Control and Design Handbook*, Part 1. Marcel Dekker, Inc., New York, 1977.
52. W. Licht, *Air Pollution Control*, Marcel Dekker, New York, 1980.
53. P. C. Reist, *Introduction To Aerosol Science*, Macmillan Publishing Co., New York, 1984.
54. R. F. Henry, S. C. Saxena, and W. F. Podolski, "Particulate Removal From High-Temperature, High-Pressure Combustion Gases." Argonne National Laboratory, Oct. 1983.
55. E. T. Mchale, R. S. Scheffe and N. P. Rossmeiss, *Combustion and Flame*, 45: 121-135 (1982).
56. D. Leith and W. Licht, A.I.Ch.E. Symp. Ser. 126, 196 (1972).
57. C. E. Lapple, *Industrial Hygiene Quarterly*, 5 (11): 40-48 (1950).
58. G. B. Shepherd and C. E. Lapple, *Ind. Eng. Chem.*, 31, 972 (1939).
59. R. H. Essenhigh, Z. Li and Kyu-il Han, "Performance Characteristics of A Hot-Wall Furnace Fired With Coal Water Slurry (CWS) Using Gas." American Flame Research Committee, Akron, Ohio, Oct. 1983.
60. A.E.S. Green, (This secondary atomization attempt was an outgrowth of personal communications with S. Ehrlich, EPRI, Palo Alto, CA, 14 April 1983).
61. T. U. Yu, S. W. Kang, M. A. Togan, P. M. Walsh, J. M. Beer and A. F. Sarofim, "Secondary Atomization of Coal-Water Slurry Fuels." Seventh International Symposium on Coal Combustion and Technology, New Orleans, LA, May 1985.
62. J. A. Fay, D. S. Golomb and Kumar, "Controlling Acid Rain Without Scrubbers," E-Lab, Massachusetts Institute of Technology, Cambridge, MA (1984). University Monograph American Gas Association, Vol. 1, No. 1. Feb. 1985.
63. B. Schlesinger, "Natural Gas is Gaining Growing Approval As An Air Pollution Control Device," Proceedings 1st International Conference on Combined Combustion of Coal and Gas, Cleveland, OH (1982).
64. J. G. Singer, *Combustion, Fossil Power Systems*, Combustion Engineering, Inc., Windsor, CT (1981).

Appendices

Appendix A

Natural Gas Supplies

GEORGE H. LAWRENCE
MICHAEL I. GERMAN
American Gas Association

The U.S. natural gas production level in 1984 was roughly 17 Tcf. The nation has discovered—and has available to be produced—166 Tcf of proven reserves in the lower 48 states. These 166 Tcf have been booked as corporate assets of the companies that own them and 88 percent of this gas has already been "sold" (i.e., committed) to a purchaser.

The Potential Gas Committee—the industry group that evaluates gas resources—has estimated that, with currently foreseeable technology and economics, there are 647 Tcf of potential gas resources in the lower 48 states which are still awaiting discovery. This estimate of potential resources is in addition to the 166 Tcf of proved reserves. Therefore, estimated *conventional* gas resources total 813 Tcf—or the equivalent of 50 years of supply at the 1984 level of gas production. This 813 Tcf is referred to as lower 48 states "remaining recoverable resources."

The perception that the nation is "playing out" the remaining recoverable resource base over time has become the principal rationale for projections that gas production will steadily decline and that this decline will in turn force higher gas prices. This pivotal perception is mistaken because it considers the calculation of the gas resource base as a static—rather than a dynamic—assessment. History has proven that the gas resource base is indeed dynamic, changing radically with technology development.

The assertion that gas supplies must eventually decline—because of the finite nature of the recoverable resource base—is incorrect. The recoverable resource base is just the "tip of the iceberg" when assessing total gas resources *in-place* in the United States.

Near-term supplemental sources of gas currently in production have identified in-place gas at least twice the size of the recoverable resource base. These include gas from Western tight sands, Devonian shale, coal seams and enhanced gas recovery.

Long-term supplemental gas sources can be virtually inexhaustible. Overall, the gas resources in-place in the U.S. are potentially several thousand times our current consumption level.

Some advocates of energy sources that compete with gas have incorrectly used past gas shortages (in the *price-controlled* interstate market) as proof that: (a) we are "running out" of gas; and (b) the United States should therefore redirect energy production and consumption patterns away from natural gas.

However, recent gas exploration, production, and research activities, as well as activities that are in progress today, have characterized and quantified the gas resource base (both conventional and unconventional) in a manner which clearly disproves this theory.

Gas Resources

To examine the gas supply issue, it is necessary to understand how gas resource estimates are collected. As previously noted, the U.S. has about 50 years of conventional gas resources available in the lower-48 states—with another 10 years in Alaska. As technology advances, additional quantities of gas not presently included in this remaining resource estimate will be added to this nearly 1000 Tcf.

Over the next 20 years we will continue to see new technological developments increase the availability of gas resources. This in turn will stabilize the remaining recoverable gas resource base in the U.S. The incorrect perception of "playing out" over time of the remaining recoverable resource base is the principal rationale for projections of declining gas production and higher prices.

Technological developments in the gas supply area have been dramatic. One key area is the ability to drill for gas in offshore waters. The record in 1965 was 632 feet of water. In 1974 it was 2,097 feet, 5,624 feet in 1982 and in 1984 Shell set a new record of 6,448 feet of water. The theoretical limit today is around 10,000 feet.

According to the estimates of the Potential Gas Committee, the increases in water depth capability from 1965 to the late 1970s increased our offshore recoverable resource base by 50 percent. At this rate this technology growth has major implications for future additions to the recoverable resource base.

In the area of drilling capability onshore, similar statistics are available. In 1938 the record depth for a well was 15,004 feet. Today the record depth is 31,441 feet—more than double the depth. From a recoverable resource base view in the lower 48 states, the increase in depth capability from 15,000 to 30,000 feet increased the recoverable gas resources by 86 percent.

The payoff of past technological developments is clear. This technological growth primarily was the result of the oil industry trying to develop oil resources. If the government, together with the Gas Research Institute (GRI) and private sector gas companies, developed the needed technology today, similar, if not greater, technological advances could be made on an accelerated timetable.

There is tremendous potential for increased gas supply from sources that are *not now in the recoverable resource base.* Some of these sources are being developed today—some are clearly long-term investments.

But the U.S. can develop these resources through technological advancements, in the same manner deep water gas and deep well gas were added to our recoverable resources. To believe otherwise is to ignore the history of gas resource development in the U.S.

A key area for gas research is tight sands—those gas-bearing formations characterized by low permeability. Research in tight formations can be categorized as between short- and intermediate-term. It therefore should be the responsibility of both the private sector and the government.

Estimates of the tight sands resource in-place range as high as 900 Tcf. However, under foreseeable economics and technology only 172 Tcf of tight sands gas currently is included in the recoverable resource base. Although there is tight formations production today, it is from the easily accessible portion of the resource and thus is limited. Research is needed to improve the production rates for tight sands.

Another major research area is the Devonian shales. Devonian shale is estimated to contain upwards of 2,000 Tcf of gas in-place, but the amount presently counted as remaining recoverable gas resources is about one hundredth this amount. Improved techniques for "fracturing" the shale need to be developed before the Devonian shale resource can be fully exploited. Like tight sands, the great majority of these gas resources are *not* counted in the recoverable resource base.

In the case of both tight formation and Devonian shale gas, the existence—even the location—of these large resource bases is not speculative. Future research breakthroughs will be directed strictly at achieving marketability of gas production.

While the risk will be significant, the payback from successful research efforts would be very large. The respective locations of both tight sands gas in the West and Devonian shale gas in the East also offer transportation advantages that ensure subsequent production will benefit different consumers and indeed different segments of the U.S. economy.

Methane occurs naturally in coalbeds, being held in the micropores of the coal or in the natural fractures of the coalbed. Despite its broad potential, there were relatively few efforts to recover natural gas from coal seams on a commercial scale before 1975.

These efforts have, however, increased in recent years in response to the removal of price controls on coal seam gas and an increasing understanding of the recovery technology. For example, in the Black Warrior Basin in Alabama alone, there are more than 100 coalbed methane wells on line today. With a total resource in-place estimated as high as 800 Tcf there are vast portions of the resource which will not be accessible without significant technology development.

This coal-seam gas is often found close to the national pipeline and distribution network. It could be transported efficiently and economically to every area of the continental United States—needing minimal processing and no conversion before use by consumers. At present, there is practically no coal-seam gas in the remaining recoverable resource base estimates by the Potential Gas Committee and the U.S. Geological Service.

In addition to naturally occurring gas, gas can be produced synthetically from either coal, municipal and agricultural wastes or biomass. There are over 30 active waste gas projects and one major high-Btu coal gasification project underway in the U.S. These projects have, in general, proven to be technically more successful and have superior economics compared to other synthetic gas projects.

Geopressured gas is methane which is found either in a free state or dissolved in water and trapped in sedimentary rock at higher-than-normal pressures. Research efforts in geopressured gas can be categorized between the intermediate- and long-term.

One of the largest geopressured areas underlies a portion of the Texas and Louisiana Gulf Coast. Estimates of the total resource in-place range up to 100,000 Tcf; however, estimates of the amounts which might ultimately be recovered range between 25 and 1,200 Tcf.

As with coal-seam gas, there is practically no geopressured gas included in the recoverable resource base estimate. Given the large in-place resource, it could become an essentially inexhaustible source of domestic energy.

Long-term gas supply research efforts are those that have little likelihood of contributing to U.S. energy needs this century. These are high risk, high payoff technologies, which if proven, have the potential of making natural gas a virtually inexhaustible energy source.

Gas hydrates are solid, ice-like compounds in which gas molecules are entrapped and bound to water molecules. Recently, it has been recognized that methane hydrates occur naturally in some permafrost zones and ocean sediments. While most of the research on gas hydrates in sediments has been conducted in the Soviet Union, drilling in the North American Arctic and the offshore Atlantic have reportedly confirmed their existence.

Recently, samples were recovered by an expedition of a deep sea drilling vessel, Glomar Challenger, off the Pacific Coast of Central America. Typical estimates of the total world resource of gas in hydrates range to over 270 million Tcf. Again, this entire potential resource is totally excluded from current recoverable resource estimates.

There are a number of theories which suggest the existence of large volumes of methane deep within the earth's crust and upper mantle. One theory concludes that massive amounts of abiogenic methane, that is, methane whose origin is of a nonbiological nature, were derived from a primordial supply of hydrocarbons trapped during the earth's formation.

There is now sufficient evidence available to conclude that at least small amounts of the natural gas being produced today are the product of outgassing of methane from deep within the earth. Many traditional geologists now believe that both biological and abiogenic theories of gas formation can coexist. Scientific evaluation of this theory could lead to a new and more complete understanding of methane production—opening up new drilling prospects that formerly were overlooked.

Resource Summary

The gas resources in-place in the U.S. potentially are several thousand times our current consumption level. To the extent that research and development programs are directed into gas supply areas, the economy of transportation to ultimate consumers is assured by the million-mile pipeline and distribution system, plus thousands of miles of field gathering systems. This transportation network is energy efficient, readily expandable, and has environmental and public policy attributes unique to gas energy.

Conclusion

Natural gas supplies need never be a problem again. There are adequate gas resources in the United States (and Canada) to guarantee supplies for not only the 20th century, but also the 21st. This fact is recognized by all segments of the gas industry—with differences of opinion between sellers and buyers of gas focusing on the cost of production, not the resource base or the capability to produce.

A number of independent government agencies have also recognized the large supply potential of our domestic gas resource base. For example, the Congressional Budget Office recently issued a report that states that future gas supplies could be assured if a diversified gas sources strategy is followed.

This gas supply *fact* is important to all people interested in the environment because it means that domestic energy policy can be made compatible with environmental goals. The future availability of gas supplies should finally debunk the mistaken notion that energy and environmental objectives must necessarily be in conflict.

Appendix B

Bibliography

American Gas Association, *An Economic and Environmental Comparison of Natural Gas and Coal Use for Large-Scale Industrial Cogeneration*, October, 1984, Energy Analysis 1984-13.

Identifies broad prospects for gas-fired cogeneration by industry, region, and size of facility; provides a guide to methodology for assessing cogeneration benefits.

American Gas Association, *Fact Book: Energy, the Environment and Natural Gas*, October, 1983.

Analyzes environmental benefits of natural gas use in comparison with other fuels.

American Gas Association, *U.S. National Energy Research Policy* (1984)

Describes natural gas supply and prospects in the context of the role of natural gas in achieving national energy priorities.

Bardin, David J., "Combined Gas/Coal Burning—A New Way to Satisfy Both Environmental and Fuel Use Regulations," *Energy Economics, Policy and Management*, Vol. 2, No. 4, Spring 1983, pp. 19-29.

Describes Fuel Use Act and Clean Air Act constraints and ways around them.

Bardin, David J., "Real and Imagined Fuel Use Restrictions on Electric Utility Use of Natural Gas," *Public Utilities Fortnightly*, March 21, 1985, p. 27.

Describes extent of FUA restrictions that apply to electric utilities in operating old units, reconstructing old units and constructing new units.

Blodgett, J., Parker, L., and Backiel, A., "Acid Rain: Current Issues," Congressional Research Service Issue Brief Number IB83016, July 10, 1984.

Report summarizes recent scientific studies of acid precipitation and reviews each of the many proposals for acid rain legislation that have been introduced in the 98th Congress. Such reports are prepared periodically by the Congressional Research Service and can be obtained from your Congressman.

Chemical Week (2/2/83), "Mixing Gas with Coal or Oil can Pay Off," Vol. 132, No. 6, p. 26.

Environmental Research and Technology, Inc., *Evaluation of Environmental and Other Benefits of the Select Use of Natural Gas,* November, 1983.

Study, funded by the American Gas Association and the Gas Research Institute, examines technical, environmental, and economic aspects of selectively firing natural gas with coal and oil feedstocks at utility powerplants.

Green, A. E., *An Alternative to Oil: Burning Oil with Gas,* (1981), University Presses of Florida, Gainesville, Florida.

Summarizes an ongoing program of combustion research and energy analysis involving natural gas with coal, undertaken over the past several years at the University of Florida.

Hay, Nelson E., ed., *Guide to New Natural Gas Utilization Technologies,* (1984), Fairmont Press, Atlanta, Georgia.

Presents a compendium of economic and environmental analyses relating to cogeneration select gas use, methane-powered vehicles, and other potential new uses for natural gas.

National Coal Association, *Steam Electric Plant Factors,* 1983.

Pinkham, Douglas, "GAS: The Pollution Solution," *American Gas Association Monthly*, October, 1984, p. 19.

Describes select gas use at the Bellefield Boiler Plant in Pittsburgh, PA; the facility burns coal and natural gas.

Proceedings: International Conference on Combined Combustion of Coal and Gas, October 12-13, 1982.

Includes a variety of papers on technical, economic, and regulatory issues associated with select gas use.

Radak, Les J., "In-Furnace Control of No Formation in Gas and Oil-fired Utility Boilers," Southern California Edison, Rosemead, CA (1983).

Describes an advanced combustion process entailing burning both oil and gas in the same boiler, that has been demonstrated to achieve significant reductions in emissions of nitrous oxides.

Raufer, Roger K., and Stephen L. Feldman, "Emissions Trading and What It May Mean For Acid Deposition Control," *Public Utilities Fortnightly*, August 16, 1984, pp. 17-25.

Schlesinger, Benjamin, "Natural Gas Can Help Coal Burn Cheaper," *Environmental Science and Technology*, Vol. 14, No. 9, September 1980, pp. 1067-1071.

Assesses, on a national basis, the potential benefits of select use of natural gas with coal.

U.S. Department of Energy, *Inventory of Power Plants in the United States*, July, 1984, DOE/EIA-0095(83), GPO Stock No. 061-003-00391-5.

Provides information about electric generating units operated by or planned to be operated by electric utilities.

U.S. Environmental Protection Agency, *Compliance Data System Users Guide*, August, 1983, EPA-340-1-83-020.

Document is a users manual for the EPA's Compliance Data System. It provides detailed information about the CDS database and the retrieval of information from the system.

U.S. Environmental Protection Agency, *Maps Depicting Non-attainment Areas Pursuant to Section 107 of the Clean Air Act—1983*, EPA-450/2-83-004.

Mapping and listing of counties (or parts of counties) in non-attainment status for each of five of the categorical pollutants (TSP, SO_2, CO, NO_X, Ozone). Organized by nation, region and state. Easy to use. (Check for an update.)

Wilkinson, Paul, and Bruce McDowell, "An Economic and Environmental Comparison of Natural Gas and Coal Use for Large-Scale Cogeneration," *Gas Energy Review*, Vol. 12, No. 12, December, 1984, pp. 2-8.

Presents results of an examination of the relative economics and environmental impacts for coal- and gas-fired large-scale cogeneration systems.

Wilkinson, Paul, and Nelson Hay, "Select-Use Sites," *Gas Energy Review*, Vol. 12, No. 10, pp. 13-15.

Provides information on facilities where the select gas use concept has been employed.

Wilkinson, Paul, "Seasonal Fuel-Switching to Natural Gas: A Logical Approach to Reduce Sulfur Dioxide and Other Air Emissions," *Gas Energy Review*, Vol. 12, No. 9, pp. 11-17.

A method for reducing sulfur dioxide and other air pollutant emissions by substituting natural gas for higher sulfur fuels on a seasonal basis is analyzed in this report.

Appendix C

Environmental Glossary

Actual Emissions

The level of pollution emitted by a source. Actual emissions may differ from "allowable" emissions, which is the level specified in a source's permit or in the State Implementation Plan (SIP). Whether allowable or actual emissions are used in determining the base line against which emission reductions are measured will depend on the manner in which the SIP was developed. In almost every state, actuals will form the base line for measuring emission reductions.

Air Pollution Control Authority (APCA)/or District (APCD)

The public agency at the state and/or local level which has primary responsibility for implementing the Clean Air Act. The APCA is the most likely institution to implement a banking system. It is possible, however, for a distinct body, e.g., the Chamber of Commerce, air quality planning organization, or an economic development agency, to develop a banking program but only with the close cooperation and support of the APCA which will be responsible for allowing implementation of the program. *Note:* The national association of air pollution control officials is also abbreviated APCA.

Air Quality Control Region (AQCR)

The Environmental Protection Agency (EPA) has divided the nation into 247 air quality control regions which are so designated on the basis of geographical, topological, and meterological conditions.

Allowable Emissions

The level of emissions permitted by the terms of a source's permit or in the SIP. The question is whether allowable emissions is the appropriate base line for measuring emissions reductions. Only if current (i.e., actual) emissions are less than allowable emissions *and* the SIP was designed using the emission levels specified in permits, would a source be allowed credit for this difference.

Allowable Ratio

This ratio is applied during the process by which banked Emission Reduction Credits (ERCs) are converted for use to partially satisfy the terms of a permit requirement. The allowance ratio is a reduction in the quantity of ERCs to satisfy ambient air quality requirements.

Attainment Area: see **Non-Attainment Area; Prevention of Significant Deterioration.**

Bank

The term and its derivatives, banked and banking, describe regulatory procedures where a firm acquires or uses Emission Reduction Credits (ERC). These terms are not used to describe an actual institution where deposits, withdrawals, and other transactions are consummated. Therefore, there is not any one institution which can be described as the ERC "bank."

1. Banking describes the process by which a firm initially reduces its emissions and applies for ERCs. The *banking* process continues until ERCs are extinguished through use.
2. *Banked* refers to the status of an ERC after it has been certified, but before it has been used.
3. Bank refers to the pool of ERCs currently registered.

See: **Certificate; Confirmation; Emissions Reduction Credits; Emissions Trading and Banking Programs; Registry.**

Base Line

The level of emissions below which a source must reduce its emission to constitute an "emission reduction." Generally, it is the more stringent requirement of actual or allowable emissions, but this will depend on how the SIP was developed and the specific policy of that locale in satisfying the requirements of the Clean Air Act.

Best Available Control Technology (BACT)
BACT refers to the degree of pollution control required of a new source locating in an area that has attained the national ambient air quality standard; it is therefore the standard that must be met to obtain a PSD permit.

Bubble
EPA's alternative emission reduction option which, when incorporated into a SIP, allows a source to reduce control requirements at one point by increasing control correspondingly at another. The bubble may be applicable either within a single plant or between different plants in the same area.

Capital Costs
Costs associated with initiating a particular project or operation, e.g., acquisition of land, construction, or purchase of vehicles.

Certificate
In ERC banking, the air pollution control agency issues certificates representing ownership of specific ERCs which appear on the register and, thus, are banked. These certificates are for recordkeeping purposes and are not legally transferable. See: **Bank.**

Clean Air Act
The Clean Air Act was passed by Congress in 1971 to protect the nation's health and welfare. The Act requires that EPA set minimally acceptable levels of air quality. It designates much of the responsibility for attaining and maintaining air quality to the states. The Act governs air pollution emitted from both stationary and mobile sources.

Clean Air Act Amendments (1977)
These amendments clarify many provisions of the 1971 Act. Specifically, the amendments addressed the issues of economic growth in both non-attainment and Prevention of Significant Deterioration (PSD) areas, strengthened the New Source Performance Standards (NSPS) provisions, and provided a new timetable for attaining the National Ambient Air Quality Standards (NAAQS).

Confirmation
Air pollution control agency's verification given to a source's creation of an emission reduction. Confirmation is the second component of the banking process and comes only after all ques-

tions about the emission reduction (i.e., quantity, permanence, how created, etc.) have been resolved. A confirmed reduction in emissions would result in a change in the terms of a source's permit. See: **Bank.**

Control Methods (Controls)

Any means by which an emission reduction is achieved. Commonly, the term is used in reference to add-on technological controls installed by a source—scrubbers, electrostatic precipitators, or other abatement equipment. However, the term also includes modified processes to reduce emissions—shutdown, cutbacks, altered work practices, alteration of fuel or raw material inputs or production processes, etc.

Control Techniques Guidelines (CTG)

Supplied by EPA specifying the degree of emissions controls required of stationary point source emissions.

Conversion Ratio

The ratio is applicable to the process whereby an emission reduction is converted to an ERC. It is the fraction or percentage used to determine the number of ERCs which will be credited to an account. The ratio will generally be 1.0 and is determined after considering many factors including any classification schemes applied to sources and pollutants, the characteristics of the ERC, measurement certainty, etc.

Emission Offset

A regulatory device designed to allow economic growth in an area where a NAAQS has not been attained. The actual offset is obtained by securing a decrease in an existing source's emissions to compensate for emissions of a new or expanding source seeking to locate in a non-attainment area (see Section 173 of the Clean Air Act).

Emission Reduction Credits (ERCs)

The value which is "banked" and can later be used by a source to satisfy the required emission limits contained in its permit. The ERC is the end product of the conversion of the emission reductions. ERCs are *used* by being converted back into physical pollution units using an allowance ratio in a manner consistent with other requirements (e.g., new or expanding source emission offsets, bubble). It is crucial that the distinction between emission reductions and ERCs be maintained—they are *not* synonymous. See: **Bank; Conversion Ratio.**

Emissions Inventory

A list of large (heavy) pollution emitters in an air quality control region, maintained by the agency responsible for air quality in that region.

Emissions Trading and Banking Programs

Under these programs, emissions reductions in excess of EPA requirements, achieved by one source, can be "banked" by them for future expansion, or "traded" to another firm for their use. See: **Bank.**

Federal Motor Vehicle Control Program (FMVCP)

The Federal program under Title II of the Clean Air Act which requires auto manufacturers to produce cars which will achieve progressively lower exhaust emissions.

Fugitive Dust Emissions

Loose dust, grit, soot and other such emissions produced in a region, but not attributable to a particular source.

Hydrocarbons (HC)

Non-methane hydrocarbons. Precursors of photochemical smog. Subject of NAAQS.

Lowest Achievable Emission Rate (LAER)

LAER refers to the level of emissions that a new source locating in a non-attainment area must achieve. States determine LAER based on either (a) the lowest level contained in the implementation plan of any state, unless the permit applicant can demonstrate that such a level cannot be achieved, or (b) the lowest emission level which is achieved in practice within a relevant industrial category, whichever is lower.

Minor Source

A subcategory of sources with emissions below some threshold defined by states in their SIPs (e.g., 25 tons per year). This subcategory of existing sources is typically excluded from permit requirements and lacks a baseline against which emission reductions can be ascertained.

Mobile Source Emissions
Emissions from highway vehicle exhaust.

Monitoring
The measurement and recording of emissions which occurs over time. The purpose of monitoring is both to obtain a measurement and to ensure the permanency of the emission reductions. Monitoring can involve in-stack devices which measure emissions or devices which measure input or output parameters.

National Ambient Air Quality Standards (NAAQS)
Standards for amounts of specific pollutants allowable in the atmosphere, established by EPA.

New Source Performance Standards (NSPS)
EPA sets NSPS for all major industrial categories of stationary sources of air pollution that must be met by any new (or modified) source in that category. NSPS reflect a balance of the environmental and health effects of air pollution with the economic and energy impacts of a proposed level of control.

Non-Attainment Area
A geographic area designated by EPA to be in violation of NAAQS. A major new or expanding source seeking to locate in a non-attainment area must arrange for sufficient offsets to ensure that Reasonable Further Progress (RFP) toward attainment of NAAQS is achieved and must meet LAER.

Nonconventional Sources (Non-Permitted Sources)
A subcategory of sources characterized by emissions which are not directly measurable. Nonconventional sources include roads, storage piles, and the like. Because of this characteristic, these sources are generally not subject to permits and an APCA may decide to exclude them from the banking system.

Nonpermitted Sources
Industrial plants which do not require an air quality permit to operate.

Offset
EPA requires all major new sources of pollution in non-attainment areas to offset any increase in emissions with a commensurate decrease in emissions from any existing source in the same plant or area. See: **Emission Offset.**

Ozone (O_3)
A pungent, colorless, toxic gas composed of three atoms of oxygen. Although it occurs naturally in the stratosphere where it serves as a protective shield against radiation, it is dangerous when found in the air we breathe. Ozone may react with non-methane hydrocarbons or VOCs to form photochemical smog.

Permit
The emission restrictions placed by the APCA on a specific source. The permit may specify a specific emissions limit, require a percent removal of a pollutant, or dictate a particular work practice. Where possible, the permit conditions should be used as the base line for evaluating emission reductions. The permit terms are also generally related to the SIP and, thus, any change in a permit requires a corresponding change in the SIP.

Point Source Emissions
Emissions created by a specific industrial process.

Prevention of Significant Deterioration (PSD)
PSD requirements are applicable in areas that meet the NAAQS. The PSD regulations restrict the amount of growth in such areas and require that all new major sources of pollution in the area meet the BACT requirements.

Reasonably Available Control Technology (RACT)
RACT refers to the level of technology required of an existing source of pollution. In non-attainment areas, RACT is the minimum level of control. RACT is set by the state and reflects technology that is reasonably available considering economic and technological feasibility. For many categories of sources, EPA has assisted states by providing guidance for technical definitions of RACT.

Reasonable Further Progress (RFP)
The requirement under the Clean Air Act that areas designated non-attainment achieve annual incremental steps toward satisfying ambient air quality standards by the designated deadlines.

Registry
The books in which the banking system's activities are recorded and which serve as the accounting record for the issuance and use of ERCs. In a banking system, these books generally will be main-

tained by the APCA. They clarify ownership issues and facilitate the search process for needed reductions. See: **Bank.**

Source

A source is any building, structure, facility, or installation which emits any air pollutant. A source may include several specific emitting points but is limited to those owned by a single legal entity.

Standard Industrial Classification (SIC)

A nationwide standard breakdown or classification of light and heavy industries and services, each with a number.

Standard Metropolitan Statistical Area (SMSA)

A large population nucleus and nearby communities which have a high degree of economic and social integration with that nucleus. Each SMSA consists of one or more entire counties that meet specified standards pertaining to population, commuting ties, and metropolitan character. SMSA's are designated by the Office of Management and Budget.

State Implementation Plan (SIP)

The legal mechanism, subject to EPA approval, by which a state proposes to achieve and maintain the ambient air quality requirements of the Clean Air Act.

Total Suspended Particulates (TSP)

Pollutant, consisting of nonvolatile substances such as grit and dust, suspended in the atmosphere. TSP is one of six pollutants for which "criteria" and NAAQS have been set.

Trade

The transfer or sales of ERCs from one legal entity to another in some kind of market situation subject to APCA review and approval. See: **Bank.**

User Fees

Charges levied against sources that make use of the banking system. The charges can be used to defray operating expenses or to fund the operation of the banking program.

Volatile Organic Compounds (VOCs)

Precursors of photochemical smog. Subject of emission restrictions.

Appendix D

List of State and Federal Air Quality Control Agencies

The annual Directory of Governmental Air Pollution Control Agencies is prepared by the Air Pollution Control Association, P.O. Box 2861, Pittsburgh, PA 15230. This appendix reproduces the 1985 Directory with that Association's permission. The Directory is updated annually in the April issue of the Journal of the Air Pollution Control Association.

UNITED STATES ENVIRONMENTAL PROTECTION AGENCY

Administrator .... **Lee M. Thomas**
Telephone: (202) 382-4700
Deputy Administrator .... **A. James Barnes**
Telephone: (202) 382-4711
Associate Administrator for
International Affairs .... **Fitzhugh Green**
Telephone: (202) 382-4870
Associate Administrator for
Regional Operation .... **Carolyn M. Clinton**
Telephone: (202) 382-4709

Assistant Administrator for
Administration and Resource
Management .... **Howard Messner**
Telephone: (202) 382-4600
Enforcement & Compliance Monitoring ...... **Courtney Price**
Telephone: (202) 382-4137
General Counsel (Acting) .... **Gerald H. Yamada**
Telephone: (202) 382-4134
Policy Planning & Evaluation .... **Milton Russell**
Telephone: (202) 382-4332
External Affairs .... **Josephine Cooper**
Telephone: (202) 382-5659
Inspector General (Acting) .... **John C. Martin**
Telephone: (202) 382-3137
Water .... **(Not known at press time)**
Telephone: (202) 382-5100
Solid Waste & Emergency Response
(Acting) .... **Jack W. McGraw**
Telephone: (202) 382-4711
Pesticides & Toxic Substances .... **John Moore**
Telephone: (202) 382-2902
Air & Radiation (Acting) .... **Charles L. Elkins**
Telephone: (202): 382-7400
Research & Development .... **Bernard Goldstein**
Telephone: (202) 382-7676

401 M Street, SW, Washington, DC 20460

Office of Air Quality Planning and Standards
Telephone: (919) 541-plus indicated extension
Director .... **Gerald A. Emison(5615)**

Division Directors
Control Program Development .... **Darryl Tyler(5551)**
Emissions Standards & Engineering ... **Jack R. Farmer(5571)**
Monitoring and Data Analysis ..... **Richard G. Rhoads(5536)**
Strategies and Air Standards .... **John R. O'Connor(5504)**
Stationary Source Compliance ..... **Edward E. Reich in D.C. (202) 382-2807**

Research and Development—Associated Laboratories
Environmental Monitoring Systems ... **Thomas Hauser(2106)**
Atmospheric Sciences Research ...... **Alfred H. Ellison(2191)**
Health Effects Research .... **Gordon Hueter(2281)**
Air & Energy Engineering
Research .... **Frank Princiotta(2821)**

Research Triangle Park, NC 27711

Office of Mobile Source Air Pollution Control
Telephone: (202) 382-plus indicated extension
Director .... **Richard D. Wilson(7645)**
Science Advisor .... **Stanley M. Blacker(7676)**
Senior Technical Advisor .... **George Kittredge(4990)**

401 M Street, SW, Washington, DC 20460

Division Directors
Telephone: (313) 688-plus indicated extension
Certification .... **Robert E. Maxwell(4402)**
Emission Control Technology .... **Charles M. Gray(4404)**
Engineering Operations .... **Glenn Reschke(4243)**

2565 Plymouth Road, Ann Arbor, MI 48105

EPA REGIONAL OFFICES

Region I—Connecticut, Maine, Massachusetts, New Hampshire, Rhode Island, Vermont

Regional Administrator .... **Michael R. Deland**
Telephone: (617) 223-7210
Deputy Regional Administrator .... **Paul G. Keough**
Telephone: (617) 223-5424
Director, Air Management Division .... **Harley F. Laing**
Telephone: (617) 223-2226
Director of Waste Management
Division .... **Merril S. Hohman**
Telephone: (617) 223-5186
Chief, State Air Programs Branch .... **Linda M. Murphy**
Telephone: (617) 223-5134
Chief, Technical Support Branch .... **Wallace E. Stickney**
Telephone: (617) 223-5633
Director, Environmental Services
Division .... **Edward V. Fitzpatrick**
Telephone: (617) 861-6700, 60 Westview Street, Lexington, MA 02173
Chief, Control Technology & Compliance
Section .... **Robert A. O'Meara**
Telephone: (617) 223-5137
Director, Office of Public Affairs .... **Brook Cook**
Telephone: (617) 223-7223

John F. Kennedy Federal Building, Boston, MA 02203

Region II—New Jersey, New York, Puerto Rico, Virgin Islands

Regional Administrator .... **Christopher J. Daggett**
Telephone: (212) 264-2525
Deputy Regional Administrator .... **Vacant**
Telephone: (212) 264-0396
Director, Air & Waste Management Division ..... **Conrad Simon**
Telephone: (212) 264-2302
Chief, Air Programs Branch .... **William S. Baker**
Telephone: (212) 264-2517
Chief, Air Compliance Branch .... **Francis W. Giaccone**
Telephone: (212) 264-9627
Director, Environmental Services Division .... **Barbara Metzger**
Telephone: (202) 321-6754, Woodbridge Avenue, Edison, NJ 08817
Director, Office of Public Affairs .... **James Marshall**
Telephone: (212) 264-2515

26 Federal Plaza, New York, NY 10278

Region III—Delaware, District of Columbia, Maryland, Pennsylvania, West Virginia, Virginia

Regional Administrator .... **James M. Seif**
Telephone: (215) 597-9814
Deputy Regional Administrator .... **Stanley L. Laskowski**
Telephone: (215) 597-9812
Director, Hazardous Waste Management
Division .... **Stephen R. Wassersug**
Telephone: (215) 597-8131
Director, Air Management Division .... **W. Ray Cunningham**
Telephone: (215) 597-9390
Director, Environmental Services
Division .... **Robert J. Mitkus**
Telephone: (215) 597-4532
Acting, Air Enforcement Branch .... **Thomas Maslany**
Telephone: (215) 597-3989
Chief, Air Programs Branch .... **James A. Sydnor**
Telephone: (215) 597-4713
Acting Director, Office of
Public Affairs .... **Janet Luffy**
Telephone: (215) 597-9370

841 Chestnut St., Philadelphia, PA 19107

Region IV—Alabama, Florida, Georgia, Kentucky, Mississippi, North Carolina, South Carolina, Tennessee

Regional Administrator .... **Jack E. Ravan**
Telephone: (404) 881-4727
Deputy Regional Administrator .... **John A. Little**
Telephone: (404) 881-4727
Director, Air, Pesticides & Toxic
Management Division .... **Winston A. Smith**
Telephone: (404) 881-3785

Director, Waste Management Division ...... **Thomas W. Devine**
Telephone: (404) 881-3454
Chief, Air Management Branch .... **James T. Wilburn**
Telephone: (404) 881-3043
Director, Environmental Services Division ..... **James H. Finger**
Telephone: (404) 546-3136, College Station Road, Athens, GA 30613
Chief, Compliance Section .... **Richard S. Dubose**
Telephone: (404) 881-4298
Director, Office of Public Affairs .... **Hagen Thompson**
Telephone: (404) 881-3004
Director, Office of External Affairs .... **Frank Redmond**
Telephone: (404) 881-3004

345 Courtland Street, NE, Atlanta, GA 30365

Region V—Illinois, Indiana, Michigan, Minnesota, Ohio, Wisconsin

Regional Administrator .... **Valdas V. Adamkus**
Telephone: (312) 353-2000
Deputy Regional Administrator .... **Alan Levin**
Telephone: (312) 353-2000
Director, Air Management Division .... **David Kee**
Telephone: (312) 353-2212
Director, Waste Management Division ... **Basil G. Constantelos**
Telephone: (312) 886-7579
Chief, Air and Radiation Branch .... **Steven Rothblatt**
Telephone: (312) 353-2211
Chief, Air Compliance Branch .... **Larry Kertcher**
Telephone: (312) 353-2086
Director, Environmental Services Division .... **William H. Sanders, III**
Telephone: (312) 353-3808, 536 South Clark Street, Chicago, IL 60605
Director, Office of Public Affairs .... **Jon Grand**
Telephone: (312) 353-2072

230 South Dearborn Street, Chicago, IL 60604

Region VI—Arkansas, Louisiana, New Mexico, Okalhoma, Texas

Regional Administrator .... **Dick Whittington**
Telephone: (214) 767-2600
Deputy Regional Administrator .... **Francis E. Phillips**
Telephone: (214) 767-2600
Director, Air & Waste Management Division .... **Allyn M. Davis**
Telephone: (214) 767-2730
Chief, Air Branch .... **Jack S. Divita**
Telephone: (214) 767-2746
Director, Environmental Services Division .. **Russell F. Rhoades**
Telephone: (214) 767-2697
Air Compliance Contact .... **Jack S. Divita**
Telephone: (214) 767-2746
Acting Director, Office of Public Affairs .... **Roger Meachum**
Telephone: (214) 767-2630

First International Building, 1201 Elm Street, Dallas, TX 75270

Region VII—Iowa, Kansas, Missouri, Nebraska

Regional Administrator .... **Morris Kay**
Telephone: (913) 236-2800
Deputy Regional Administrator .... **William W. Rice**
Telephone: (913) 236-2800
Director, Waste Management Division ...... **David A. Wagoner**
Telephone: (816) 374-6539, 324 E. 11th St., Kansas City, MO 64106 (until 5-15-85)
Director, Air & Waste Toxics Management Division .... **William A. Spratlin**
Telephone: (913) 236-2834
Chief, Air Branch .... **Carl M. Walter**
Telephone: (816) 374-3791, 324 E. 11th St., Kansas City, MO 64106
Director, Environmental Services Division ... **John C. Wicklund**
Telephone: (913) 236-3720, 25 Funston Road, Kansas City, KS 66115
Air Compliance Contact .... **Charles Whitmore**
Telephone: (816) 374-6525, 324 E. 11th St., Kansas City, MO 64106
Director, Office of Public Affairs .... **Rowena Michaels**
Telephone: (913) 236-2803

726 Minnesota Ave., Kansas City, KS 66101

Region VIII—Colorado, Montana, North Dakota, South Dakota, Utah, Wyoming

Regional Administrator .... **John G. Welles**
Telephone: (303) 293-1603
Deputy Regional Administrator .... **Alexandra B. Smith**
Telephone: (303) 293-1603
Director, Air & Toxics Management Division .... **Irwin L. Dickstein**
Telephone: (303) 293-1437
Director, Waste Management Division .... **Robert L. Duprey**
Telephone: (303) 293-1720
Chief, Air Programs Branch .... **Robert S. DeSpain**
Telephone: (303) 293-1753
Director, Environmental Services Division .... **James B. Lehr**
Telephone: (303) 236-5061, Denver Federal Center, Lakewood, CO
Air Compliance Contact .... **John Philbrook**
Telephone: (303) 293-1720
Director, Office of Public Affairs .... **Doris Sanders**
Telephone: (303) 837-5927

1860 Lincoln Street, Denver, CO 80295

Region IX—Arizona, California, Hawaii, Nevada, Guam, American Samoa

Regional Administrator .... **Judith E. Ayres**
Telephone: (415) 974-8153
Deputy Regional Administrator .... **John C. Wise**
Telephone: (415) 974-8153
Director, Air Management Division .... **David P. Howekamp**
Telephone: (415) 974-8201
Acting Deputy Director, Waste Management Division .... **David Mowday**
Telephone: (415) 974-8201
Chief, Air Programs Branch .... **David L. Calkins**
Telephone: (415) 974-8058
Acting Chief, Air Operations Branch .... **Carl C. Kohnert**
Telephone: (415) 974-8201
Director, Office of Technical & Scientific Assistance .... **David S. Mowday**
Telephone: (415) 974-8147
Director, Office of External Affairs .... **Deanna Wieman**
Telephone: (415) 974-8023

215 Fremont Street, San Francisco, CA 94105

Region X—Washington, Oregon, Idaho, Alaska

Regional Administrator .... **Ernesta Ballard Barnes**
Telephone: (206) 442-5810
Deputy Regional Administrator .... **L. Edwin Coate**
Telephone: (206) 442-1220
Director, Air & Toxics Division .... **Gary L. O'Neal**
Telephone: (206) 442-4762
Acting Director, Waste Management Division .... **Charles E. Findley**
Telephone: (206) 442-1906
Chief, Air Programs Branch .... **Randy Smith**
Telephone: (206) 442-1941
Director, Environmental Services Division .... **Dick Bower**
Telephone: (206) 442-1295
Air Compliance Contact .... **Randy Smith**
Telephone: (206) 442-1941
Public Relations Officer .... **Bob Jacobson**
Telephone: (206) 442-1465

1200 6th Avenue, Seattle, WA 98101

UNITED STATES
STATE AND LOCAL AGENCIES

■ ALABAMA

STATE AGENCY

Alabama Department of Environmental Management, State Capitol, Montgomery, AL 36130 — Telephone: (205) 271-7861

Director .... **Joe B. Broadwater**
Chief, Air Division .... **Richard E. Grusnick**

Supplementary personnel — full time: 2 administrators, 3 chemists, 2 data processing specialists, 15 engineers, 15 inspectors, 1 meteorologist, 5 office staff, 3 technicians; part time: 1 attorney

Alabama Environmental Management Commission (same address and phone as above)

Chairman .... **Dr. Dewey White**

BIRMINGHAM (See Jefferson County)

CULLMAN COUNTY (See TriCounty)

HUNTSVILLE

Air Pollution Control Department, City of Huntsville, 2033 C Airport Road, Huntsville, AL 35802 — Telephone: (205) 881-7803

Air Pollution Control Officer .... **I. N. Vaughan**

Supplementary personnel — full time: 1 administrator, 1 chemist, 1 office staff, 2 specialists

Air Pollution Control Board (same address and phone as above)

Chairperson .... **Robert Willisson**

JEFFERSON COUNTY (Includes Birmingham)

Jefferson County Department of Health, 1400 Sixth Avenue South, Birmingham, AL 35233 — Telephone: (205) 933-9110

Acting Health Officer .... **Carole W. Samuelson, M.D.**
Director, Bureau of Environmental Health .... **Paul Pate**

Supplementary personnel — full time: 1 administrator, 1 chemist, 1 data processing specialist, 5 engineers, 1 meteorologist, 3 office staff, 6 sanitarians, 1 technician, 1 air monitoring supervisor, 2 senior sanitarians, 1 senior maintenance repair worker; part time: 1 attorney, 1 public information specialist

Jefferson County Board of Health (same address and phone as above)

Chairman .... **Doris Phillips, M.D.**

LAWRENCE COUNTY (See TriCounty)

LIMESTONE COUNTY (See TriCounty)

MORGAN COUNTY (See TriCounty)

TRICOUNTY (Includes Cullman, Lawrence, Limestone and Morgan Counties)

The TriCounty District Health Service, Division of Air Pollution Control, 510 Cherry St., N.E., P.O. Box 1628, Decatur, AL 35602 — Telephone: (205) 353-7021

District Medical Director .... **Betty W. Vaughan, M.D.**
Air Program Director .... **Danny R. Seaver**

Supplementary personnel — full time: 1 administrator, 2 inspectors, 1 office staff; part time; 10 inspectors

Boards of Health: Cullman, Lawrence, Limestone and Morgan Counties.

Chairman, Cullman County .... **J. W. David**
Chairperson, Lawrence County .... **Garland Hall**
Chairperson, Limestone County .... **C. D. Roberson**
Chairperson, Morgan County .... **Paul McCain**

■ ALASKA

STATE AGENCY

State of Alaska, Department of Environmental Conservation, Pouch O, Juneau, AK 99811 — Telephone: (907) 465-2666

Commissioner .... **Richard A. Nevé**
Director, Environmental Quality Management .... **Keith Kelton**
Chief, Air and Solid Waste Management ...... **Stan Hungerford**

Supplementary personnel — full time: 6 engineers, 1 transportation planner, 1 office staff, 1 instrument technician, 1 ecologist; part time: 1 administrator, 1 chemist, 3 inspectors, 1 management analyst

FAIRBANKS NORTH STAR BOROUGH

Fairbanks North Star Borough, P.O. Box 1267, Fairbanks, AK 99707 — Telephone: (907) 452-4761

Public Works Director .... **Donald L. Moore**
Environmental Services Director .... **Heather J. Stockard**

Supplementary personnel — full time: 1 administrator, 1 engineer, 1 I/M program manager, 6 technicians; part time: 1 attorney, 1 data processing specialist, 1 meteorologist

Pollution Control Commission (same address and phone as above)

Chairperson .... **John Hargesheumer**

SOUTH CENTRAL CLEAN AIR AUTHORITY (Includes Anchorage Air Pollution Control Agency)

Anchorage Air Pollution Control Agency, 825 L Street, Anchorage, AK 99503 — Telephone: (907) 264-4713

Director, DEHP .... **Rodman Wilson, M.D.**
Air Program Manager .... **George La More**

Supplementary personnel — full time: 1 office staff, 4 air resource specialists; part time: 1 attorney

South Central Clean Air Authority Commission (same address and phone as above)

Chairperson .... **Dorothy Jones**

■ AMERICAN SAMOA

Environmental Quality Commission, Governor's Office, Pago Pago, American Samoa 96799 — Telephone: 633-4116, 633-4398

Chairperson .... **Lyle Richmond**
Executive Secretary .... **Pati Faiai**

Supplementary personnel — full time: 6 sanitarians, 4 technicians; part time: 1 administrator, 1 attorney, 1 data processing specialist, 1 engineer, 1 public information specialist, 1 chemist, 1 inspector, 1 office staff

■ ARIZONA

STATE AGENCY

Arizona Department of Health Services, 1740 West Adams Street, Phoenix, AZ 85007 — Telephone: (602) 255-1140

Director .... **Lloyd F. Novick, M.D.**
Assistant Director for Environmental
Health Services .... **Charles D. Anders**
Manager, Permits & Compliance .... **Arthur A. Aymar, P.E.**

Supplementary personnel — full time: 1 administrator, 1 chemist, 2 data processing specialists, 12 engineers, 1 meteorologist, 7 office staff, 5 technicians; part time: 1 attorney

Arizona State Air Pollution Control Hearing Board (same address as above)

Chairman ....

COCONINO COUNTY

Coconino County Air Pollution Control District, 2500 North Fort Valley Road, Flagstaff, AZ 86001 — Telephone: (602) 779-5164

Director, Health Department .... **William J. Thomas, M.D.**
Director, Environmental Health .... **Dan Smith**

Supplementary personnel — full time: 5 sanitarians; part time: 1 administrator, 1 attorney, 1 data processing specialist, 1 office staff

Coconino County Air Pollution Control Hearing Board (same address and phone as above)

Chairperson .... **William J. Thomas, M.D.**

MARICOPA COUNTY

Maricopa County Department of Health Services, 1825 E. Roosevelt St., Phoenix, AZ 85006 — Telephone: (602) 258-6381

Director, Dept. of Health Services .... **Adolfo Echeveste**
Assistant Director,
Environmental Services Division .... **John A. Power, P.E.**
Chief, Bureau of Air Pollution Control .... **Robert W. Evans**

Supplementary personnel — full time: 4 administrators, 1

chemist, 1 data processing specialist, 2 engineers, 10 inspectors, 3 office staff, 3 technicians, 1 planner; part time: 1 attorney, 1 public information specialist

Maricopa County Air Pollution Control Advisory Council and Maricopa County Air Pollution Control Hearing Board (same address and phone as above)

Chairperson, Advisory Council .... **Jan Hunsker**
Chairperson, Hearing Board .... **Donna Hahn**

PIMA COUNTY

Pima County Health Department — Air Quality Control District, 151 West Congress Street, Tucson, AZ 85701 — Telephone: (602) 792-8803

Director, Pima County Health Department .... **Patricia A. Nolan, M.D.**
Director, Air Quality Control District .... **C. Lee Fox**

Supplementary personnel — full time: 1 administrator, 3 air quality analysts, 1 engineer, 6 inspectors, 2 supervisors, 1 technician, 3 office staff, 1 senior air quality analyst; part time: 1 attorney

Pima County Air Quality Advisory Council (same address and phone as above)

Chairperson .... **Raymond White, Jr., Ph.D.**

PINAL-GILA COUNTIES

Pinal-Gila Counties Air Quality Control District, P.O. Box 1076, Florence, AZ 85232 — Telephone (602) 868-5801, ext. 486

Director and Air Pollution Control Officer .... **Dorothy M. Rankin**

Supplementary personnel — full time: 1 administrator, 1 office staff, 1 technician; part time: 2 attorneys, 1 engineer

Pinal-Gila Counties Air Pollution Advisory Council, 5161 E. Shiprock St., Apache Junction, AZ 85220

Chairperson .... **Larry Kluth**

■ ARKANSAS

STATE AGENCY

Arkansas Department of Pollution Control and Ecology, 8001 National Dr., Little Rock, AR 72209 — Telephone (501) 562-7444

Director .... **Phyllis Garnett, Ph.D.**
Chief, Division of Air Pollution Control .... **Wilson Tolefree**

Supplementary personnel — full time: 4 administrators, 4 attorneys, 5 chemists, 2 data processing specialists, 4 engineers, 14 inspectors, 1 meteorologist, 2 office staff, 1 public information specialist, 3 technicians

Arkansas Commission of Pollution Control and Ecology (same address and phone as above)

Chairman .... **Clyde Broyles**

■ CALIFORNIA

STATE AGENCY

California Air Resources Board, 1102 Q St., Sacramento, CA 95814 — Telephone: (916) 322-2990

Chairperson, ARB/Secretary of Envir. Affairs .... **Gordon Duffy**
Executive Officer .... **James D. Boyd**

Supplementary personnel — full time, 20 administrators, 4 attorneys, 5 chemists, 18 data processing specialists, 111 engineers, 9 inspectors, 8 meteorologists, 43 office staff, 3 public information specialists, 37 technicians, 24 research specialists, 141 air pollution specialists; part time: 3 attorneys, 1 data processing specialist, 8 engineers, 7 office staff, 3 technicians, 2 research specialists, 12 air pollution specialists

ALAMEDA COUNTY (See Bay Area)

ALPINE COUNTY (See Great Basin Valley)

AMADOR COUNTY

Amador County Air Pollution Control District, 108 Court Street, Jackson, CA 95642 — Telephone: (209) 223-3230, Ext. 292

Air Pollution Control Officer.......... **Dr. James McClenahan**
Deputy APCO .... **Rod Schyler**
Assistant APCO .... **Raymond Kruth**

Supplementary personnel — part time: 2 administrators, 1 engineer, 1 office staff

Air Pollution Control Board, Amador County Board of Supervisors, (same address and phone as above)

Chairperson .... **Kenneth H. Deaver**

BAY AREA (Includes Counties of Alameda, Contra Costa, Marin, Napa, San Francisco, San Mateo, Santa Clara, Sonoma and Solano)

Bay Area Air Quality Management District, 939 Ellis St., San Francisco, CA 94109 — Telephone: (415) 771-6000

Air Pollution Control Officer................. **Milton Feldstein**

Supplementary personnel — full time: 24 administrators, 3 attorneys, 4 chemists, 3 data processing specialists, 33 engineers, 46 inspectors, 4 meteorologists, 46 office staff, 3 public information specialists, 25 technicians

Bay Area Air Quality Management District Board of Directors (same address and phone as above)

Chairperson .... **Ray Diridon**

BUTTE COUNTY

Butte County Air Pollution Control District, 316 Nelson Avenue, P.O. Box 1229, Oroville, CA 95965 — Telephone: (916) 534-4383

Air Pollution Control Officer.................... **Joe E. Bandy**
Deputy APCO .... **Craig L. Gilmour**

Supplementary personnel — full time: 2 inspectors, 1 administrative engineering technician, 1 office staff; part time; 2 administrators, 1 attorney

Butte County Board of Supervisors, Administrative Building, Oroville, CA 95965 — Telephone: (916) 534-4780

Chairperson .... **Len Fulton**

CALAVERAS COUNTY

Calaveras County Air Pollution Control District, Government Center, San Andreas, CA 95249 — Telephone: (209) 754-3742

Air Pollution Control Officer......... **Robert E. Marshall, M.D.**
Deputy Air Pollution Control Officer...... **Michael P. Redgrave**

Supplementary personnel — part time: 1 administrator, 1 office staff, 1 engineer, 1 sanitarian

Calaveras County Board of Supervisors (same address as above) — Telephone; (209) 754-3241

Chairperson .... **Tom Taylor**

COLUSA COUNTY

Colusa County Air Pollution Control District, 751 Fremont St., Colusa, CA 95932 — Telephone: (916) 458-5891

Agricultural Commissioner .... **Wilber C. Disney, Jr.**

Supplementary personnel — part time: 1 administrator, 1 office staff

Colusa County Board of Supervisors, 546 Jay St., Colusa, CA 95932 — Telephone: (916) 458-4516

Board Chairperson .... **Vacant**

CONTRA COSTA COUNTY (See Bay Area)

DEL NORTE COUNTY (See North Coast Unified)

EL DORADO COUNTY

El Dorado County Air Pollution Control District, 360 Fair Lane, Placerville, CA 95667 — Telephone: (916) 626-2407

Air Pollution Control Officer.............. **Vernon C. Peterson**

Supplementary personnel — full time: 1 technician; part time: 1 administrator, 1 engineer, 1 office staff

El Dorado County Air Pollution Control Board (same address and phone as above)

Chairperson .... **Patricia Lowe**

FRESNO COUNTY

Fresno County Air Pollution Control District, 1221 Fulton Mall, Fresno, CA 93721 — Telephone: (209) 445-3239

Director of Health.......................... **Donald R. Rowe**
Assistant Director, Envir. Health .... **Louis A. Dooley**

Supplementary personnel — full time: 2 administrators, 4 engineers, 1 inspector, 2 office staff, 4 sanitarians, 3 technicians; part time; 1 attorney, 1 chemist

Fresno County Board of Supervisors, Hall of Records, Third Floor, 2281 Tulare St., Fresno, CA 93721

Chairperson .... **Sharon Levy**

GLENN COUNTY

Glenn County Air Pollution Control District, 720 North Colusa Street, Willows, CA 95988 — Telephone: (916) 934-4651

Air Pollution Control Officer..................... **Ed Romano**
Air Pollution Technician II .... **Joseph McIntyre**

Supplementary personnel — part time: 1 administrator, 3 inspectors, 2 office staff, 1 technician

Glenn County Air Pollution Control Board, 526 W. Sycamore St., Willows, CA 95988

Chairperson .... **Joe Williams**

GREAT BASIN VALLEY (Includes Counties of Alpine, Inyo and Mono)

Great Basin Unified Air Pollution Control District, 157 Short Street, Suite 6, Bishop, CA 93514 — Telephone: (619) 872-8211

Air Pollution Control Officer............... **Charles L. Fryxell**

Supplementary personnel — full time: 1 administrator, 1 office staff; part time: 1 attorney, 1 inspector, 1 sanitarian, 1 technician

Great Basin Unified Air Pollution Control Board (same address and phone as above)

Chairperson .... **Paul F. Johnson**

HUMBOLDT COUNTY (See North Coast Unified)

IMPERIAL COUNTY

Imperial County Air Pollution Control District, 150 South Ninth St., El Centro, CA 92243-2801 — Telephone (619) 339-4341

Air Pollution Control Officer............... **Claude M. Finnell**

Supplementary personnel — full time: 2 engineers, 1 deputy, 1 office staff; part time: 1 administrator

Air Pollution Control Board, Imperial County Board of Supervisors, 940 W. Main St., El Centro, CA 92243

Chairperson .... **Abe F. Seabolt**

INYO COUNTY (See Great Basin Valley)

KERN COUNTY

Kern County Air Pollution Control District, 1601 H Street — Suite 150, Bakersfield, CA 93301 — Telephone: (805) 861-3682

Air Pollution Control Officer........ **Leon M. Herbertson, M.D.**
Chief Air Sanitation Officer.................. **Citron Toy, R.S.**

Supplementary personnel — full time: 1 administrator, 3 program managers, 8 engineers, 16 inspectors, 3 office staff, 1 technician; part time: 1 administrator, 1 attorney, 1 chemist, 1 data processing specialist, 37 fire department personnel, 1 meteorologist, 1 public information specialist

Kern County Air Pollution Control Board, 1415 Truxtun Ave., Bakersfield, CA 93301

KINGS COUNTY

Kings County Air Pollution Control District, 330 Campus Drive, Hanford, CA 93230 — Telephone: (209) 584-1411

Air Pollution Control Officer..........**Sheldon R. Minkin, D.O.**
Asst. Air Pollution Control Officer .... **Dennis Otani, R.S.**

Supplementary personnel — full time: 1 engineer; part time: 1 office staff, 2 sanitarians

Kings County Air Pollution Control Board, Kings County Government Center, 1400 W. Lacey Blvd., Hanford, CA 93230

Chairman ....**Joseph Hammond**

LAKE COUNTY

Lake County Air Pollution Control District, 255 N. Forbes St., Lakeport, CA 95453 — Telephone: (707) 263-2391

Air Pollution Control Officer.............. **Robert L. Reynolds**

Supplementary personnel — full time: 1 administrator, 1 meteorologist, 1 office staff, 2 technicians; part time: 1 chemist

Lake County Board of Supervisors (same address and phone as above)

Chairperson ....**Walter Wilcox**

LASSEN COUNTY

Lassen County Air Pollution Control District, 175 Russell Ave., Susanville, CA 96130 — Telephone: (916) 257-8311, ext. 110

Air Pollution Control Officer................ **Fredric L. Evans**

Supplementary personnel — full time: 1 administrator, 2 inspectors, 1 office staff; part time: 1 attorney, 2 engineers

Lassen County Air Pollution Hearing Board (same address and phone as above)

Chairperson .... **Paul Simpson**

LOS ANGELES COUNTY (See South Coast AQMD)

MADERA COUNTY

Madera County Air Pollution Control District, 135 W. Yosemite Ave., Madera, CA 93637 — Telephone: (209) 675-7823

Director, Environmental Health .... **David W. Fishel**
Air Pollution Control Officer................. **David W. Fishel**

Supplementary personnel — full time: 1 administrator; part time: 1 attorney, 1 engineer, 1 office staff, 4 sanitarians

Madera County Air Pollution Control Board, 209 W. Yosemite Ave., Madera, CA 93637

Chairperson .... **J. Gordon Kennedy**

MARIN COUNTY (See Bay Area)

MARIPOSA COUNTY

Mariposa County Air Pollution Control District, P.O. Box 5, Mariposa, CA 95338 — Telephone: (209) 966-3689

County Health Officer .... **Avery E. Sturm, M.D.**

Supplementary personnel — full time: 1 administrator, 1 attorney, 1 inspector, 1 office staff

Mariposa County Board of Supervisors, P.O. Box 247, Mariposa, CA 95338

Chairman .... **Gene Dalton**

MENDOCINO COUNTY

Mendocino County Air Pollution Control District, Courthouse Square, Ukiah, CA 95482 — Telephone: (707) 463-4391

Air Pollution Control Officer.........**Craig M. McMillan, M.D.**
Deputy Director, APC........................**Robert F. Swan**

Supplementary personnel — full time: 1 administrator/inspector; part time: 1 office staff

Air Pollution Control Board (same address and phone as above)

Chairperson .... **Marilyn Butcher**

MERCED COUNTY

Merced County Air Pollution Control District, 210 E. 15th Street, P.O. Box 471, Merced, CA 95341 — Telephone: (209) 385-7391

Health Officer .... **Vacant**
Manager, Air Pollution
Control District.............. **Roland D. Brooks, MPA, R.S.**

Supplementary personnel — full time: 1 administrator, 1 engineer; part time: 1 inspector, 2 office staff, 8 sanitarians

Merced County Air Pollution Control Board, 2222 M St., Merced, CA 95340

Chairperson .... **Harvey Fonseca**

MODOC COUNTY

Modoc County Air Pollution Control District, 202 W. Fourth St., Alturas, CA 96101 — Telephone: (916) 233-2225

Air Pollution Control Officer............**Clinton B. Greenbank**

Supplementary personnel — part time: 1 administrator, 1 office staff, 2 technicians

Modoc County Air Pollution Control Board, P.O. Box 1130, Alturas, CA 96101

Chairperson .... **John L. Coulson**

MONO COUNTY (See Great Basin Valley)

MONTEREY BAY UNIFIED (Includes Counties of Monterey, Santa Cruz and San Benito)

Monterey Bay Unified Air Pollution Control District, 1164 Monroe Street, Suite 10, Salinas, CA 93906 — Telephone: (408) 443-1135

Air Pollution Control Officer............... **Lawrence D. Odle**

Supplementary personnel — full time: 2 administrators, 1 chemist, 3 engineers, 3 inspectors, 4 office staff, 3 technicians; part time: 1 attorney

Monterey Bay Unified Air Pollution Control District Board (same address and phone as above)

Chairperson ....**Gary Patton**

NAPA COUNTY (See Bay Area)

NEVADA COUNTY

Nevada County Air Pollution Control District, 10433 Willow Valley Road, Nevada City, CA 95959 — Telephone: (916) 265-1452

Air Pollution Control Officer............ **Jerry Zarriello, M.D.**
Assistant Air Pollution Control Officer .... **Ronald E. Earles**

Supplementary personnel — part time: 1 office staff

NORTH COAST UNIFIED (Includes Del Norte, Humboldt, and Trinity Counties)

North Coast Unified Air Quality Management District, 5630 S. Broadway, Eureka, CA 95501 — Telephone: (707) 443-3093

Air Pollution Control Officer.............. **Charles Sassenrath**

Supplementary personnel — full time: 1 administrator, 1 engineer; part time: 1 office staff

North Coast Air Quality Management District Board (same address and phone as above)
Chairperson .... **Henry Pritchard**

NORTHERN SONOMA COUNTY

Northern Sonoma County Air Pollution Control District, 118 North Street, Healdsburg, CA 95448 — Telephone: (707) 433-5911
Air Pollution Control Officer .... **Michael W. Tolmasoff**
Supplementary personnel — full time: 1 engineer, 3 inspectors; part time: 2 office staff
Northern Sonoma County Air Pollution Control District Board of Directors, 2555 Mendocino Ave., Santa Rosa, CA 95401
Chairperson .... **Ernest Carpenter**

ORANGE COUNTY (See South Coast AQMD)

PLACER COUNTY

Placer County Air Pollution Control District, 11484 B Ave., Auburn, CA 95603 — Telephone: (916) 823-4443
Air Pollution Control Officer .... **Kenneth F. Selover**
Supplementary personnel — full time: 1 administrator, 1 engineer, 2 inspectors, 1 office staff; part time: 1 attorney
Placer County Board of Supervisors, 175 Fulweiler Avenue, Auburn, CA 95603

PLUMAS COUNTY

Plumas County Air Pollution Control District, P.O. Box 480, Quincy, CA 95971 — Telephone: (916) 283-1256
Air Pollution Control Officer .... **Mike Murray**
Supplementary personnel — part time: 1 administrator/sanitarian/inspector, 1 office staff
Plumas County Board of Supervisors, P.O. Box 207, Quincy, CA 95971
Chairperson .... **Vacant**

RIVERSIDE COUNTY (See South Coast AQMD)

SACRAMENTO COUNTY

Sacramento County Air Pollution Control District, 9323 Tech Center Drive, Suite 800, Sacramento, CA 95827 — Telephone: (916) 366-2107
Air Pollution Control Officer .... **Norman Covell**
Supplementary personnel — full time: 3 administrators, 2 engineers, 5 inspectors, 2 office staff, 3 technicians; part time: 1 attorney, 1 agricultural commissioner, 5 agricultural biologists
Sacramento County Air Pollution Control District Board, 700 H Street, Sacramento, CA 95814
Chairperson .... **Sandra Smoley**

SAN BENITO COUNTY (See Monterey Bay Unified)

SAN BERNARDINO COUNTY

San Bernardino County Air Pollution Control District, 15505 Civic Drive, Victorville, CA 92392 — Telephone: (619) 245-4247
Executive Officer .... **Walter R. Mook**
Assistant Executive Officer .... **Oscar P. Hellrich**
Supplementary personnel — 1 administrator, 1 air pollution specialist, 4 engineers, 2 inspectors, 2 technicians, 3 office staff
San Bernardino County Board of Supervisors, County Government Center, 385 North Arrowhead Ave., Fifth Floor, San Bernardino, CA 92415-0110
Board Chairperson .... **Robert O. Townsend**

SAN DIEGO COUNTY

San Diego County Air Pollution Control District, 9150 Chesapeake Dr., San Diego, CA 92123 — Telephone: (619) 565-5901
Air Pollution Control Officer .... **R. J. Sommerville**
Supplementary personnel — full time: 6 administrators, 6 chemists, 3 data processing specialists, 15 engineers, 15 inspectors, 3 meteorologists, 13 office staff, 1 public information specialist, 13 technicians, 4 air resource specialists
Air Pollution Control Board, 1600 Pacific Highway, San Diego, CA 92101
Chairperson .... **Tom Hamilton**

SAN FRANCISCO COUNTY (See Bay Area)

SAN JOAQUIN COUNTY

San Joaquin County Air Pollution Control District, 1601 E. Hazelton Ave., P.O. Box 2009, Stockton, CA 95201 — Telephone: (209) 466-6781 and (209) 462-8526
Air Pollution Control Officer .... **Jogi Khanna, M.D.**
Air Pollution Program Supervisor .... **Lakhmir Grewal**
Supplementary personnel — full time: 2 engineers, 3 inspectors, 2 office staff; part time: 2 administrators, 1 attorney
San Joaquin County Air Pollution Control Board, 222 E. Weber St., Stockton, CA 95202
Chairperson .... **Evelyn Costa**

SAN LUIS OBISPO COUNTY

San Luis Obispo Air Pollution Control District, 2156 Sierra Way, Suite B, San Luis Obispo, CA 93401 — Telephone: (805) 549-5912
Air Pollution Control Officer .... **George Rowland, M.D.**
Supplementary personnel — full time: 1 administrator, 1 engineer, 1 inspector, 1 office staff, 2 technicians; part time: 1 attorney
Air Pollution Control Board, County Government Center, San Luis Obispo, CA 93408
Chairperson .... **William Coy**

SAN MATEO COUNTY (See Bay Area)

SANTA BARBARA COUNTY

Santa Barbara County Air Pollution Control District, 315 Camino del Remedio, Santa Barbara, CA 93110 — Telephone: (805) 964-8658
Air Pollution Control Officer ..... **Lawrence Hart, M.D., M.P.H.**
Director, Air Pollution Control .... **John B. English**
Supplementary personnel — full time: 1 administrator, 7 engineers, 5 inspectors, 7 office staff, 3 technicians, 2 air quality inspectors
Air Pollution Control Board (Board of Supervisors), 105 E. Anapamu, Santa Barbara, CA 93101
Chairperson .... **William Wallace**

SANTA CLARA COUNTY (See Bay Area)

SANTA CRUZ COUNTY (See Monterey Bay Unified)

SHASTA COUNTY

Shasta County Air Pollution Control District, 1615 Continental Street, Rm. 120, Redding, CA 96001 — Telephone: (916) 225-5674
Public Health Officer .... **Stephen J. Plank, M.D.**
Air Pollution Control Officer .... **Richard B. Booth**
Supplementary personnel — full time: 1 administrator, 2 inspectors, 1 office staff; part time: 1 administrator
Air Pollution Control Board, County Courthouse, Redding, CA 96001
Chairman .... **Roy F. Peters**

SIERRA COUNTY

Sierra County Air Pollution Control District, County Courthouse, Downieville, CA 95936 — Telephone: (916) 289-3274
Air Pollution Control Officer .... **Thomas Koch, M.D.**
Air Basin Engineer .... **Earl Withycombe**
Supplementary personnel — part time: 1 administrator
Sierra County Air Pollution Control Board, P.O. Drawer D, Downieville, CA 95936
Chairperson .... **Earl Withycombe**

SISKIYOU COUNTY

Siskiyou County Air Pollution Control District, 525 So. Foothill Dr., Yreka, CA 96097 — Telephone: (916) 842-3906
Air Pollution Control Officer .... **Edmond W. Hale**
Air Pollution Specialist .... **Kenneth L. Corbin**
Supplementary personnel — full time: 1 engineer; part time: 1 administrator, 1 office staff
Siskiyou County Air Pollution Control Board, Courthouse, Yreka, CA 96097
Chairperson .... **Jim Steinhaus**

SOLANO COUNTY (See Bay Area, Yolo-Solano)

SOUTH COAST AIR QUALITY MANAGEMENT DISTRICT (Includes Los Angeles, Orange and Riverside Counties)

(Headquarters) 9150 Flair Drive, El Monte, CA 91731 — Telephone: (818) 572-6200
Executive Officer .... **Joseph A. Stuart**
Supplementary personnel — full time: 4 administrators, 5 at-

torneys, 24 chemists, 8 data processing specialists, 111 engineers, 126 inspectors, 3 meteorologists, 120 office staff, 6 public information specialists, 50 technicians

South Coast Air Quality Management District Board

Chairman .... **Thomas F. Heinsheimer**

STANISLAUS COUNTY

Stanislaus County Air Pollution Control District, 1716 Morgan Road, Modesto, CA 95351 — Telephone: (209) 571-6908

Deputy Air Pollution Control Officer .... **Wayne Morgan**

Supplementary personnel — full time: 1 deputy air pollution control officer, 3 air pollution control specialists, 1 secretary

Stanislaus County Air Pollution Control Board, 1100 H St., Modesto, CA 95352

Chairperson .... **Sal Cannella**

SUTTER COUNTY

Sutter County Air Pollution Control District, 142 Garden Way, Yuba City, CA 95991 — Telephone: (916) 674-2851

Air Pollution Control Officer .... **Alfred C. Perrin, Jr.**

Supplementary personnel — full time: 2 administrators, 7 inspectors, 3 office staff; part time: 2 attorneys, 1 chemist, 1 engineer

Sutter County Air Pollution Control District Board, 463 Second Street, Yuba City, CA 95991

Chairperson .... **Joe Benatar**

TEHAMA COUNTY

Tehama County Air Pollution Control District, 1760 Walnut Street (P.O. Box 38), Red Bluff, CA 96080 — Telephone: (916) 527-4504

Air Pollution Control Officer .... **Donald R. Hill**

Supplementary personnel — full time: 1 administrator, 1 inspector, 1 office staff

Tehama County Air Pollution Control Board, P.O. Box 250, Red Bluff, CA 96080

Chairperson .... **Russ Frey**

TRINITY COUNTY (See North Coast Unified)

TULARE COUNTY

Tulare County Air Pollution Control District, Health Building, County Civic Center, Visalia, CA 93291 — Telephone: (209) 733-6438

Air Pollution Control Officer .... **Ronald W. Probasco**
Environmental Health Engineer .... **Gary D. Criscione**

Supplementary personnel — full time: 2 engineers, 1 inspector, 1 office staff; part time: 7 sanitarians

Air Pollution Control Board (same address and phone as above)

Chairperson .... **LeRoy Swiney**

TUOLUMNE SONORA COUNTY

Tuolumne Sonora County Air Pollution Control District, 9 North Washington St., Sonora, CA 95370 — Telephone: (209) 533-5693

Air Pollution Control Officer .... **Gerald A. Benincasa**

Supplementary personnel — part time: 1 administrator, 1 inspector, 1 office staff, 1 attorney, 1 engineer

Tuolumne County Air Pollution Control Board, 2 South Green Street, Sonora, CA 95370

Chairperson .... **Greg Hurt**

VENTURA COUNTY

Ventura County Air Pollution Control District, 800 South Victoria Avenue, Ventura, CA 93009 — Telephone: (805) 654-2806

Air Pollution Control Officer .... **Richard H. Baldwin**

Supplementary personnel — full time: 1 administrator, 8 engineers, 5 inspectors, 1 meteorologist, 3 office staff, 7 technicians, 7 air quality specialists; part time: 2 attorneys, 1 accountant, 2 air quality specialists

Ventura County Air Pollution Control Board (Board of Supervisors) (same address and phone as above)

Chairman .... **Edwin A. Jones**

YOLO-SOLANO

Yolo-Solano Air Pollution Control District, 323 First St., Suite 5, P.O. Box 1006, Woodland, CA 95695 — Telephone: (916) 666-8433

Air Pollution Control Officer .... **James A. Koslow, P.E.**

Supplementary personnel — full time: 1 administrator/APCO, 1 engineer, 1 inspector, 2 office staff, 3 technicians; part time: 1 attorney

Yolo-Solano Air Pollution Control Board (same address and phone as above)

Chairperson .... **Don Pippo**

YUBA COUNTY

Yuba County Air Pollution and Control District, 938 14th St., Marysville, CA 95901 — Telephone: (916) 741-6484

Air Pollution Control Officer .... **Earnie E. Vickrey**

Supplementary personnel — full time: 1 administrator; part time: 1 engineer, 5 inspectors, 2 office staff

Office of the Board of Supervisors, Yuba County Courthouse, 215 Fifth St., 3rd Floor, Marysville, CA 95901

Supervisor, 1st District .... **George Deveraux**
Supervisor, 2nd District .... **John R. Dower**
Supervisor, 3rd District .... **William Harper**
Supervisor, 4th District .... **Mimi Mathews**
Supervisor, 5th District .... **J. E. McGill**

■ COLORADO

STATE AGENCY

Colorado Department of Health — Air Pollution Control Division, 4210 E. 11th Ave., Denver, CO 80220 — Telephone: (303) 320-4180

Executive Director .... **Thomas M. Vernon, M.D.**
Division Director .... **James M. Lents, Ph.D.**

Supplementary personnel — full time: 8 administrators, 5 data processing specialists, 9 engineers, 15 air pollution specialists, 2 meteorologists, 9 office staff, 3 public information specialists, 17 technicians, 6 planners, 5 chemists; part time: 2 attorneys

Colorado Air Quality Control Commission (same address and phone as above)

Chairperson .... **Clemons Kopf, Jr.**

BOULDER COUNTY

Boulder County Health Department, 3450 Broadway, Boulder, CO 80302 — Telephone: (303) 441-1180

Environmental Health Director .... **Bryan Miller**
Environmental Health Sanitarian .... **George Mathews**
Air Quality Sanitarian .... **Susan Martino**

Supplementary personnel — full time: 1 sanitarian; part time: 1 administrator, 1 attorney, 1 public information specialist, 1 sanitarian

Boulder County Board of Health (same address and phone as above)

President .... **Hester McNulty**

COLORADO SPRINGS (See El Paso County)

DENVER CITY-COUNTY

Air Pollution Control Section, Department of Health and Hospitals, Environmental Health Service, 605 Bannock St., Denver, CO 80204 — Telephone: (303) 893-6243

Director, Environmental
Health Service .... **J. Douglas McCluskie, D.V.M., M.P.H.**
Chief, Air Pollution Section .... **Thomas I. Peabody, P.E.**

Supplementary personnel — full time: 1 office staff, 2 sanitarians; part time: 1 administrator, 1 engineer, 1 inspector, 1 office staff, 1 sanitarian

EL PASO AND TELLER COUNTIES (Includes Colorado Springs)

El Paso County Health Department, Air Quality Control Section, 501 North Foote Ave., Colorado Springs, CO 80909-4598 — Telephone: (303) 636-0137

Director — Medical .... **John B. Muth, M.D., M.P.H.**
Chief, Air Pollution Section .... **James P. Easton**

Supplementary personnel — full time: 1 administrator, 2 technicians; part time: 1 attorney, 1 office staff

Air Quality Technical Committee, c/o PPACG 27 East Vermijo, Colorado Springs, CO 80903

Chairperson .... **James P. Easton**

JEFFERSON COUNTY

Jefferson County Health Department, 260 S. Kipling, Lakewood, CO 80226 — Telephone: (303) 238-6301

Director .... **Dr. C. H. Miller**
Air Pollution Control Officer .... **Richard A. Bell**

Supplementary personnel — full time: 1 inspector; 1 sanitarian; part time: 1 administrator, 1 office staff, 1 technician

Jefferson County Board of Health (same address and phone as above)

Chairperson .... **Richard Newman**

LARIMER COUNTY

Environmental Health Division, Larimer County Health Department, 363 Jefferson Street, Fort Collins, CO 80524 — Telephone: (303) 221-7497

Director, Environmental Health ..... **Laurence Johnston, Ph.D.**
Air Program Sanitarian .... **Myrna Jamison Hansen**

Supplementary personnel — full time: 1 inspector; part time: 1 administrator, 1 attorney, 1 engineer, 1 inspector, 2 office staff, 1 public information specialist, 2 sanitarians

Colorado Air Quality Control Commission, 4210 East 11th Avenue, Denver, CO 80220

LAS ANIMAS-HUERFANO COUNTIES

Las Animas-Huerfano Counties District Health Department, 119 East 5th, Walsenburg, CO 81089 — Telephone: (303) 738-2650; Las Animas-Huerfano Counties District Health Department, 412 Benedicta Avenue, Trinidad, CO 81082 — Telephone: (303) 846-2213

Director .... **Gary Williams**
Chief Sanitarian .... **Ed Cruz**

Supplementary personnel — full time: 1 sanitarian

MESA COUNTY

Mesa County Health Department, 515 Patterson Road, Grand Junction, CO 81501 — Telephone: (303) 244-1750

Director .... **Kenneth J. Lampert, M.D., M.P.H.**
Air Pollution Control Specialist .... **Steven L. DeFeyter**

Supplementary personnel — full time: 5 sanitarians; part time: 1 administrator, 1 attorney, 1 office staff

PUEBLO CITY-COUNTY

Pueblo City-County Health Department, 151 Central Main St., Pueblo, CO 81003 — Telephone: (303) 544-8376

Administrator .... **Robert H. Redwine, M.D.**
Supervisor .... **Gerald Hollandsworth**

Supplementary personnel — full time: 1 sanitarian; part time: 1 administrator, 1 sanitarian

WELD COUNTY

Weld County Health Department, 1516 Hospital Rd., Greeley, CO 80631 — Telephone: (303) 353-0540

Director, Environmental Health Protection Services.. **Wes Potter**
Environmental Specialist .... **Millie Turner**

Supplementary personnel — part time: 1 administrator, 4 environmental health specialists, 1 office staff

Weld County Board of Commissioners, 915 10th St., Greeley, CO 80631

■ CONNECTICUT

STATE AGENCY

Department of Environmental Protection, Air Compliance Unit, 165 Capitol Ave., Room 144, Hartford, CT 06106 — Telephone: (203) 566-4030

Commissioner .... **Stanley J. Pac**
Director .... **Leonard Bruckman**

Supplementary personnel — full time: 4 administrators, 42 engineers, 1 inspector, 10 office staff, 1 public information specialist, 1 senior environmental analyst, 7 technicians; part time, 1 public information specialist, 1 librarian

BRIDGEPORT

Department of Air Pollution Control, Bridgeport Office of Humane Affairs, 752 E. Main St., Bridgeport, CT 06608 — Telephone: (203) 576-7451

Director, Environmental Health .... **Henry Gross**

Supplementary personnel — part time: 1 administrator, 2 engineers, 1 inspector, 1 sanitarian

Environmental Commission (same address and phone as above)

Chairperson .... **Kaye Williams**

BRISTOL-BURLINGTON

Bristol-Burlington Health Department, City Hall, 111 North Main Street, Bristol, CT 06010 — Telephone: (203) 584-7682

Director of Health .... **William E. Furniss, M.D.**

Supplementary personnel — full time: 1 administrator, 1 inspector, 2 office staff, 2 sanitarians, 7 public nurses; part time: 1 attorney, 1 inspector

District Board of Health (Same address and phone as above)

Mayor .... **Michael L. Werner**

GREENWICH

Greenwich Health Department, Town Hall, Field Point Road, Greenwich, CT 06830 — Telephone: (203) 622-7843

Director, Division of Environmental Health and Laboratory .... **Robert C. Brown**

Supplementary personnel — part time: 1 administrator, 1 attorney, 1 office staff, 1 public information specialist, 2 sanitarians, 1 lab supervisor

Board of Health (same address and phone as above except 7836)

Chairman .... **Marjorie B. Chadsey**

MERIDEN

Meriden Department of Health/Human Services, 165 Miller St., Meriden, CT 06450 — Telephone: (203) 634-0003

Director of Health/Human Services .... **Frank Grosso, Jr., M.P.H., R.S.**
Chief, Environmental Health .... **Raymond E. West, R.S.**

Supplementary personnel — full time: 1 office staff, 1 registered sanitarian, 2 sanitarians

MILFORD

Department of Public Health, Environmental Division, 2051 Bridgeport Ave., Milford, CT 06460 — Telephone: (203) 783-3287, 3288, 3289

Director of Health .... **George Kraus, M.D.**
Air Pollution Control Officer .... **Frederick C. Shumacher, R.S.**

Supplementary personnel — part time: 1 inspector, 1 office staff, 1 sanitarian

NEW HAVEN

New Haven Health Department, One State St., New Haven, CT 06511 — Telephone: (203) 787-8173

Director of Health .... **Edward R. DeLouise, M.P.H., R.S.**
Director of Environmental Health . **Paul Kowalski, M.P.H., R.S.**

Supplementary personnel — full time: 1 administrator, 1 office staff, 8 sanitarians; part time: 1 office staff

Board of Health, City of New Haven

Chairman .... **Luca Celentano, M.D.**

NORWALK

Department of Health, 137 East Ave., Norwalk, CT 06851 — Telephone: (203) 838-7531

Director of Health .... **Louise S. Leary, R.N., M.S.**
Laboratory Director .... **Alfred C. Kao, M.S.**

Supplementary personnel — full time: 1 administrator, 1 office staff, 3 sanitarians; part time: 1 attorney

Board of Health (same address and phone as above)

Chairman .... **Robert Appleby, M.D.**

STAMFORD

Stamford Health Department, 229 North St., Stamford, CT 06902 — Telephone: (203) 358-4396

Director of Health and School Services .... **Ralph M. Gofstein, M.D., M.P.H.**
Director of Environmental Health .... **Vincent J. Radke, M.P.H., R.S.**

Supplementary personnel — part time: 2 administrators, 1 chemist, 1 office staff, 8 sanitarians, 1 attorney

Stamford Health Commission (same address and phone as above)

Chairperson .... **Dr. Angelo Mastrangelo**

STRATFORD

Department of Health, 2730 Main St., Stratford, CT 06497 — Telephone: (203) 385-4090

Health Administrator .... **William J. Watson, R.S.**

Supplementary personnel — full time: 1 administrator, 1 sanitarian; part time: 1 attorney, 1 office staff

■ DELAWARE

STATE AGENCY

Delaware Department of Natural Resources and Environmental Control, P.O. Box 1401, Dover, DE 19903 — Telephone: (302) 736-4764

Acting Director, Division of Environmental Control .... **Robert J. Touhey**
Manager, Air Resources Section .... **Robert R. French**

Supplementary personnel — full time: 1 administrator, 2 chemists, 8 engineers, 3 office staff, 12 technicians

■ DISTRICT OF COLUMBIA

STATE AGENCY

Department of Consumer and Regulatory Affairs, 5010 Overlook Ave., S.W., Washington, DC 20032 — Telephone: (202) 767-7370

Director .... **Carol B. Thompson**
Program Manager .... **Anantha Padmanabha, P.E.**

Supplementary personnel — full time: 1 administrator, 1 chemist, 1 data processing specialist, 5 engineers, 6 inspectors, 1 meteorologist, 4 office staff, 4 technicians, 1 urban planner; part time: 1 attorney, 2 public information specialists

■ FLORIDA

STATE AGENCY

Florida State Agency, Department of Environmental Regulation, Twin Towers Office Building, 2600 Blair Stone Road, Tallahassee, FL 32301 — Telephone: (904) 488-1344

Secretary .... **Victoria J. Tschinkel**
Chief, Bureau of Air Quality Mgmt .... **Steven Smallwood**

Supplementary personnel — full time: 5 administrators, 1 attorney, 1 chemist, 20 engineers, 2 meteorologists, 4 office staff, 3 technicians, 7 environmental specialists; part time: 1 attorney, 10 chemists, 13 engineers, 29 office staff, 9 technicians, 21 environmental specialists

Environmental Regulation Commission (same address and phone as above)

Chairperson .... **Robert Parks**

ALACHUA COUNTY

Alachua County Dept. of Environmental Services, 825 N.W. 23rd Avenue, Building III, Gainesville, FL 32609 — Telephone: (904) 373-8509

Director .... **John Schert**
Environmental Engineering .... **Vince McLeod**

Supplementary personnel — full time: 1 administrator, 1 chemist, 2 engineers, 1 office staff, 1 technician, 1 environmental field specialist

Alachua Board of County Commissioners, P.O. Drawer CC, Gainesville, FL 32602

Chairperson .... **Leveda Brown**

BROWARD COUNTY

Broward County Environmental Quality Control Board, 500 S.W. 14th Court, Fort Lauderdale, FL 33315 — Telephone: (305) 765-5881

Pollution Control Officer .... **Victor N. Howard, P.E.**
Chief, Air Section .... **Gary D. Carlson, M.S.**

Supplementary personnel — full time: 1 chemist, 3 engineers, 3 technicians; part time: 1 administrator, 2 attorneys, 1 data processing specialist, 7 inspectors, 5 office staff

Broward County Environmental Quality Control Board (same address and phone as above)

Chairperson .... **George E. Fitzpatrick**

DUVAL COUNTY (See Jacksonville)

HILLSBOROUGH COUNTY

Hillsborough County Environmental Protection Commission, 1900-9th Ave., Tampa, FL 33605 — Telephone: (813) 272-5960

Director .... **Roger P. Stewart**
Environmental Program Manager of
Air Programs .... **Iwan Choronenko**

Supplementary personnel — full time: 2 chemists, 4 engineers, 1 environmental scientist, 7 technicians, 7 environmental specialists, 1 administrator, 1 data processing specialist; part time: 2 administrators, 1 attorney, 2 data processing specialists, 4 office staff, 1 technician, 1 accountant

Board of County Commissioners, P.O. Box 1110, Tampa, FL 33601

Commissioner .... **E. L. Bing**

JACKSONVILLE (Includes Duval County)

Bio-Environmental Services Division, 515 W. 6th St., Jacksonville, FL 32211 — Telephone: (904) 633-3303

Division Chief .... **Donald C. Bayly**
Air Pollution Engineer .... **Kunshid Mehta**

Supplementary personnel — full time: 2 chemists, 5 engineers, 2 office staff, 1 public information specialist, 12 technicians, 4 pollution control specialists; part time: 2 administrators, 1 attorney

Environmental Protection Board (same address and phone as above)

Chairman .... **Jack Milne**

MANATEE COUNTY

Manatee County Health Unit, 202 Sixth Ave., East Bradenton, FL 33508 — Telephone: (813) 748-0666

Medical Executive Director .... **John Ambrusko, M.D.**
Pollution Control Director .... **W. L. Priesmeyer**

Supplementary personnel — full time: 2 chemists, 3 engineers, 3 office staff, 7 technicians, 3 environmental specialists; part time: 1 attorney

Manatee County Pollution Control Board (same address and phone as above)

Chairperson .... **Bernie Moler**

METROPOLITAN DADE COUNTY

Metropolitan Dade County Department of Environmental Resources Management, 909 S.E. First Avenue, Miami, FL 33131 — Telephone: (305) 579-2760

Director .... **Anthony J. Clemente, P.E.**
Chief, Environmental Planning Division .... **Rafael Rodon**

Supplementary personnel — full time: 2 chemists, 2 data processing specialists, 3 engineers, 2 inspectors, 3 technicians, 1 office staff; part time: 5 administrators, 2 attorneys, 1 data processing specialist, 10 inspectors, 5 office staff

Environmental Quality Control Board (same address and phone as above)

Chairperson .... **Carlton L. Nelson, P.E.**

ORANGE COUNTY (See Reedy Creek Improvement District)

OSCEOLA COUNTY (See Reedy Creek Improvement District)

PALM BEACH COUNTY

Palm Beach County Health Department, Division of Environmental Sciences and Engineering, 901 Evernia St., West Palm Beach, FL 33402 — Telephone: (305) 837-3070

County Health Officer .... **C. L. Brumback, M.D.**
Director, Environmental
Sciences and Engineering .... **F. J. Gargiulo, P.E.**
Air Pollution Control Administrator .... **E. J. Sacco**

Supplementary personnel — full time: 1 administrator, 2 environmental specialists, 1 office staff, 2 technicians; part time: 1 attorney, 2 chemists, 4 engineers, 2 environmental specialists, 1 technician

Palm Beach County Environmental Control Board, Palm Beach County Courthouse, West Palm Beach, FL 33401

Chairperson .... **Ken Spillias**

PINELLAS COUNTY

Pinellas County Department of Environmental Management, Division of Air Quality, 16100 Fairchild Drive, Bldg. V102, Clearwater, FL 33520 — Telephone: (813) 530-6522

Director, Department of
Environmental Management .... **Jacob Stowers**
Chief, Division of Air Quality .... **Joyce M. Gibbs**

Supplementary personnel — full time: 1 administrator, 1 chemist, 1 engineer, 1 environmental specialist II, 2 environmental specialists I, 1 office staff, 1 quality assurance officer, 1 air monitoring supervisor, 1 air quality analyst; part time: 2 administrators, 1 attorney, 1 office staff

Pinellas County Board of County Commissioners, 315 Court Street, Clearwater, FL 33516

Chairperson .... **Barbara Sheen Todd**

REEDY CREEK IMPROVEMENT DISTRICT (Orange and Osceola Counties)

Reedy Creek Improvement District, P.O. Box 36, Lake Buena Vista, FL 32830 — Telephone: (305) 824-3460

Director, Pollution Control Department .... **Gary Gornto**

Supplementary personnel — full time: 3 chemists, 1 data processing specialist, 1 public information specialist, 1 technician

Reedy Creek Improvement District Board of Supervisors (same address and phone as above)

President, Board of Supervisors .... **Thomas DeWolfe**

SARASOTA COUNTY

Sarasota County Environmental Services Department/Environmental Services Laboratory Division, 2086 Main Street, Sarasota, FL 33577 — Telephone: (813) 365-1000, ext. 2064 or 2390

Environmental Services Director .... **Russell P. Klier, P.E.**
Environmental Services Laboratory
Director .... **Jose M. Guira, Ph.D.**

Supplementary personnel — full time: 1 chemist, 1 engineer, 2 office staff, 5 environmental specialists

Sarasota County Board of Commissioners, P.O. Box 8, Sarasota, FL 33578

Chairman .... **Jerry Hente**

■ GEORGIA

STATE AGENCY

Environmental Protection Division, Department of Natural Resources, 27D Washington St., S.W., Atlanta, GA 30334 — Telephone: (404) 656-4713

Assistant Director .... **Harold F. Reheis**
Chief, Air Protection Branch .... **Robert H. Collom**

Supplementary personnel — full time: 3 chemists, 2 data processing specialists, 17 engineers, 1 meteorologist, 10 office staff, 3 technicians, 4 managers, 31 environmental specialists; part time: 2 attorneys, 1 public information specialist

Board of Natural Resources (same address and phone as above)

Commissioner .... **J. Leonard Ledbetter**

■ GUAM

Guam Environmental Protection Agency, Government of Guam, P.O. Box 2999, Agana, Guam 96910 — Telephone: (671) 646-8863/64/65

Administrator .... **James B. Branch**
Head, Air and Land Program Division .... **James L. Canto**

Supplementary personnel — full time: 1 administrator, 1 inspector; part time: 1 chemist, 1 public information specialist, 1 technician

Guam Environmental Protection Agency Board of Directors (same address as above)

Chairperson .... **Antonio T. Lizama**

■ HAWAII

STATE AGENCY

Environmental Protection and Health Services Division, Hawaii State Department of Health, 1250 Punchbowl Street, Honolulu, HI 96813 — Telephone: (808) 548-6455

Director of Health .... **Leslie S. Matsubara**
Deputy Director for Environmental Health .. **Melvin K. Koizumi**

Supplementary personnel — full time: 3 chemists, 3 engineers, 4 technicians; part time: 4 administrators, 1 attorney, 2 engineers, 16 inspectors, 3 office staff, 1 public information specialist

■ IDAHO

STATE AGENCY

Idaho Department of Health and Welfare — Division of Environment, Statehouse, Boise, ID 83720 — Telephone: (208) 334-5360

Director, Department of Health and Welfare ..... **Rose Bowman**
Administrator, Division of Environment .. **Lee W. Stokes, Ph.D.**
Chief, Bureau of Air Quality .... **Kenneth D. Brooks**

Supplementary personnel — full time: 1 administrator, 1 data processing specialist, 3 engineers, 5 inspectors, 1 meteorologist, 2 office staff, 2 managers, 1 analyst

Board of Health and Welfare (same address and phone as above)

Chairperson .... **Marvin Wittman**

■ ILLINOIS

STATE AGENCY

Illinois Environmental Protection Agency, 2200 Churchill Rd., Springfield, IL 62706 — Telephone: (217) 782-7326

Director .... **Richard J. Carlson**
Acting Manager, Division of
Air Pollution Control .... **Michael J. Hayes**

Supplementary personnel — full time: 2 administrators, 7 attorneys, 2 accountants, 1 data processing specialist, 37 engineers, 2 environ. prot. associates, 23 office staff, 10 research economists, 9 technicians, 39 specialists

Department of Energy and Natural Resources, 309 West Washington Street, Chicago, IL 60606 — Telephone: (312) 793-3780

Director .... **Michael B. White**

BEDFORD PARK

Village of Bedford Park, 6701 South Archer Road, Bedford Park, IL 60501 — Telephone: (312) 458-2975

Chairperson .... **Richard W. Weeks**

Supplementary personnel — full time: 23 inspectors; part time: 1 attorney, 1 chemist, 5 engineers, 5 office staff, 2 BP-EQCB

Bedford Park Environmental Quality Control Board, P.O. Box 128, Bedford Park, IL 60501

Village President .... **Ernest B. Martin**

BENSENVILLE

Bensenville Air Pollution Control District, 1253 Roosevelt Avenue, Glenview, IL 60025 — Telephone: (312) 498-9840

Director and Chief
Pollution Control Officer .... **Richard A. Young**

Supplementary personnel — full time: 1 administrator, 1 engineer, 1 inspector; part time: 1 attorney, 1 inspector, 1 office staff, 2 technicians

CHICAGO

City of Chicago, Department of Consumer Services, 121 N. LaSalle Street, Room 810, Chicago, IL 60602 — Telephone: (312) 744-4034

Commissioner .... **Jesse D. Madison**
Director of Energy and
Environmental Programs .... **John Evanoff**

Supplementary personnel — full time: 4 administrators, 2 chemists, 4 engineers, 10 inspectors, 1 meteorologist, 8 office staff, 5 technicians; part time: 1 public information specialist

Environmental Control Appeal Board (same address and phone as above)

Chairperson .... **Norman H. Nachman**

COOK COUNTY

Cook County Department of Environmental Control, Maybrook Civic Center, 1500 Maybrook Dr., Maywood, IL 60153 — Telephone: (312) 865-6165

Director .... **Philip J. Molé**
Assistant Director .... **Mario G. Tonelli**

Supplementary personnel — full time: 6 administrators, 4 attorneys, 5 chemists, 2 data processing specialists, 15 engineers, 20 inspectors, 1 meteorologist, 20 office staff, 1 public information specialist, 15 sanitarians, 14 technicians

Cook County Board of Commissioners, Room 537, County Building, 118 N. Clark St., Chicago, IL 60602

President, Cook County
Board of Commissioners .... **George W. Dunne**

DUPAGE COUNTY

DuPage County Health Department, 111 North County Farm Road, Wheaton, IL 60187

Executive Director .... **James Paulissen, M.D., M.P.H.**
Director of Environmental Health .... **Orville L. Meyer, P.E.**
Air Pollution Sanitarian .... **Bruce E. Schroer**

Supplementary personnel — full time: 2 engineers, 28 sanitarians, 3 technicians, 6 office staff; part time: 7 technicians

EAST ST. LOUIS

City of East St. Louis, IL, 7 Collinsville Avenue, East St. Louis, IL 62201 — Telephone: (618) 874-2115

Supplementary personnel — full time: 2 administrators, 1 attorney, 1 data processing specialist, 1 engineer, 1 inspector, 2 office staff, 1 sanitarian

City Commission (same address and phone as above)

Chairperson .... **Mayor Carl E. Officer**

EVANSTON

City of Evanston — Department of Inspections and Permits, 2100 Ridge Ave., Evanston, IL 60204 — Telephone: (312) 328-2100

City Manager .... **Joel A. Asprooth**
Director, Building & Zoning .... **Reed Carlson**

Supplementary personnel — full time: 1 administrator, 1 office staff; part time: 1 inspector

Evanston Environmental Control Board (same address and phone as above)

Chairperson .... **Cheryl Wollin**

GRANITE CITY

Granite City Air Pollution Control Agency, 2060 B Delmar Avenue, Granite, IL 62040 — Telephone: (618) 452-6218

City Engineer .... **Monroe F. Brewer**

Supplementary personnel — full time: 1 inspector; part time: 1 administrator

Air Pollution Control Board (same address and phone as above)

Technical Secretary .... **Vacant**

McCOOK

Village of McCook Environmental Board, 50th and Glencoe Ave., McCook, IL 60525 — Telephone: (312) 447-1231

Chairperson .... **Jerry Bubash**

Supplementary personnel — part time: 3 administrators, 1 engineer, 4 inspectors, 1 technical advisor

■ INDIANA

STATE AGENCY

Indiana State Board of Health/Air Pollution Control Division, 1330 W. Michigan Street, Indianapolis, IN 46206 — Telephone: (317) 633-0600

State Health Commissioner . **T. S. Danielson, Jr., M.D., M.P.H.**
Division Director .... **Harry D. Williams**

Supplementary personnel — full time: 16 administrators, 2 attorneys, 4 chemists, 18 engineers, 1 grant coordinator, 1 meteorologist, 10 office staff, 2 public information specialists, 41 sanitarians, 2 technicians, 1 training officer, 4 planners, 1 maintenance repair

Air Pollution Control Board (same address as above)

Technical Secretary .... **Harry D. Williams**

ANDERSON

Anderson Air Pollution Control Department, P.O. Box 2100, Anderson, IN 46018 — Telephone: (317) 646-9835

Administrator and/or Director .... **William A. Dorff**

Supplementary personnel — full time: 1 administrator, 1 inspector/technician; part time: 1 engineer, 1 office staff

Clean Air Council (same address and phone as above)

Chairperson .... **Richard O. Wilkins**

EAST CHICAGO

Department of Air Quality Control, 4525 Indianapolis Blvd., East Chicago, IN 46312 — Telephone: (219) 392-8297

Assistant Director .... **M. M. Ali Khan, P.E.**

Supplementary personnel — full time: 1 administrator, 2 engineers, 3 inspectors, 1 data analyst, 1 office staff; part time: 1 attorney

Air Quality Control Advisory and Appeal Board (same address and phone as above)

Chairperson .... **Astor Tillotson**

EVANSVILLE

Air Pollution Control Office, City of Evansville, Room 207, Administration Building, Evansville, IN 47708 — Telephone: (812) 426-5597

Director .... **Joan D. Shelton**

Supplementary personnel — full time: 2 administrators, 2 inspectors, 1 office staff, 1 technician; part time: 1 attorney

Environmental Protection Agency Board (same address and phone as above)

Chairperson ....**Ward Peyronnin**

GARY

Gary Air Pollution Control, 3600 W 3rd Avenue, Gary, IN 46406 — Telephone: (219) 944-0667, 8 or 9

Chief .... **Dennis J. McGuire**

Supplementary personnel — full time: 1 administrator, 2 chemists, 2 engineers, 2 office staff, 1 technician

Air Pollution Control Advisory Board (same address and phone as above)

Chairperson .... **Elaine G. Beck**

HAMMOND

Hammond Air Pollution Control, 5925 Calumet Ave., Hammond, IN 46320 — Telephone: (219) 853-6306

Chief .... **Ronald L. Novak**

Supplementary personnel — full time: 1 administrator, 3 engineers, 4 inspectors, 1 office staff; part time: 1 attorney

Air Pollution Control Board (same address and phone as above)

President .... **Doyal Yaney**

INDIANAPOLIS

Air Pollution Control Division, 2700 South Belmont Avenue, Indianapolis, IN 46221 — Telephone: (317) 633-5496

Administrator .... **Walter J. Kulakowski**

Supplementary personnel — full time: 1 administrator, 1 chemist, 2 engineers, 1 meteorologist, 2 office staff, 3 technicians, 3 supervisors, 1 emissions analyst, 6 investigators, 1 assistant administrator; part time: 1 attorney

Indianapolis Air Pollution Control Board (same address and phone as above)

Chairperson .... **Arlie J. Ullrich**

LAKE COUNTY

Lake County Health Department, 2293 N. Main St., Crown Point, IN 46307 — Telephone: (219) 738-2020

Health Commissioner .... **Peter Stecy, M.D.**
Administrative Assistant .... **Nick Doffin**

Supplementary personnel — part time: 1 attorney, 1 chemist, 1 administrator, 1 office staff, 1 sanitarian

Lake County Health Board (same address and phone as above)

Chairperson .... **Adrian Bowman**

ST. JOSEPH COUNTY

St. Joseph County Health Department, 227 Jefferson Blvd., South Bend, IN 46601 — Telephone: (219) 284-9721

Health Officer .... **George B. Plain, M.D.**
Pollution Control Officer .... **Paul E. Trost**

Supplementary personnel — full time: 1 administrator, 2 sanitarians; part time: 1 attorney, 1 chemist

Air Pollution Advisory Committee (same address and phone as above)

Chairperson .... **Daniel Lawson**

VIGO COUNTY

Vigo County Air Pollution Control, 201 Cherry Street, Terre Haute, IN 47807 — Telephone: (812) 238-8433

Director .... **Robert L. Moran**

Supplementary personnel — full time: 1 administrator, 1 engineer, 2 inspectors, 1 office staff; part time: 1 attorney

Vigo County Air Pollution Control Board (same address and phone as above)

Chairperson .... **Dr. Martin Thomas**

■ IOWA

STATE AGENCY

Iowa Department of Water, Air and Waste Management, 900 East Grand Avenue, Henry A. Wallace Building, Des Moines, IA 50319 — Telephone: (515) 281-8690

Executive Director .... **Stephen W. Ballou**
Director of Program Operations .... **Allen Stokes**

Supplementary personnel — full time: 3 administrators, 1 attorney, 1 chemist, 2 engineers, 6 inspectors, 5 office staff, 4 environmental specialists; part time: 1 data processing specialist, 1 public information specialist

Iowa Water, Air and Waste Management Commission (same address and phone as above)

Chairman .... **Robert W. Schlutz**

CEDAR RAPIDS-LINN COUNTY

Linn County Health Department, 751 Center Point Rd. N.E., Cedar Rapids, IA 52402 — Telephone: (319) 398-3551

Health Department Director .... **H. Keith Erickson**
Air Pollution Control Officer .... **Gregory Slager**

Supplementary personnel — full time: 2 air pollution control officers; part time: 1 administrator, 1 attorney, 1 chemist, 2 sanitarians, 2 technicians, 2 office staff

Linn County Board of Health (same address as above)

Chairperson ....**Rita Huber**

POLK COUNTY

Polk County Health Center, 1915 Hickman, Des Moines, IA 50314 — Telephone: (515) 286-3927

Director, Physical Planning
and Health .... **Gary L. Pryor**
Supervisor .... **Gary L. Young**

Supplementary personnel — full time: 1 engineer, 1 office staff, 1 technician, 1 supervisor; part time: 1 administrator

Polk County Board of Supervisors, Second and Court Avenue, Des Moines, IA

Chairperson ....**Tom Whitney**

■ KANSAS

STATE AGENCY

Kansas Department of Health and Environment, Forbes Field, Topeka, KS 66620 — Telephone: (913) 862-9360

Director, Division of Environment .... Allan S. Abramson
Director, Bureau of Air Quality and
Radiation Control .... David J. Romano

Supplementary personnel — full time: 3 administrators, 1 chemist, 2 electronic technicians, 5 engineers, 2 office staff, 15 technicians; part time: 4 administrators, 2 attorneys, 1 data processing specialist, 1 public information specialist

KANSAS CITY-WYANDOTTE COUNTY

Kansas City, Kansas-Wyandotte County Health Department, Air Pollution Control Division, 619 Ann Ave., Kansas City, KS 66101 — Telephone: (913) 321-4803, ext. 340

Director of Health .... Darrel D. Newkirk, M.D., M.P.H.
Director, Division of Air Pollution .... Richard S. Michael

Supplementary personnel — full time: 3 engineers, 2 technicians, 1 office staff

Kansas City-Wyandotte County Joint Board of Health (same address and phone as above)

Chairperson .... J. J. Swanson, D.V.M.

SEDGWICK COUNTY (See Wichita-Sedgwick County)

TOPEKA-SHAWNEE COUNTY

Topeka-Shawnee County Health Department, 1615 W. Eighth St., Topeka, KS 66606 — Telephone: (913) 233-8961

Health Officer .... Ray D. Baker, M.D.
Environmental Engineer .... Lynn E. Couch, P.E.

Supplementary personnel — full time: 2 technicians; part time: 2 engineers, 1 office staff

Topeka City Commission, Shawnee County Commission, City Building, 215 E. Seventh St., Topeka, KS 66606

Mayor .... Douglas S. Wright

WICHITA-SEDGWICK COUNTY

Wichita-Sedgwick County Department of Community Health, 1900 E. 9th St., Wichita, KS 67214 — Telephone: (316) 268-8351

Environmental Health Director .... Michael J. Everhart

Chief, Air Quality Control .... George K. Huenergart

Supplementary personnel — full time: 1 administrator, 1 office staff, 2 technicians

WYANDOTTE COUNTY (See Kansas City-Wyandotte County)

■ KENTUCKY

STATE AGENCY

Division of Air Pollution Control, Department of Environmental Protection, Fort Boone Plaza, 18 Reilly Road, Frankfort, KY 40601 — Telephone: (502) 564-3382

Director .... Roger B. McCann
Assistant Director .... Hisham M. Sa'aid

Supplementary personnel — full time: 13 administrators, 1 data processing specialist, 20 engineers, 10 inspectors, 14 office staff, 4 technicians, 13 environmental specialists, 3 permit reviewers, 2 air monitoring specialists, 3 program coordinators

Kentucky Natural Resources and Environmental Protection, Cabinent Fifth Floor, Capital Plaza Tower, Frankfort, KY 40601

Secretary .... Mrs. Jackie Swigart

JEFFERSON COUNTY

Air Pollution Control District of Jefferson County, 914 East Broadway, Louisville, KY 40204 — Telephone: (502) 584-8151

Secretary-Treasurer .... Robert T. Offutt

Air Pollution Control Officer ... Michael T. DeBusschere, P.E.

Supplementary personnel — full time: 4 administrators, 1 attorney, 1 chemist, 2 data processing specialists, 10 engineers, 6 inspectors, 5 office staff, 1 public information specialist, 9 technicians, 4 VET info. clerks, 1 auto repair coordinator, 1 vehicle exhaust test spec.

Air Pollution Control Board of Jefferson County (same address and phone as above)

Chairperson .... Paul K. Young

■ LOUISIANA

STATE AGENCY

Department of Environmental Quality, Office of Air Quality and Nuclear Energy, 625 North Fourth Street, Baton Rouge, LA 70804 — Telephone: (504) 342-1206

Assistant Secretary .... L. Hall Bohlinger
Administrator, Air Quality Division .... Gus Von Bodungen

Supplementary personnel —full time: 2 administrators, 13 office staff, 66 environmental specialists, 1 data processing specialist, 2 administrative staff; part time: 3 attorneys

Louisiana Environmental Control Commission, P.O. Box 44066, Baton Rouge, LA 70804

Chairperson .... Frank P. Simoneaux

■ MAINE

STATE AGENCY

Department of Environmental Protection, Bureau of Air Quality Control, State House, Station 17, MDEP, Augusta, ME 04333 — Telephone: (207) 289-2437

Commissioner .... Henry E. Warren
Director .... John L. Bastey

Supplementary personnel — full time: 4 administrators, 2 chemists, 1 data processing specialist, 5 engineers, 3 inspectors, 3 meteorologists, 3 office staff, 1/2 public information specialist, 10 field office people

Board of Environmental Protection (same address and phone as above)

Chairperson .... Samuel Zaitlin

■ MARYLAND

STATE AGENCY

Air Management Administration, Office of Environmental Programs, 201 West Preston Street, Baltimore, MD 21201 — Telephone: (301) 383-2779

Assistant Secretary .... William M. Eichbaum
Director, Air Management Administration .. George P. Ferreri

Supplementary personnel — full time: 2 administrators, 2 attorneys, 1 data processing specialist, 24 engineers, 5 inspectors, 1 meteorologist, 13 office staff, 1 public information specialist, 16 sanitarians, 8 technicians, 4 division chiefs, 1 lab scientist

Air Quality Control Advisory Council (same address and phone as above)

Chairperson .... Dr. Charles Billings

ALLEGANY COUNTY

Allegany County Health Department, Willowbrook Road, Cumberland, MD 21502 — Telephone: (301) 777-5653

Deputy State Health Officer .... Jane Fiscus, M.D.
Asst. Director of Environmental
Health .... Conrad B. Zimmermann

Supplementary personnel — full time: 1 office staff, 1 sanitarian; part time: 6 sanitarians, 1 technician

Allegany County Board of Commissioners, County Office Building, 3 Pershing Street, Cumberland, MD 21502

President .... Arthur T. Bond

ANNE ARUNDEL COUNTY

Anne Arundel County Department of Health, Division of Environmental Health, Air Quality Control Section, 3 Harry S Truman Parkway, Annapolis, MD 21401 — Telephone: (301) 224-7360

Health Officer .... J. Howard Beard, M.D.
Director, Division of Environmental Health ... N. Singh Dhilon
Chief, Air Quality Control Section .... Joseph G. Abey

Supplementary personnel — full time: 1 engineer, 2 environmental health aides, 1 office staff, 4 sanitarians

BALTIMORE

Baltimore City Health Department, Bureau of Industrial Hygiene, Air Pollution Control, Room S-219, 111 N. Calvert Street, Baltimore, MD 21202 — Telephone: (301) 396-4427

Commissioner of Health .... Susan R. Guarnieri, M.D., M.P.H.
Director, Bureau of
Industrial Hygiene .... Elkins W. Dahle, Jr., P.E.

Supplementary personnel — full time: 2 engineers, 1 office staff, 4 sanitarians

BALTIMORE COUNTY

Baltimore County Division of Air Pollution Control, 300 E. Towsontown Blvd., Towson, MD 21204 — Telephone: (301) 494-3775

Director .... David Filbert

Supplementary personnel — full time: 1 administrator, 5 engineers, 2 office staff, 10 sanitarians

Baltimore County Department of Health, 401 Bosley Avenue, Tow-

son, MD 21204
Deputy State and County
Health Officer .... **Donald J. Roop, M.D., M.P.H.**

FREDERICK COUNTY

Frederick County Health Department, 12 East Church Street, Frederick, MD 21701 — Telephone: (301) 694-1700

Health Officer .... **Martha T. Schipper, M.D., M.P.H.**
Air Quality Sanitarian .... **Charles L. Gillis, Jr.**
Director, Environmental Health Services .. **Edgar A. Woods, R.S.**

Supplementary personnel — full time: 1 inspector, 1 office staff, 1 sanitarian; part time: 1 administrator

Frederick County Board of County Commissioners, 12 East Church Street, Frederick, MD 21701, Telephone: (301) 694-1100

President .... **Galen R. Claggett**

HARFORD COUNTY

Harford County Health Department, 119 Hays St., Bel Air, MD 21014 — Telephone: (301) 838-6000, ext. 7249

Health Officer .... **Thomas M. Thomas**
Sanitarian .... **Thomas Kusterer**

Supplementary personnel — full time: 1 inspector, 1 sanitarian

HOWARD COUNTY

Howard County Health Department, P.O. Box 476, Ellicott City, MD 21042 — Telephone: (301) 461-9955

Health Officer .... **Joyce M. Boyd, M.D.**
Director, Bureau of Envir. Health .... **Robert W. Sheesley**
Director, Technical Services Program .... **Thomas S. Ogle**

Supplementary personnel — full time: 2 sanitarians, 2 technicians

MONTGOMERY COUNTY

Montgomery County Department of Environmental Protection, Division of Environmental Planning and Monitoring, 101 Monroe Street, Executive Office Building, Rockville, MD 20850 — Telephone: (301) 251-2380

Director, Dept. of Environmental
Protection .... **John L. Menke**
Director, Division of Environmental
Planning & Monitoring .... **David G. Sobers**
Environmental Protection Manager for
Air Pollution and Noise Control .... **Eric S. Mendelsohn**

Supplementary personnel — full time: 1 administrator, 1 environmental investigator; part time: 1 attorney, 3 inspectors, 1 meteorologist, 2 office staff, 2 technicians

PRINCE GEORGE'S COUNTY

Prince George's County Health Department, Hospital Road, Cheverly, MD 20785 — Telephone: (301) 794-6800

Health Officer .... **Helen B. McAllister, M.D.**
Chief, Division of Air Quality .... **Donald A. Nork**

Supplementary personnel — full time: 1 office staff, 7 sanitarians

WASHINGTON COUNTY

Washington County Health Department, 1302 Pennsylvania Ave., Hagerstown, MD 21742 — Telephone: (301) 791-3270

Director, Division of Environmental Health .. **James A. Rooney**

Supplementary personnel — part time: 1 administrator, 1 office staff, 2 sanitarians

■ MASSACHUSETTS

STATE AGENCY

Division of Air Quality Engineering, One Winter Street, Boston, MA 02108 — Telephone: (617) 292-5593

Commissioner .... **S. Russell Sylva**
Director, Division of Air Quality Control ..... **Kenneth A. Hagg**

Supplementary personnel — full time: 3 administrators, 1 attorney, 4 data processing specialists, 18 engineers, 14 office staff, 14 technicians, 1 hearings officer, 16 planners, 1 statistician, 1 economist, 7 consultants

BOSTON

Boston Air Pollution Control Commission, Room 805, Boston City Hall, Boston, MA 02201 — Telephone: (617) 725-4416

Executive Director .... **Geoffrey M. Boehm**

Supplementary personnel — full time: 1 administrator, 1 inspector; part time: 1 attorney, 1 office staff, 1 public information specialist

Air Pollution Control Commission (same address and phone as above)

Chairperson .... **Philip J. Caruso**

CENTRAL MASSACHUSETTS

Central Massachusetts Air Pollution Control District, 75 Grove Street, Worcester, MA 01605 — Telephone: (617) 792-7650

Chief, Air Quality .... **Thomas P. Cusson**

Supplementary personnel — full time: 8 engineers, 3 inspectors, 2 office staff

FITCHBURG

Fitchburg Board of Health, 718 Main St., Fitchburg, MA 01420 — Telephone: (617) 342-9843

Chairman .... **Frederick P. Ross, M.D.**
Director, Public Health .... **Jeffrey P. Jerszyk, R.S.**

Supplementary personnel — part time: 1 administrator, 1 sanitarian, 2 office staff, 2 inspectors

Central Massachusetts Air Pollution Control District, 75 Grove St., Worcester, MA 01605 — Telephone: (617) 791-3672

METROPOLITAN BOSTON-NORTHEAST REGION

Merrimack Valley and Metropolitan Boston Air Pollution Control Districts, 323 New Boston Street, Woburn, MA 01801 — Telephone: (617) 935-2160

Air Quality Section Chief .... **Michael Maher**

Supplementary personnel — full time: 3 clerical help, 18 engineers, 12 inspectors

SOUTHEAST REGION

Southeastern Massachusetts Air Pollution Control District, Lakeville Hospital, Lakeville, MA 02346 — Telephone: (617) 947-1231, ext. 680

Commissioner .... **J. Russell Sylva**
Regional Air Quality Chief .... **Vaughan M. Steeves**

Supplementary personnel — full time: 7 engineers, 3 inspectors, 1 office staff, 1 environmental analyst

Division of Air Quality Control (same address as above)

WESTERN MASSACHUSETTS

Berkshire and Pioneer Valley Air Pollution Control Districts, State House West, Fourth Floor, 436 Dwight Street, Springfield MA 01103 — Telephone: (413) 727-8640

Air Quality Chief .... **David Howland**

■ MICHIGAN

STATE AGENCY

Air Quality Division, Michigan Department of Natural Resources, P.O. Box 30028, Lansing, MI 48909 — Telephone: (517) 322-1330

Chief, Air Quality Division .... **Robert P. Miller**

Supplementary personnel — full time: 1 administrator, 4 chemists, 3 data processing specialists, 48 engineers, 3 meteorologists, 26 resource specialists, 17 office staff, 7 technicians; part time: 2 attorneys, 1 public information specialist

Michigan Air Pollution Control Commission (same address and phone as above)

Chairperson .... **Lee E. Jager**

GRAND RAPIDS

City of Grand Rapids Environmental Services Department, Air Pollution Control Division, 1300 Market, S.W., Grand Rapids, MI 49503 — Telephone: (616) 456-3158

Environmental Services Director .... **Daryle Smith**
Air Pollution Control Officer .... **Richard Olson**

Supplementary personnel — full time: 1 administrator, 1 inspector; part time: 1 office staff

MUSKEGON COUNTY

Muskegon County Health Department, 1611 Oak Street, Muskegon, MI 49442 — Telephone: (616) 724-6266

Director .... **Anita M. Herald, M.D.**
Environmental Specialist .... **Richard O'Neal**

Supplementary personnel — full time: 1 administrator, 1 inspector, 1 office staff, 1 technician; part time: 1 attorney, 1 chemist, 1 data processing specialist, 1 public information specialist, 8 sanitarians

Muskegon County Board of Health (same address and phone as above)

Chairperson .... **Joseph Jurick**

WAYNE COUNTY

Wayne County Air Pollution Control Division, Wayne County Health Department, 1311 E. Jefferson St., Detroit, MI 48207 — Telephone: (313) 224-4650

County Director .... **Vernice Davis-Anthony**
Director .... **Allen M. Greenberg**

Supplementary personnel — full time: 1 administrator, 2 attorneys, 4 chemists, 2 data processing specialists, 11 engineers, 12 inspectors, 1 meteorologist, 5 technicians, 1 public information specialist, 1 accountant, 1 technical services director, 9 office staff

■ MINNESOTA

STATE AGENCY

Minnesota Pollution Control Agency, 1935 W. County Rd., B-2 Roseville, MN 55113 — Telephone: (612) 296-7301

Executive Director .... **Tom Kalitowski**
Director, Division of Air Quality .... **J. Michael Valentine**

Supplementary personnel — full time: 4 administrators, 6 data processing specialists, 13 engineers, 7 inspectors and enforcers, 1 meteorologist, 8 office staff; part time: 2 attorneys, 1 public information specialist, 10 technicians

Minnesota Pollution Control Agency Board (same address and phone as above)

Chairperson .... **Cynthia C. Jepsen**

BLOOMINGTON

City of Bloomington, Environmental Pollution Section, 2215 W. Old Shakopee Rd., Bloomington, MN 55431 — Telephone: (612) 881-5811

Manager, Building and
Environmental Inspections .... **Robert A. Mood**
Environmental Specialist .... **John Nelson**

Supplementary personnel — full time: 3 sanitarians; part time: 1 administrator, 1 attorney, 1 chemist, 1 data processing specialist, 1 office staff

MINNEAPOLIS

Pollution Control Division, Department of Inspections, Public Health Center, Room 300, 250 South 4th St., Minneapolis, MN 55415 — Telephone: (612) 348-7897

Director, Department of Inspections .... **Sol J. Jacobs**
Supervisor, Pollution Control Division .... **Robert L. Lines**

Supplementary personnel — full time: 1 inspector; part time: 1 administrator, 1 attorney, 1 office staff

RICHFIELD

City of Richfield, 6700 Portland Avenue, Richfield, MN 55423 — Telephone: (612) 869-7521

Supervisor, Environmental Health Division .... **Pat Coughlin**

Supplementary personnel — full time: 1 administrator, part time: 1 attorney

ST. CLOUD (Includes all Stearns County)

City-County Air Pollution Control Agency, 315 St. Germain St., St. Cloud, MN 56301 — Telephone: (612) 251-5541, ext. 160

Air Pollution Control Officer .... **Duane M. Beckstrom**

Supplementary personnel — full time: 1 air pollution control specialist; part time: 1 administrator, 1 attorney, 1 data processing specialist, 1 office staff

St. Cloud Board of Health (same address and phone as above)

Health Officer .... **Roger Slanga, M.D.**

ST. LOUIS PARK (Suburb of Minneapolis)

St. Louis Park Inspectional Services, 5005 Minnetonka Blvd., St. Louis Park, MN 55416 — Telephone: (612) 920-3000

Director of Inspectional Services .... **Harvey J. McPhee**

Supplementary personnel — full time: 1 sanitarian; part time: 1 attorney, 1 chemist

STEARNS COUNTY (See St. Cloud)

■ MISSISSIPPI

STATE AGENCY

Mississippi Department of Natural Resources, Bureau of Pollution Control, 2380 Highway 80 West, Jackson, MS 39209 — Telephone: (901) 961-5171 or 961-5099

Director .... **Charles H. Chisolm**
Chief, Air Quality Division .... **Dwight K. Wylie**

Supplementary personnel — full time: 1 attorney, 3 chemists, 15 engineers, 3 office staff, 14 technicians; part time: 2 administrators, 3 data processing specialists, 8 office staff, 6 technicians

Mississippi Commission on Natural Resources, P.O. Box 20305, Jackson, MS 39209

Commission Chairman .... **Bob Travis**

■ MISSOURI

STATE AGENCY

Missouri Department of Natural Resources, Division of Environmental Quality, Air Pollution Control Program, 1101 Rear Southwest Blvd., Jefferson City, MO 65101 — Telephone: (314) 751-4817

Director, Department of Natural Resources ...... **Fred A. Lafset**
Director, Division of
Environmental Quality .... **Robert J. Schreiber, Jr., P.E.**
Staff Director, Air Pollution Control .... **Nick Nikkila**

Supplementary personnel — full time: 2 administrators, 11 engineers, 1 meteorologist, 5 office staff, 4 research analysts, 6 planners, 2 environmental specialists; part time: 1 attorney

Missouri Air Conservation Commission (same address and phone as above)

Chairman .... **William A. Vaughn**

GREENE COUNTY-CITY OF SPRINGFIELD

Greene County-City of Springfield Air Pollution Control Authority, 227 East Chestnut Expressway, Springfield, MO 65802 — Telephone: (417) 864-1622

Director of Health .... **Harold K. Bengsch**
Chief, Air Pollution Control .... **Ronald Boyer**

Supplementary personnel — full time: 3 air pollution control specialists, 1 office staff

INDEPENDENCE

City of Independence Planning Dept., Air Pollution Control Division, 111 E Maple Avenue, Independence, MO 64050 — Telephone: (816) 836-8300, ext. 363

Planning Director .... **William C. Bullard**
Air Pollution Control Officer .... **Randy Clark**

Supplementary personnel: full time: 1 air pollution control officer; part time: 1 administrator, 1 attorney, 1 engineer, 1 inspector, 1 office staff

Air Pollution Board of Appeals (same address as above)

Chairman .... **John Lamkin**

KANSAS CITY

Kansas City Air Quality Program, 414 E. 12th Street, Kansas City, MO 64106 — Telephone: (816) 274-2501

Acting Program Manager .... **Paul F. Stablein, P.E.**

Supplementary personnel — full time: 1 administrator, 1 data processing specialist, 2 inspectors, 1 office staff; part time: 1 attorney

Air Pollution Control Advisory Board (same address and phone as above)

Chairperson .... **Joseph Ochoa**

ST. LOUIS CITY

Division of Air Pollution Control, Room 419, City Hall, St. Louis, MO 63103 — Telephone: (314) 622-3334

Supplementary personnel — full time: 1 administrator, 3 chemists, 3 engineers, 8 inspectors, 6 office staff, 1 technician

Board of Air Pollution Ap (same address and phone as above)

Chairperson .... **Michael Stahl**

ST. LOUIS COUNTY

St. Louis County Air Pollution Control Branch, 801 South Brentwood Blvd., Clayton, MO 63106 — Telephone: (314) 854-6921

Director, Environmental Care Services .... **Conn B. Roden**
Program Manager .... **Blaine J. Rhoades**

Supplementary personnel — full time: 1 administrator, 2 chemists, 4 engineers, 4 inspectors, 1½ office staff, 1 technician, 3 instrument specialists, 1 data processing specialist

Air Pollution Control Appeal Board, 801 S. Brentwood, Clayton, MO 63106

Chairperson .... **Raymond T. Gusnar**

SPRINGFIELD (See Greene County-City of Springfield)

MONTANA

STATE AGENCY

Montana State Department of Health and Environmental Sciences, Air Quality Bureau, Cogswell Building, Helena, MT 59620 — Telephone: (406) 444-3454

Director .... **John J. Drynan, M.D.**
Chief .... **Harold W. Robbins**

Supplementary personnel — full time: 2 administrators, 1 attorney, 2 chemists, 3 data processing specialists, 3 engineers, 2 inspectors, 1 meteorologist, 1 office staff, 4 technicians; part time: 1 office staff

Board of Health and Environmental Sciences (same address and phone as above)

Chairperson .... **John M. McGregor, M.D.**

CASCADE COUNTY

Cascade City-County Air Pollution Control Program, 1130 17th Ave., S., Great Falls, MT 59405 — Telephone: (406) 761-1190

Health Officer .... **Donald E. Pizzini**
Assistant Administrator .... **Samuel R. Kalafat**

Supplementary personnel — full time: 2 administrators, 1 office staff; part time: 1 attorney, 1 chemist, 1 data processing specialist, 1 public information specialist, 4 sanitarians, 2 technicians

City-County Board of Health (same address and phone as above)

Chairperson .... **John H. Stone, M.D.**

MISSOULA COUNTY

Missoula City-County Health Department, 301 West Alder St., Missoula, MT 59802 — Telephone: (406) 721-5700

Health Officer .... **Gary Boe**
Environmental Health Director .... **Elaine Bild**

Supplementary personnel — full time: 1 meteorologist, 1 office staff, 2 sanitarians; part time: 2 administrators, 1 attorney, 1 data processing specialist, 4 work study students

Missoula City-County Air Pollution Control Board (same address and phone as above)

Chairperson .... **Ann Mary Dussault**

YELLOWSTONE COUNTY

Yellowstone County Air Pollution Control Agency, 3306 2nd Avenue North, Billings, MT 59101 — Telephone: 1-(406) 252-0756

Program Director .... **Steven A. Duganz**

Supplementary personnel — full time: 1 administrator, 1 chemist, 1 office staff

Yellowstone County Air Pollution Control Board (same address and phone same as above)

Chairperson .... **L. D. Collins**

NEBRASKA

STATE AGENCY

Department of Environmental Control, 301 Centennial Mall South, P.O. Box 94877, Lincoln, NE 68509 — Telephone: (402) 471-2186

Director .... **Dennis Grams, P.E.**
Chief, Air Pollution Control Div. .... **Gene Robinson, P.E.**

Supplementary personnel — full time: 1 administrator, 1 engineer, 5 inspectors; part time: 1 attorney, 1 office staff, 1 public information specialist

Environmental Control Council (same address and phone as above)

Chairperson .... **Homer Loutzenheiser**

LANCASTER COUNTY (See Lincoln-Lancaster County)

LINCOLN-LANCASTER COUNTY

Lincoln-Lancaster County Air Pollution Control Agency, 2200 St. Marys Ave., Lincoln, NE 68502 — Telephone: (402) 471-7800

Air Pollution Control Officer .... **M. Jane Ford**
Supervisor, Air, Noise & Occupational Health .... **Otis L. Sunderman, R.S.**

Supplementary personnel — full time: 1 engineer, 1 air pollution control specialist; part time: 2 administrators, 1 attorney, 1 chemist, 1 office staff, 1 sanitarian, 1 air pollution control specialist

Lincoln-Lancaster County Air Pollution Control Advisory Board (same address and phone as above)

President .... **Joe Schneider**

OMAHA CITY

Air Quality Control Division, 1819 Farnam Street, Omaha, NE 68183 — Telephone: (402) 444-5378

Superintendent .... **Joseph A. Rogers**
Chief, Air Quality Control Inspector .... **R. L. Timmerman**

Supplementary personnel — full time: 1 administrator, 1 chemist, 3 inspectors, 2 office staff, 3 sanitarians; part time: 1 administrator, 1 attorney, 2 office staff

Omaha City Council (same address and phone as above)

President .... **Bernie Simon**

NEVADA

STATE AGENCY

Division of Environmental Protection, 201 South Fall Street, Carson City, NV 89710— Telephone: (702) 885-4670

Administrator .... **Lew Dodgion**
Air Quality Officer .... **Dick Serdoz**

Supplementary personnel — full time: 1 administrator, 1 engineer, 3 inspectors, 5 technicians; part time: 1 administrator, 1 chemist, 1 attorney, 1 data processing specialist, 5 office staff

State Environmental Commission (same address and phone as above)

Chairperson .... **John Fransway**

CLARK COUNTY

Clark County Health District, Air Pollution Control Division, 625 Shadow Lane, Las Vegas, NV 89106 — Telephone: (702) 383-1276

Chief Health Officer .... **Otto Ravenholt, M.D.**
Director .... **Michael H. Naylor, P.E.**

Supplementary personnel — full time: 1 administrator, 1 chemist, 1 engineer, 4 inspectors, 3 office staff, 6 technicians; part time: 1 attorney, 1 data processing specialist, 1 public information specialist

Clark County District Board of Health (same address and phone as above)

Chairperson .... **Thalia Dondero**

WASHOE COUNTY

Washoe County District Health Department, 1001 E. Ninth, Reno, NV 89520 — Telephone: (702) 785-4290

Health Officer .... **Michael Ford**
Director .... **David J. Minedew**

Supplementary personnel — full time: 1 administrator, 3 sanitarians; part time: 2 administrators, 1 attorney, 1 engineer, 3 office staff, 1 public information specialist

Washoe County District Board of Health

Chairperson .... **V. A. Salvadorini, M.D.**

NEW HAMPSHIRE

STATE AGENCY

New Hampshire Air Resources Agency, Hazen Dr., Concord, NH 03301 — Telephone: (603) 271-4582

Director .... **Dennis R. Lunderville**

Supplementary personnel — full time: 2 administrators, 1 chemist, 6 engineers, 4 office staff, 7 technicians, 2 environmentalists; part time: 1 attorney, 2 chemists

New Hampshire Air Resources Commission (same address and phone as above)

Chairperson .... **William C. Stephan, M.D.**

NEW JERSEY

STATE AGENCY

New Jersey Bureau of Air Pollution Control, Division of Environmental Quality, Department of Environmental Protection, CN027, Trenton, NJ 08625 — Telephone: (609) 292-5450

Director, Division of Environmental Quality .. **Steven G. Kuhrtz**
Chief, Bureau of Air Pollution Control ... **Herbert I. Wortreich**

Supplementary personnel — full time: 4 administrators, 3 chemists, 32 engineers, 21 office staff, 27 technicians, 44 environmental specialists, 1 maintenance worker, 1 mechanic, 2 environmental scientists; part time: 1 administrator, 10 office staff

Clean Air Council (same address and phone as above)

Chairperson .... **Dr. Paul Lioy**

ELIZABETH

City of Elizabeth City Hall, 60 Winfield Scott Plaza, Elizabeth, NJ 07201 — Telephone: (201) 820-4000

Director of Health, Welfare & Housing .... **John N. Surmay**
Health Officer .... **Kenneth J. Sandor**
Chief Sanitary and Air Pollution Inspector .... **Joseph Faccone**

Supplementary personnel — full time: 5 inspectors, 2 office staff; part time: 1 attorney, 1 sanitarian, 1 office staff

HUDSON COUNTY

Hudson Regional Health Commission, 313 Harrison Avenue, Harrison, NJ 07029 — Telephone: (201) 485-7001 (2)

Executive Director .... **Robert R. Ferraiuolo**

Supplementary personnel — full time: 1 administrator, 2 data processing specialists, 4 inspectors, 1 occupational hygienist; part time: 1 attorney, 1 engineer, 1 inspector

Board of Commissioners (same address and phone as above)

President .... **Richard D. Manney, H.O.**

MIDDLESEX COUNTY (Includes Middlesex, Linden, Rahway and Union Counties)

Middlesex County Air Pollution Control Program, 280 Hobart Street, Room 518, Perth Amboy, NJ 08861 — Telephone: (201) 826-3100

Program Coordinator .... **Richard J. Hills**

Supplementary personnel — full time: 2 administrators, 5 inspectors, 2 office staff, 1 emergency response supervisor

Middlesex County Health Department, 417 Dennison Avenue, Highland Park, NJ 08904

Director .... **Laszlo Szabo**

SUBURBAN (Includes Belleville, East Orange, Irvington, Livingston, Montclair, Newark, Orange, Union, West Caldwell and West Orange)

Suburban Regional Health Commission, 377 S. Harrison St., Suite 1F, East Orange, NJ 07018 — Telephone: (201) 675-1774

Executive Director .... **John J. Farley**

Supplementary personnel — full time: 1 administrator, 3 inspectors, 1 office staff; part time: 1 attorney

East Orange Health Dept., 143 New Street, East Orange, NJ 07017

President .... **David C. Byrnes**

■ NEW MEXICO

STATE AGENCY

Environmental Improvement Division, P.O. Box 968, Santa Fe, NM 87503 — Telephone: (505) 984-0020

Director .... **Denise Fort**
Bureau Chief .... **Gordon Venable**

Supplementary personnel — full time: 1 data processing specialist, 7 engineers, 16 environmental scientists, 1 meteorologist, 5 office staff, 4 technicians, 4 administrators, 1 chemist; part time: 1 attorney

Environmental Improvement Board (same address and phone as above)

Chairperson .... **William Atkins**

ALBUQUERQUE-BERNALILLO COUNTY

Albuquerque Environmental Health and Energy Department, 924 Park Avenue, SW, Albuquerque, NM 87102 — Telephone: (505) 766-7290

Director .... **Larry J. Gordon**
Manager, Air Pollution Control
Division .... **Steven W. Walker**

Supplementary personnel — full time: 1 administrator, ½ attorney, ½ data processing specialist, 2 engineers, ½ meteorologist, 1 office staff, 3 technicians, 4 environmental scientist, 1 environmental planner

Albuquerque-Bernalillo County Air Quality Board, P.O. Box 1293, Albuquerque, NM 87103

Chairperson .... **Robert Luna, Ph.D.**

BERNALILLO COUNTY (See Albuquerque, etc.)

■ NEW YORK

STATE AGENCY

New York Department of Environmental Conservation, Division of Air, 50 Wolf Road, Albany, NY 12223 — Telephone: (518) 457-7230

Commissioner of Environmental
Conservation .... **Henry G. Williams**
Director, Division of Air .... **Harry H. Hovey, Jr., P.E.**
Associate Director, Division of Air ...... **Thomas M. Allen, P.E.**
Attorney for Division of Air .... **Vacant**
(Telephone: (518) 457-3468)
Bureau of Abatement Planning .... **E. Davis**
(Telephone: (518) 457-6379)
Bureau of Air Quality Surveillance .... **D. Gower**
(Telephone: (518) 457-3676)
Bureau of Technical Services .... **G. Blanchard**
(Telephone: (518) 457-5385)
Bureau of Impact Assessment and Meteorology ...... **S. Marlow**
(Telephone: (518) 457-7688)
Bureau of Source Control .... **J. Davis**
(Telephone: (518) 457-5618)
Bureau of Air Toxics .... **A. Fossa**
(Telephone: (518) 457-7454)
Special Assistant, Clean Air Act and Acid Rain .... **D. Shaw**
(Telephone: (518) 457-2823)

Supplementary personnel — full time: 9 administrators, 3 chemists, 1 data processing specialist, 78 engineers, 6 meteorologists, 25 office staff, 2 public information specialists, 51 technicians, 13 research specialists

State Environmental Board (same address as above) — Telephone: (518) 457-3446

Commissioner .... **Henry G. Williams**

ALBANY COUNTY

Albany County Department of Health, S. Ferry and Green Streets, Albany, NY 12201 — Telephone: (518) 445-7811

Commissioner of Health .... **William A. Grattan, M.D.**
Director, Division of Environmental
Health Services .... **Stephen S. Lukowski, P.E.**

Supplementary personnel — part time: 1 engineer, 2 technicians

Albany County Board of Health (same address and phone as above)

President .... **Richard J. Conners**

COLUMBIA COUNTY (See State Agency)

ERIE COUNTY

Erie County Department of Environment and Planning, 95 Franklin Street, Buffalo, NY 14202 — Telephone: (716) 846-6370

Commissioner .... **Joan E. Loring**
Deputy Commissioner, Division of
Environmental Control .... **Anthony T. Voell, P.E.**
Air Quality Section, Operating Division,
Senior Environmental
Quality Engineer .... **Eugene J. Sciascia, P.E.**

Supplementary personnel — part time: 4 engineers, 5 inspectors, 2 office staff

INTERSTATE (See New York Metropolitan Area)

NASSAU COUNTY

Department of Health, Division of Environmental Health Services, 240 Old County Rd., Mineola, NY 11501 — Telephone: (516) 535-3232

Commissioner of Health .... **John J. Dowling, M.D., M.P.H.**
Director, Bureau of Air
Quality Management .... **Robert V. Close, P.E., M.Ch.E.**

Supplementary personnel — full time: 2 chemists, 5 engineers, 3 office staff, 9 sanitarians; part time: 1 administrator, 1 attorney

Nassau County Board of Health (same address as above) — Telephone: (516) 535-2037

Chairperson .... **Bruce A. Lister**

NEW ROCHELLE (See Westchester)

NEW YORK

Department of Environmental Protection, City of New York, 1 Centre Street, Room 2358, New York, NY 10007 — Telephone: (212) 566-4124

Commissioner .... **Joseph T. McGough, Jr.**
Deputy Commissioner, Planning &
Enforcement .... **Richard G. Mendes**
Deputy Commissioner, Intergovernmental &
Public Affairs .... **Martha Holstein**

Supplementary personnel — full time: 13 administrators, 8 analysts, 7 chemists, 49 engineers, 3 highway transportation specialists, 40 inspectors, 68 office staff, 3 physicists, 11 technicians, 4 computer specialists, 2 attorneys, 1 public information specialist, 3 project coordinators, 1 research scientist

Environmental Control Board, Department of Environmental Protection, 333 West 34th Street, New York, NY 10001
Executive Director .... **George Aridas**

NEW YORK METROPOLITAN AREA

Interstate Sanitation Commission, 311 W. 43 St., New York, NY 10036 — Telephone: (212) 582-0380
Director and Chief Engineer .... **Alan I. Mytelka, Ph.D.**

Supplementary personnel — full time: 1 engineer, 1 sanitarian; part time: 2 administrators, 1 attorney, 3 chemists, 1 data processing specialist, 4 office staff, 1 technician

Interstate Sanitation Commission (same address and phone as above)
Chairperson .... **Natale Colosi, Ph.D.**

NIAGARA COUNTY

Niagara County Health Department, Air Resources Bureau, 10th and East Falls Street, Niagara Falls, NY 14302 — Telephone: (716) 284-3128
Asst. Commissioner of Environmental Health .... **John A. Kehoe, P.E.**
Associate Sanitarian .... **Maurice Vaughan**

Supplementary personnel — full time: 1 engineer, 1 office staff, 2 sanitarians, 3 technicians; part time: 1 administrator, 1 attorney, 1 technician

Niagara County Board of Health (same address and phone as above)
President .... **Dr. Theodore Gerwig**

ONONDAGA COUNTY

Division of Environmental Sanitation, Onondaga County Department of Health, P.O. Box 190, Syracuse, NY 13215 — Telephone: (315) 469-6955
Director, Division of Environmental Sanitation .... **Peter Guala, P.E.**
Chief, Bureau of Air Resources .... **S. M. Burdick, P.E.**

Supplementary personnel — full time: 2 engineers, 2 inspectors, 1 office staff, 1 technician

RENSSELAER COUNTY

Rensselaer County Department of Health, Division of Environmental Health, County Office Building, 1600 Seventh Ave., Troy, NY 12180 — Telephone: (518) 270-2664
Public Health Director .... **Kenneth Van Praag**

Supplementary personnel — part time: 1 administrator, 1 attorney, 1 engineer, 2 sanitarians, 2 technicians

ROCKLAND COUNTY

Rockland County Department of Health, Sanatorium Rd., Pomona, NY 10970 — Telephone: (914) 354-0200
Commissioner of Health .... **Dr. Israel Praiss**
Asst. Commissioner for Environmental Health .... **George E. O'Keefe, P.E.**
Assoc. Public Health Engineer .... **Thomas M. Micelli, P.E.**

Supplementary personnel — full time: 1 administrator, 1 engineer, 1 technician, 1 office staff

Rockland County Board of Health (same address and phone as above)

SUFFOLK COUNTY

Suffolk County Department of Health Services, 65 Jetson Lane, Hauppauge, NY 11790 — Telephone: (516) 234-2622
Commissioner .... **David Harris, M.D., M.P.H.**
Chief, Air Pollution/Solid Waste Control .... **James C. Maloney, P.E.**

Supplementary personnel — full time: 1 data processing specialist, 3 engineers, 3 inspectors, 1 office staff; part time: 1 attorney, 1 chemist, 1 public information specialist

WESTCHESTER COUNTY (Includes City of New Rochelle)

Westchester County Department of Health, Bureau of Environmental Quality, 112 E. Post Road, White Plains, NY 10601 — Telephone: (914) 285-5010
Commissioner of Health .... **Anita S. Curran, M.D., M.P.H.**
Assistant Commissioner of Health .... **Calvin E. Weber, P.E.**

Supplementary personnel — full time: 3 engineers, 1 sanitarian, 1 technician; part time: 1 administrator, 1 office staff

Westchester County Board of Health (same address and phone as above)
President .... **John A. Ramsdell, M.D.**

■ NORTH CAROLINA

STATE AGENCY

Division of Environmental Management, P.O. Box 27687, Raleigh, NC 27611 — Telephone: (919) 733-7015
Director .... **Robert F. Helms**
Chief, Air Quality Section .... **W. Fin Johnson**

Supplementary personnel — full time: 1 administrator, 5 chemists, 1 data processing specialist, 27 engineers, 2 meteorologists, 5 office staff, 1 quality assurance specialist, 1 statistical research assistant, 1 environmental scientist, 1 social research associate, 37 technicians; part time: 2 attorneys

Environmental Management Commission (same address and phone as above)
Chairperson .... **Thomas W. Bradshaw, Jr.**

BUNCOMBE-HAYWOOD COUNTIES (See Western North Carolina Region)

CLEVELAND COUNTY

Cleveland County Health Department, 315 Grover Street, Shelby, NC 28150 — Telephone: (704) 487-1131
Health Director .... **Richard G. Steeves**
Sanitarian Supervisor .... **Jerry R. Freeman**

Supplementary personnel — part time: 1 administrator, 1 attorney, 1 sanitarian

Cleveland County Board of Health (address and phone same as above)
Chairperson .... **Charles Rhoden, Jr.**

CUMBERLAND COUNTY

E. Newton Smith Public Health Center, 227 Fountainhead Lane, Fayetteville, NC 28301 — Telephone: (919) 483-9046
Health Director .... **Jesse F. Williams, M.D.**
Director, Environmental Health .... **Lacy Williams, Jr., R.S.**

Supplementary personnel — full time: 1 administrator, 2 sanitarians, 2 office staff, 1 attorney

Cumberland County Board of Health (same address and phone as above)
Chairperson .... **James R. Hickmon, RPh**

DURHAM COUNTY (See State Agency)

FORSYTH COUNTY

Forsyth County Environmental Affairs Department, 537 N. Spruce Street, Winston-Salem, NC 27101 — Telephone: (919) 727-8064
Director .... **Robert R. Fulp**

Supplementary personnel — full time: 2 administrators, 3 chemists, 2 engineers, 4 inspectors, 2 office staff, 2 technicians

Forsyth County Environmental Affairs Board (same address and phone as above)
Chairperson .... **Dr. Ruth P. Phillips**

GASTON COUNTY (See State Agency)

GUILFORD COUNTY

Guilford County Department of Public Health, 301 N. Eugene St., Greensboro, NC 27401 — Telephone: (919) 373-3771
Health Director .... **Joseph L. Holliday, M.D., M.P.H.**
Director of Environmental Health ...... **Thomas F. Owens, R.S.**

Supplementary personnel — full time: 1 sanitarian; part time: 1 administrator, 1 sanitarian

Guilford County Board of Health (same address and phone as above)
Chairman .... **Dr. Eugene Grubb**

HAYWOOD COUNTY (See Western North Carolina Region)

MECKLENBURG COUNTY

Mecklenburg County Department of Environmental Health, 1200 Blythe Blvd., Charlotte, NC 28203 — Telephone: (704) 376-4649
Director .... **John Barry**
Manager, Air Quality .... **Bobby Cobb**
Laboratory Director .... **Tom Ward**

Supplementary personnel — full time: 2 administrators, 3 chemists, 2 engineers, 5 sanitarians, 1 office staff

WESTERN NORTH CAROLINA REGION (Includes

Buncombe and Haywood Counties)

Western North Carolina Region Air Pollution Control, 189 College St., P.O. Box 7215, Asheville, NC 28807 — Telephone: (704) 255-5655

Director .... Ronald Boone

Supplementary personnel — full time; 1 administrator, 1 chemist, 3 engineers, 3 inspectors, 1 office staff; part time; 1 attorney

Western North Carolina Regional Air Pollution Control Board, P.O. Box 7215, Asheville, NC 28807

Chairperson .... Roy Roberts, M.D.

■ NORTH DAKOTA

STATE AGENCY

North Dakota State Department of Health, 1200 Missouri Ave., Bismarck, ND 58501 — Telephone: (701) 224-2348

Chief, Environmental Health Section .... Gene A. Christianson

Director, Division of Environmental Engineering................. Dana K. Mount

Supplementary personnel — full time: 2 attorneys, 1 data processing specialist, 7 engineers, 4 office staff, 3 sanitarians, 2 environmental quality specialists, 4 technicians; part time: 2 administrators, 2 attorneys, 4 chemists, 2 meteorologists, 1 office staff

State Air Pollution Control Advisory Council (same address and phone as above)

Chairperson .... O. P. Madhock

■ OHIO

STATE AGENCY

Ohio Environmental Protection Agency, 361 East Broad St., Columbus, OH 43215 — Telephone: (614) 466-6166

Director .... Robert Maynard

Chief, Division of Air Pollution Control...... Charles M. Taylor

Supplementary personnel — full time: 5 administrators, 1 attorney, 7 chemists, 4 data processing specialists, 24 engineers, 36 inspectors, 10 office staff, 1 meteorologist, 10 technicians; part time: 1 attorney, 2 data processing specialists, 2 public information specialists

AKRON, BARBERTON, KENT, RAVENNA, MEDINA, SUMMIT AND PORTAGE COUNTIES

Akron Regional Air Pollution Control Agency, Akron City Health Department, 177 S. Broadway, Akron, OH 44308 — Telephone: (216) 375-2480

Director of Health-Akron ...... C. William Keck, M.D., M.P.H.

Administrator, Air Pollution Control .. Norman F. Keckler, P.E.

Air Monitoring Officer .... Jerry J. Garro

Supplementary personnel — full time: 1 administrator, 1 chemist, 1 monitoring officer, 3 engineers, 3 inspectors, 2 office staff, 2 technicians; part time: 1 accountant, 1 attorney

BARBERTON (See Akron, etc.)

BELMONT COUNTY (See Steubenville)

BUTLER COUNTY (See Cincinnati)

CANTON

Canton City Health Department, Air Pollution Control Division, City Hall, Canton, OH 44702 — Telephone: (216) 489-3385

Health Commissioner .... Robert E. Pattison, M.P.A.

Assistant Director .... Bruce E. Blankenship

Supplementary personnel — full time: 1 administrator, 2 inspectors; part time: 1 administrator

Canton City Board of Health (same address and phone as above)

President Pro Tem .... M. L. Wilen, D.V.M.

CARROLL COUNTY (See Steubenville)

CINCINNATI (Includes Butler, Clermont, Hamilton and Warren Counties)

Hamilton County-Southwestern Ohio Air Pollution Control Agency, 2400 Beekman St., Cincinnati, OH 45214 — Telephone: (513) 251-8777

Director .... Charles E. Schumann

Administrator .... Michael J. Maloney

Supplementary personnel — full time: 2 administrators, 3 chemists, 6 engineers, 6 inspectors, 4 office staff, 9 technicians, 1 accountant

Hamilton County Board of Commissioners, 1000 Main Street, Cincinnati, OH 45202 (Hamilton County Courthouse)

Chairperson .... Norman A. Murdock

CLARK COUNTY (See Montgomery)

CLERMONT COUNTY (See Cincinnati)

CLEVELAND

Division of Air Pollution Control, Department of Public Health and Welfare, 2735 Broadway Avenue, Cleveland, OH 44115 — Telephone: (216) 664-3500

Commissioner .... Gary J. Nied

Bureau of Field Services .... Edward J. Fasko

Bureau of Engineering Services .... Ikram Habib

Bureau of Technical Services .... Richard A. Dell

Supplementary personnel — full time: 4 administrators, 1 attorney, 3 chemists, 5 engineers, 11 inspectors, 8 office staff, 8 technicians

Board of Building Standards and Appeals, Room 15, City Hall, Cleveland, OH 44114

Chairperson .... J. L. Denk

Secretary .... Frank Eppolito

COLUMBIANA COUNTY (See Steubenville)

DARKE COUNTY (See Montgomery)

GREENE COUNTY (See Montgomery)

HAMILTON COUNTY (See Cincinnati)

HARRISON COUNTY (See Steubenville)

IRONTON (See Portsmouth)

JEFFERSON COUNTY (See Steubenville)

KENT (See Akron, etc.)

LAKE COUNTY

Lake County General Health District, Administration Center, 105 Main St., Painesville, OH 44077 — Telephone: (216) 352-0766

Health Commissioner..................... Joel Lucia, M.P.H.

Air Pollution Control Supervisor .... Leon A. Weitzel

Supplementary personnel — full time: 1 chemist, 1 engineer, 2 sanitarians; part time: 2 administrators, 1 attorney, 3 office staff

Board of Health for Lake County General Health District (same address and phone as above)

President .... Charles Magee

LUCAS COUNTY (See Wood and Lucas Counties)

MAHONING-TRUMBULL (Includes Youngstown)

Mahoning-Trumbull Air Pollution Control Agency, 9 West Front Street, Youngstown, OH 44503 — Telephone: (216) 744-1928

Director .... Robert R. Ramhoff

Administrator .... Larry C. Himes

Supplementary personnel — full time: 1 administrator, 1 chemist, 2 engineers, 1 office staff, 3 technicians

Youngstown Board of Health, City Hall, 26 S. Phelps, Youngstown, OH 44503

President .... Patrick J. Ungaro

MEDINA COUNTY (See Akron, etc.)

MIAMI COUNTY (See Montgomery)

MIDDLETOWN (See Cincinnati)

MONROE COUNTY (See Steubenville)

MONTGOMERY COUNTY (Includes Clark, Darke, Greene, Miami, Montgomery and Preble Counties)

Regional Air Pollution Control Agency, 451 W. Third St., P.O. Box

972, Dayton, OH 45422 — Telephone: (513) 225-4435

Acting Supervisor .... **D. David Redic**
Health Commissioner .... **Morton Nelson, M.D., M.P.H.**

Supplementary personnel — full time: 1 administrator, 1 lab specialist, 3 air quality analysts, 1 data processing specialist, 9 engineers, 5 inspectors, 4 office staff, 4 technicians; part time: 1 attorney, 1 meteorologist, 1 public information specialist

Montgomery County Board of Health (same address as above)
President .... **Herman I. Abromowitz, M.D.**

OREGON (See Toledo)

PORTAGE COUNTY (See Akron, etc.)

PORTSMOUTH (Includes Ironton)

Portsmouth Local Air Agency, 740 Second St., Portsmouth, OH 45662 — Telephone: (614) 353-5153

Director-Commissioner,
Air Pollution Control .... **Donald R. Cavote, P.E.**

Supplementary personnel — full time: 1 administrator, 1 engineer, 1 inspector, 1 office staff, 3 technicians

Portsmouth Board of Health (same address and phone as above)
President-Mayor, City of Portsmouth .... **Richard Grimm**

PREBLE COUNTY (See Montgomery)

RAVENNA (See Akron, etc.)

ROSSFORD (See Toledo)

STEUBENVILLE (Includes Jefferson, Belmont, Columbiana, Monroe, Carroll and Harrison Counties)

North Ohio Valley Air Authority, 814 Adams Street, Steubenville, OH 43952 — Telephone: (614) 282-3908

Director .... **Pat J. DeLuca**

Supplementary personnel — full time: 1 administrator, 3 chemists, 2 engineers, 3 inspectors, 1 office staff, 3 technicians; part time: 1 attorney, 1 public information specialist

North Ohio Valley Air Authority Board of Control (same address and phone as above)
Chairperson .... **Ed Swiger**

SUMMIT COUNTY (See Akron, etc.)

TOLEDO (Includes Lucas County and City of Rossford)

Environmental Services Agency, 26 Main Street, Toledo, OH 43605 — Telephone: (419) 693-0350

Administrator .... **Donald M. Moline**

Supplementary personnel — full time: 1 administrator, 4 chemists, 4 engineers, 7 inspectors, 4 office staff, 4 technicians; part time: 2 technicians

WARREN COUNTY (See Cincinnati)

YOUNGSTOWN (See Mahoning-Trumbull)

■ OKLAHOMA

STATE AGENCY

Air Quality Service, Environmental Health Services, Oklahoma State Department of Health, 1000 Northeast 10th Street, P.O. Box 53551, Oklahoma City, OK 73152 — Telephone: (405) 271-5220

Commissioner of Health................ **Joan K. Leavitt, M.D.**
Chief, Air Quality Service .... **John W. Drake**

Supplementary personnel — full time: 4 administrators, 1 attorney, 3 engineers, 4 office staff, 13 technicians, 4 first level supervisors

Air Quality Council (same address and phone as above)
Chairperson .... **William B. Breisch, P.E.**

OKLAHOMA CITY-COUNTY

Oklahoma City-County Health Department, Air Quality Control Section, 921 N.E. 23rd Street, Oklahoma City, OK 73105 — Telephone: (405) 427-8651

Director .... **George Dellaportas, M.D., M.P.H.**
Supervisor, Air Quality Control .... **Jim Armstrong, M.E.S.**

Supplementary personnel — full time: 1 administrator, 1 data processing specialist, 1 engineer, 2 inspectors, 1 office staff, 1 technician, 2 monitoring personnel; part time: 1 administrator, 1 public information specialist

Air Quality Advisory Board (same address and phone as above)
Chairperson .... **Keith L. Stanley, Ph.D., R.P.E.**

TULSA CITY-COUNTY

Tulsa City-County Health Department, 4616 E. 15th St., Tulsa, OK 74112 — Telephone: (918) 744-1000

Director .... **Edgar M. Cleaver, M.D.**
Supervisor .... **Roger D. Randolph**

Supplementary personnel — full time: 1 administrator, 1 chemist, 2 engineers, 2 inspectors, 1 office staff, 4 technicians; part time: 1 attorney, 1 data processing specialist, 1 public information specialist

Environmental Advisory Council (same address and phone as above)
Chairperson .... **John W. Moody**

■ OREGON

STATE AGENCY

State of Oregon, Department of Environmental Quality, 522 S.W. Fifth Avenue, P.O. Box 1760, Portland, OR 97204/97207 — Telephone: (503) 229-5696

Director .... **Fred Hansen**
Administrator .... **Thomas R. Bispham**

Supplementary personnel — full time: 1 administrator, 5 chemists, 3 data processing specialists, 20 engineers, 55 inspectors, 1 meteorologist, 16 office staff, 1 public information specialist, 22 technicians, 4 supervisors; part time: 1 attorney

Environmental Quality Commission (same address and phone as above)
Chairperson .... **James E. Petersen**

LANE COUNTY

Lane Regional Air Pollution Authority, 1244 Walnut Street, Eugene, OR 97403 — Telephone: (503) 686-7618

Director .... **Donald R. Arkell**

Supplementary personnel — full time: 1 administrator, 1 data processing specialist, 2 engineers, 1 meteorologist, 2 office staff, 1 public information specialist, 2 technicians; part time: 1 attorney, 1 financial aid

Board of Directors, Lane Regional Air Pollution Authority (same address and phone as above)
Chairperson .... **Richard Hansen**

■ PENNSYLVANIA

STATE AGENCY

Bureau of Air Quality Control, 200 N. Third Street, P.O. Box 2063, Harrisburg, PA 17120 — Telephone: (717) 787-9702

Secretary, Department of
Environmental Resources .... **Nicholas DeBenedictis**
Director, Bureau of Air Quality Control ... **James K. Hambright**

Supplementary personnel — full time: 2 administrators, 74 engineers, 3 meteorologists, 11 office staff, 44 specialists, 29 technicians

Environmental Quality Board, P.O. Box 2063, 9th Fl., Fulton Building, Harrisburg, PA 17120 — Telephone: (717) 787-4526
Chairperson .... **Nicholas DeBenedictis**

ALLEGHENY COUNTY

Health Department, Bureau of Air Pollution Control, 301 39th St., Pittsburgh, PA 15201 — Telephone: (412) 578-8111

Director-Health Department........... **N. Mark Richard, M.D.**
Deputy Director-Air Pollution Control .... **Ronald J. Chleboski**

Supplementary personnel — full time: 6 administrators, 3 attorneys, 5 chemists, 4 coke oven process technicians, 5 data processing specialists, 8 engineers, 6 inspectors, 1 meteorologist, 7 office staff, 13 technicians; part time: 1 public information specialist

Board of County Commissioners, 119 Court House, Grant Street, Pittsburgh, PA 15219
Chairperson .... **Thomas J. Forester**

PHILADELPHIA

Air Management Services, Philadelphia Department of Public Health, 500 South Broad St., Philadelphia, PA 19146 — Telephone: (215) 686-7840

Commissioner, Department
of Public Health .... **Stuart H. Shapiro, M.D.**
Deputy Commissioner for Environmental
and Specialized Health Services .... **A. J. Henley**

Assistant Health Commissioner for
Air Management Services .... **William Reilly**
Supplementary personnel — full time: 1 administrator, 12 chemists, 1 draftsman, 1 enforcement officer, 25 engineers, 18 inspectors, 1 meteorologist, 10 office staff, 1 public information specialist, 8 technicians; part time: 2 attorneys

Philadelphia Air Pollution Control Board, 540 Municipal Services Building, Philadelphia, PA 19107
Chairperson .... **Dr. I. M. Levitt**

■ PUERTO RICO

COMMONWEALTH AGENCY

Environmental Quality Board of Puerto Rico, Del Parque Street, #204 Corner Pumarada Street, Santurce, Puerto Rico 00910 — Telephone: (809) 725-5140 or (809) 722-0077

Chairman .... **Santos Rohena**
Director .... **Juan Merced Mateo, P.E.**
Supplementary personnel — full time: 1 administrator, 1 attorney, 4 chemists, 6 engineers, 12 inspectors, 3 statisticians, 2 programmers, 1 programmer assistant, 11 office staff, 1 planning technician, 1 quality assurance officer, 2 specialists in environmental sciences, 1 electrician, 2 environmental scientists, 4 electronic technicians, 1 statistician/clerk

■ RHODE ISLAND

STATE AGENCY

Rhode Island Division of Air and Hazardous Materials, 204 Cannon Building, Davis Street, Providence, RI 02908 — Telephone: (401) 277-2808

Chief .... **Barry W. Muller**
Supplementary personnel — full time: 1 administrator, 5 chemists, 6 engineers, 7 inspectors, 3 office staff, 1 planner, 2 technicians; part time: 2 attorneys

Department of Environmental Management (same address and phone as above)
Director .... **Robert L. Bendick**

■ SOUTH CAROLINA

STATE AGENCY

South Carolina Department of Health and Environmental Control, Bureau of Air Quality Control, 2600 Bull St., Columbia, SC 29201 — Telephone: (803) 758-5406

Commissioner .... **Robert S. Jackson, M.D.**
Deputy Commissioner for
Environmental Quality Control .... **R. Lewis Shaw, P.E.**
Chief, Bureau of Air Quality Control ...... **Otto E. Pearson, P.E.**
Supplementary personnel — full time: 6 administrators, 9 chemists, 14 engineers, 30 environmental quality managers, 7 office staff, 1 statistician, 8 technicians; part time: 1 attorney, 5 data processing specialists, 1 meteorologist, 1 public information specialist

Board of Health and Environmental Control (same address and phone as above)
Chairman .... **Moses H. Clarkson, Jr.**

COLUMBIA

City of Columbia, 1225 Laurel Street, Columbia, SC 29217 — Telephone: (803) 733-8320

Chief Building Officer .... **Edward C. Hornsby**
Pollution Control Officer .... **Darlene Boyd**
Supplementary personnel — full time: 1 inspector; part time: 1 administrator, 1 attorney

Pollution Control, Appeals Board (same address and phone as above)
Chairperson .... **William Otis**

■ SOUTH DAKOTA

STATE AGENCY

Department of Water and National Resources, Office of Air Quality and Solid Waste, Joe Foss Building, Pierre, SD 57501 — Telephone: (605) 773-3329

Secretary .... **Warren R. Neufield**
Administrator .... **Joel C. Smith**
Supplementary personnel — full time: 3 inspectors, 3 monitoring specialists; part time: 1 administrator, 1 attorney, 1 chemist, 1 engineer, 3 inspectors, 2 office staff, 1 data processing specialist

Board of Minerals and Environment (same address and phone as above)
Chairperson .... **Richard C. Sweetman**

■ TENNESSEE

STATE AGENCY

Tennessee Division of Air Pollution Control, TERRA Bldg., 150 Ninth Avenue, North, Nashville, TN 37203 — Telephone: (615) 741-3931

Commissioner, Department of Health and
Environment .... **James E. Word**
Director .... **Harold E. Hodges, P.E.**
Supplementary personnel — full time: 2 administrators, 16 engineers, 1 meteorologist, 10 office staff, 11 technicians, 5 environmental managers, 26 environmental specialists

Tennessee Air Pollution Control Board (same address and phone as above)
Chairperson .... **James E. Word**

CHATTANOOGA-HAMILTON COUNTY

Chattanooga-Hamilton County Air Pollution Control Bureau, 3511 Rossville Boulevard, Chattanooga, TN 37407 — Telephone: (615) 867-4321

Director .... **J. Wayne Cropp**
Supplementary personnel — full time: 1 administrator, 1 attorney, 1 chemist, 4 engineers, 2 inspectors, 3 office staff, 1 operations manager (finance and budget), 1 public information specialist, 3 technicians

Chattanooga-Hamilton County Air Pollution Control Board (same address and phone as above)
Chairperson .... **Robert D. Britt**

DAVIDSON COUNTY (See Nashville-Davidson County)

HAMILTON COUNTY (See Chattanooga-Hamilton County)

KNOX COUNTY

Knox County Department of Air Pollution Control, 400 Main Avenue, City-County Bldg., Rm. L222, Knoxville, TN 37902 — Telephone: (615) 521-2488

County Executive .... **W. Dwight Kessel**
Director .... **James E. Lovett**
Supplementary personnel — full time: 1 administrator, 1 engineer, 1 inspector, 2 office staff, 2 technicians, 3 environmentalists

Knox County Air Pollution Control Board (same address and phone as above)
Chairman .... **John R. McDowell**

MEMPHIS-SHELBY COUNTY

Memphis and Shelby County Health Department, 814 Jefferson Avenue, Memphis, TN 38105 — Telephone: (901) 528-3828

Acting Health Officer .... **Sherman Kahn, M.D.**
Manager, Pollution Control .... **Tommie A. Gibbs**
Supervisor, Air Pollution Control Branch .... **Denis Fritchie**
Supplementary personnel — full time: 2 administrators, 1 chemist, 1 data processing specialist, 3 engineers, 3 inspectors, 3 office staff, 1 sanitarian, 3 technicians; part time: 2 attorneys, 1 engineer, 1 meteorologist, 1 public information specialist, 1 law clerk

Environmental Services Bureau
Administrator .... **Norman L. Chapelle**

NASHVILLE-DAVIDSON COUNTY

Metropolitan Health Department of Nashville and Davidson County, 311 23rd Ave., N., Nashville, TN 37203 — Telephone: (615) 327-9313

Director of Health .... **Joseph M. Bistowish, M.D.**
Director, Pollution Control Division .... **Paul J. Bontrager**
Supplementary personnel — full time: 2 administrators, 1 chemist, 1 engineer, 3 environmentalists, 2 office staff, 3 technicians, 4 vehicle inspectors

Metropolitan Board of Health for Nashville and Davidson County (same address and phone as above)
Chairperson .... **Harold W. Jordan, M.D.**

SHELBY COUNTY (See Memphis-Shelby County)

■ TEXAS

STATE AGENCY

Texas Air Control Board, 6330 Highway 290 East, Austin, TX 78723 — Telephone: (512) 451-5711

Executive Director .... **Bill Stewart, P.E.**

Supplementary personnel — full time: 22 administrators, 7 attorneys, 9 chemists, 14 data processing specialists, 78 engineers, 78 office staff, 1 public information specialist, 42 technicians, 3 administrative assistants, 21 staff services officers, 78 environmental quality specialists; part time: 7 data processing specialists, 10 office staff

Texas Air Control Board (same address and phone as above)

Chairperson .... **John L. Blair**
Deputy Executive Director .... **Allen Eli Bell**
Director, Administration and Research .... **Alex Opiela, Jr., P.E.**
Director, Central Regulatory Operations .... **Steve Spaw, P.E.**
Director, Policy Analysis .... **Roger Wallis**
General Counsel .... **John B. Turney**

BEXAR COUNTY (See San Antonio-Bexar County)

CORPUS CHRISTI-NUECES COUNTY

Corpus Christi-Nueces County Department of Public Health and Welfare, 1702 Horne Rd., Corpus Christi, TX 78469 — Telephone: (512) 855-4051

Director .... **C. M. G. Buttery, M.D., M.P.H.**
Director, Air Pollution Control .... **Albert L. Cock, P.E.**

Supplementary personnel — full time: 1 chemist, 1 engineer, 1 office staff, 1 environmental health specialist

DALLAS

City of Dallas, Health and Human Services, Environmental Health Division, 1500 W. Mockingbird Lane, Dallas, TX 75235 — Telephone: (214) 670-6283

Director, Health & Human Services .... **Callie F. Struggs**
Assistant Director .... **Vacant**
Manager, Environmental Control .... **Fred M. Barnes**
Manager, Air Pollution Control .... **H. Gary Burlbaw**

Supplementary personnel — full time: 1 administrator, 3 chemists, 1 engineer, 3 inspectors, 1 office staff, 1 sanitarian; 2 technicians, 3 samplers

EL PASO CITY-COUNTY

El Paso County Health Unit, 222 S. Campbell St., El Paso, TX 79901 — Telephone: (915) 541-4262

Director .... **Laurance N. Nickey, M.D., F.A.A.P.**
Chief, Environmental Health Services .... **Raul V. Muñoz, Jr.**

Supplementary personnel — full time: 1 administrator, 1 chemist, 6 technicians

El Paso City County Board of Health (same address as above)

Chairman .... **David Wilbanks, D.D.S.**

FORT WORTH

Department of Public Health, 1800 University Dr., Fort Worth, TX 76107 — Telephone: (817) 870-7280

Director .... **B. Brooks Taylor, M.D., M.P.H.**
Supervisor .... **James C. Webb**

Supplementary personnel — full time: 1 chemist, 4 sanitarians; part time: 1 administrator

GALVESTON COUNTY

Galveston County Health District, 1207 Oak Street, P.O. Box 939, La Marque, TX 77568 — Telephone: (409) 938-7221

Acting Director .... **C. M. Porter, M.D.**
Air Control Director .... **Edward R. Ibert, Ph.D.**

Supplementary personnel — full time: 1 administrator, 2 inspectors, 1 office staff, 1 chemist

Galveston County United Board of Health (same address and phone as above)

Chairperson .... **Bernard A. Milstein, M.D.**

HARRIS COUNTY

Harris County Pollution Control District, 107 N. Munger, Pasadena, TX 77506 — Telephone: (713) 221-5224

Director .... **A. R. Peirce**

Supplementary personnel — full time: 6 administrators, 9 chemists, 1 engineer, 23 inspectors, 6 office staff, 2 janitors

Harris County Commission Court, 1001 Preston St., Houston, TX 77002

County Judge .... **The Hon. Jon Lindsay**

HOUSTON

City of Houston Bureau of Air Quality Control, 7411 Park Place Blvd., Houston, TX 77087 — Telephone: (713) 645-5316

Chief .... **Dallas E. Evans**

Supplementary personnel — full time: 2 administrators, 7 engineers, 1 meteorologist, 7 office staff, 1 public information specialist, 50 technicians, 3 instrument technicians

Department of Public Health, 1115 N. MacGregor, Houston, TX 77030

Director .... **James G. Haughton, M.D., M.P.H.**

LUBBOCK CITY

Lubbock City Health Department, 1001 Texas Avenue, P.O. Box 2548, Lubbock, TX 79408 — Telephone: (806) 762-6411, ext. 2918

Health Director .... **Anthony Way, M.D.**
Public Health Administrator,
Chief Sanitarian .... **R. D. Goodman, R.S.**

Supplementary personnel — part time: 1 administrator, 1 attorney, 1 office staff, 1 sanitarian

Lubbock Board of Health (same address and phone as above)

Chairperson .... **John Brown, D.D.S.**

NUECES COUNTY (See Corpus Christi, etc.)

ODESSA-ECTOR COUNTY

Odessa-Ector County Health Department, 221 North Texas, Odessa, TX 79761 — Telephone: (915) 332-4261

Air Pollution Program Director .... **Rick Costa**
County Health Officer .... **Robert Rowntree, III, M.D.**

Supplementary personnel — full time: 1 public information specialist, 1 county health officer, 1 sanitarian

SAN ANTONIO-BEXAR COUNTY

San Antonio Metropolitan Health District, 322 W. Commerce Street, San Antonio, TX 78285 — Telephone: (512) 299-8853

Director of Health .... **C. N. Rothe, M.D.**
Air Pollution Control Officer .... **G. Walter Harrison.**

Supplementary personnel — full time: 1 chemist, 1 office staff, 3 sanitarians

San Antonio Metropolitan Health District Advisory Board (same address and phone as above)

■ UTAH

STATE AGENCY

Utah Department of Health, Division of Environment, Bureau of Air Quality, 150 West North Temple, Salt Lake City, UT 84110 — Telephone: (801) 533-6108

Director, Division of Environmental Health . **Kenneth L. Alkema**
Director, Bureau of Air Quality .... **Brent C. Bradford**

Supplementary personnel — full time: 2 administrators, 5 engineers, 1 meteorologist, 4 office staff, 1 data processing specialist, 23 scientists; part time: 1 attorney

Utah Air Conservation Committee (same address and phone as above)

Chairperson .... **J. Ralph Macfarlane, M.D.**

■ VERMONT

STATE AGENCY

Agency of Environmental Conservation, Air Pollution Control, State Office Building, Montpelier, VT 05602 — Telephone: (802) 828-3395

Secretary, A.E.C. .... **Leonard U. Wilson**
Air Pollution Control Officer .... **Richard A. Valentinetti**

Supplementary personnel — full time: 3 administrators, 1 chemist, 1 data processing specialist, 3 engineers, 2 inspectors, 2 office staff, 6 technicians; part time: 1 attorney

Solid Waste and Air Quality Variance Board (same address and phone as above)

Chairperson .... **Carl H. Lisman**

■ VIRGIN ISLANDS

Department of Conservation & Cultural Affairs, Division of Natural Resources Management, P.O. Box 4399, St. Thomas, VI 00801-4399 — Telephone: (809) 774-3320

Commissioner .... **Angel Luis LeBron**
Director of Air Pollution
Program .... **Vernon Richards**

Supplementary personnel — full time: 1 administrator, 1 chemist, 2 inspectors, 1 engineer, 1 office staff

■ VIRGINIA

STATE AGENCY

State Air Pollution Control Board, Room 801, Ninth Street Office Building, Richmond, VA 23219 — Telephone: (804) 786-2378

Executive Director .... **William R. Meyer**

Supplementary personnel — full time: 18 administrators, 7 chemists, 3 data processing specialists, 25 engineers, 22 inspectors, 1 meteorologist, 19 office staff, 2 public information specialists, 9 technicians, 1 lab instrument supervisor, 1 personnel technician, 1 fiscal technician; part time: 1 attorney

Chairperson .... **Elizabeth H. Haskell**

Regional Directors

Southwest Virginia, Abingdon .... **M. D. Overstreet**
Valley of Virginia, Roanoke .... **Donald L. Shepherd**
Central Virginia, Lynchburg .... **T. L. Henderson**
State Capitol, Richmond .... **H. A. Moss**
Hampton Roads, Virginia Beach .... **R. P. Minx**
Northern Virginia, Springfield .... **J. C. Doherty**

ALEXANDRIA

Alexandria Health Department, Office of Environmental Quality, 517 N. Saint Asaph St., Alexandria, VA 22314 — Telephone: (703) 838-4850 or 838-4865

Health Director .... **A. J. Albertson, M.D., M.P.H.**
Chief, Office of Environmental Quality .... **James T. Alexander, Jr.**

Supplementary personnel — full time: 1 administrator, 1 administrative aide, 2 engineers; part time: 1 industrial hygienist, 2 office staff, 3 administrators

Environmental Policy Commission, City Hall, 301 King Street, Alexandria, VA 22313

Chairperson .... **Harry A. Wolf**

FAIRFAX COUNTY

Fairfax County Air Pollution Control, 4080 Chain Bridge, Rd., Fairfax, VA 22030 — Telephone: (703) 691-2541

Director, Health Services..... **Richard K. Miller, M.D., M.P.H.**
Director, Air Pollution Control .... **Edgar M. Chase**

Supplementary personnel — full time: 1 administrator, 1 data processing specialist, 8 engineers, 1 meteorologist, 1 office staff; part time: 1 attorney, 1 public information specialist, 1 planner, 1 chemist

ROANOKE

Roanoke City Engineering Department, 215 West Campbell Avenue, Roanoke, VA 24011 — Telephone: (703) 981-2731

City Engineer .... **Charles M. Huffine**
Air Pollution Control Officer............. **Richard V. Hamilton**

Supplementary personnel — part time: 1 administrator, 1 engineer, 1 office staff

ROANOKE COUNTY/CITY OF SALEM

Roanoke County/City of Salem Air Pollution Control, P.O. Box 1074, Salem VA 24153 — Telephone: (703) 387-6224

Medical Director .... **Nancy Welch, M.D.**
Air Pollution Control Director .... **W. Carter Bradley**

Supplementary personnel — full time: 1 air pollution control specialist, 1 inspector; part time: 1 bookkeeper specialist, 2 office staff

City of Salem Advisory and Appeal Board (same address and phone as above)

Chairperson .... **B. T. Arnold**

Roanoke County Advisory and Appeal Board (same address and phone as above)

Chairperson .... **David G. Dillon**

SALEM (See Roanoke, etc.)

■ WASHINGTON

STATE AGENCY

Washington State Department of Ecology, M.S. PV-11, Olympia, WA 98504 — Telephone: (206) 459-6253

Director .... **Donald W. Moos**
Assistant Director, Office of Hazardous Materials & Air Pollution Control .... **Lynda Brothers**

Supplementary personnel — full time: 2 administrators, 1 attorney, 2 chemists, 2 data processing specialists, 5 engineers, 1 meteorologist, 4 office staff, 1 public information specialist, 16 technicians, 9 air quality control specialists; part time: 1 chemist, 2 engineers, 1 office staff, 2 air quality control specialists, 2 environmentalists

Ecological Commission (same address and phone as above)

Chairperson .... **Janet Chalupnik**

BENTON-FRANKLIN-WALLA WALLA COUNTIES

Air Pollution Authority, 650 George Washington Way, Richland, WA 99352 — Telephone: (509) 545-2354

Chairperson .... **Lorn D. Proctor**
Control Officer .... **J. Philip Cooke**

Supplementary personnel — full time: 1 administrator, 1 office staff; part time: 1 attorney, 1 inspector, 1 meteorologist

Board of Directors (same address and phone as above)

Board Chairperson .... **Lorn D. Proctor**

NORTHWEST (Includes Island, Skagit and Whatcom Counties)

Northwest Air Pollution Authority, 207 Pioneer Building, Mount Vernon, WA 98273 — Telephone: (206) 336-5705

Air Pollution Control and Executive Officer .... **Glen F. Hallman**

Supplementary personnel — full time: 1 administrator, 1 chemist, 1 office staff, 2 technicians; part time: 1 attorney, 1 technician, 1 inspector

NWAPA Board of Directors (same address and phone as above)

Chairman .... **E. Duane Kemp**

OLYMPIC REGION (Includes Clallam, Grays Harbor, Jefferson, Mason, Pacific and Thurston Counties)

Olympic Air Pollution Control Authority, 120 E. State Ave., Olympia, WA 98501 — Telephone: (206) 352-4881

Control Officer .... **Charles E. Peace**

Supplementary personnel — full time: 1 administrator, 1 inspector, 2 technicians, 1 office staff; part time: 1 attorney

Olympic Air Pollution Control Authority (same address and phone as above)

Chairperson .... **William O. Hunter**

PUGET SOUND (Includes King, Kitsap, Pierce and Snohomish Counties)

Puget Sound Air Pollution Control Agency, 200 West Mercer Street, P.O. Box 9863, Seattle, WA 98109 — Telephone: (206) 344-7330

Air Pollution Control Officer......... **Arthur R. Dammkoehler**

Supplementary personnel — full time: 4 administrators, 1 data processing specialist, 7 engineers, 11 inspectors, 7 office staff, 1 sanitarian, 7 technicians, 1 public information specialist, 1 meteorologist, 2 chemists

Puget Sound Air Pollution Control Board of Directors (same address and phone as above)

Chairperson .... **Harvey S. Poll**

SOUTHWEST (Includes Clark, Cowlitz, Lewis, Skanamia and Wahkiakum Counties)

Southwest Air Pollution Control Authority, 7601 N.E. Hazel Dell Ave., Vancouver, WA 98665 — Telephone: (206) 696-2508

Executive Director .... **Edward K. Taylor**

Supplementary personnel — full time: 1 administrator, 1 chemist, 1 engineer, 3 inspectors, 1 office staff; part time: 1 attorney

Southwest Air Pollution Control Board (same address and phone as above)

Chairperson .... **Luther Peek**

SPOKANE COUNTY

Spokane County Air Pollution Control Authority, West 1101 College Ave., Room 230, Spokane, WA 99201 — Telephone: (509) 456-4727

Director .... **Fred O. Gray**

Supplementary personnel — full time: 1 administrative assistant, 1 chemist, 1 inspector, 1 office staff; part time: 1 attorney

Spokane County Air Pollution Control Authority Board of Directors (same address and phone as above)

Chairperson .... **Grant D. Peterson**

YAKIMA COUNTY

Yakima County Clean Air Authority, Room 212, County Courthouse, Yakima, WA 98901 — Telephone: (509) 575-4116

Director .... **Robert L. Crossland**

Supplementary personnel — full time: 1 administrator, 1 office staff; part time: 1 engineer

Yakima County Clean Air Authority Board of Directors (same address and phone as above)

Chairperson .... **Marvin Sundquist**

■ WEST VIRGINIA

STATE AGENCY

West Virginia Air Pollution Control Commission, 1558 Washington St., East, Charleston, WV 25311 — Telephone: (304) 348-3286

Director .... **Carl G. Beard, II**

Supplementary personnel — full time: 2 administrators, 5 air pollution specialists, 7 chemists, 2 data processing specialists, 17 engineers, 3 laboratory assistants, 1 meteorologist, 20 office staff, 1 public information specialist, 10 technicians; part time: 1 attorney

West Virginia Air Pollution Control Commission (same address and phone as above)

Chairperson .... **Don R. Richardson**

WHEELING

Wheeling Air Pollution Control Department, City County Building, 1500 Chapline Street, Wheeling, WV 26003 — Telephone: (304) 234-3604

Director .... **Tim Carroll**
City Manager .... **F. Wayne Barte**

Supplementary personnel — full time: 1 attorney

■ WISCONSIN

STATE AGENCY

Wisconsin Department of Natural Resources, Bureau of Air Management, Box 7921, Madison, WI 53707 — Telephone: (608) 266-7718

Secretary .... **Carroll Besadny**
Director .... **Donald Theiler**

Supplementary personnel — full time: 10 administrators, 2 attorneys, 11 chemists, 34 engineers, 14 inspectors, 6 planning analysts, 4 office staff, 2 public information specialists, 14 technicians; part time: 1 engineer, 1 planning analyst

Natural Resources Board (same address and phone as above)

Chairperson .... **John A. Lawton**

EAU CLAIRE CITY-COUNTY

Eau Claire City-County Health Department, 720 2nd Ave., Eau Claire, WI 54701 — Telephone: (715) 839-4718

Director .... **John A. Bacharach, M.P.H.**
Director, Environmental Health .... **Darryll D. Farmer, M.P.H., R.S.**

Supplementary personnel — part time: 2 administrators, 1 chemist, 6 sanitarians, 2 technicians

Eau Claire Board of Health (same address and phone as above)

President .... **Robert Shaw**

FOND DU LAC

Fond du Lac City Health Department, 160 South Macy Street, Fond du Lac, WI 54935-0150 — Telephone: (414) 929-3290

City Health Officer .... **Harold E. Hillmer, M.P.A.**
Environmental Health Sanitarian ...... **Dennis A. Hibray, R.S.**

Supplementary personnel — full time: 1 administrator, 1 attorney, 2 sanitarians; part time: 1 office staff

MADISON

Madison Department of Public Health, 210 Monona Ave., Room 507, Madison, WI 53710 — Telephone: (608) 266-4843

Director of Public Health.......... **Patricia J. Natzke, M.P.H.**
Director, Environmental Health & Laboratory .... **Jill Schmidt, Ph.D.**

Supplementary personnel — full time: 2 administrators, 2 chemists, 6 sanitarians, 2 microbiologists, 2 office staff

Public Health Commission (same address and phone as above)

Chairperson .... **Jim Tillotson**

MILWAUKEE COUNTY

Milwaukee County DPW, Professional Services Division, Engineering, Environmental and Energy Services Section, 901 N. 9th St., Milwaukee, WI 53233 — Telephone: (414) 278-4874

Director of Transportation/Public Works .... **Gerald Scherm**
Acting Director .... **Richard A. Keyes**

Supplementary personnel — full time: 1 administrator, 1 engineer, 1 office staff

■ WYOMING

STATE AGENCY

Air Quality Division, Department of Environmental Quality, 122 W. 25th Street, Cheyenne, WY 82002 — Telephone: (307) 777-7391

Administrator .... **Randolph Wood, P.E.**

Supplementary personnel — full time: 1 administrator, 2 chemists, 1 data processing specialist, 10 engineers, 4 office staff, 1 quality assurance specialist, 2 technicians; part time: 1 attorney, 1 public information specialist

Air Quality Advisory Board (same address and phone as above)

CANADIAN FEDERAL AGENCIES
DEPARTMENT OF THE ENVIRONMENT

Minister .... **Suzanne Blais-Grenier**
Deputy Minister .... **J. Gérin**

ENVIRONMENTAL PROTECTION SERVICE

Department of the Environment, Environmental Protection Service, Ottawa, Ontario, K1A 1C8 — Telephone: (819) 947-1575

Assistant Deputy Minister.............................. **R. W. Slater**
Director General, Environmental Protection Programs Directorate .... **P. M. Higgins**
Director, Industrial Programs Branch .... **P. J. Choquette**

Federal/Provincial Advisory Committee on Air Quality (same address and phone as above)

Secretary .... **R. J. Powell**

REGIONAL OFFICES

Atlantic Region

Environmental Protection Service, Environment Canada, 5th Floor, 45 Alderney Drive, Dartmouth, Nova Scotia, B2Y 2N6 — Telephone: (902) 426-3593

Atlantic Regional Director.............................. **E. J. Dorrena**
Chief, Air Pollution Control .... **J. H. Kozak**

Supplementary personnel — full time: 2 engineers, 2 technicians

Western and Northern Region

Environmental Protection Service, Environment Canada, 804, 9942-108th Street, Edmonton, Alberta T5K 2J5 — Telephone: (403) 420-2570

Western and Northern Regional Director .... **R. K. Lane**
Chief, Air Pollution Control .... **A. C. Edwards**

Supplementary personnel — full time: 4 engineers; part time: 5 administrators, 1 chemist, 4 engineers, 4 technicians

Ontario Region

Environmental Protection Service, Environment Canada, 25 St. Clair Avenue, East, Toronto, Ontario, M4T 1M2 — Telephone: (416) 966-5840

Regional Director .... **K. Shikaze**

Manager, Pollution Control .... S. Llewelyn
Supplementary personnel — part time: 1 administrator, 4 engineers, 1 technician

Pacific Region

Environmental Protection Service, Environment Canada, Kapilano 100, West Vancouver, BC, V7T 1A2 — Telephone: (604) 666-6711

Regional Director .... B. A. Heskin
Manager, Air Programs .... E. P. Witushek
Supplementary personnel — full time: 1 engineer

Quebec Region

Environment Canada, Environmental Protection Service, 1550 De Maisonneuve West, Suite 410, Montreal, Quebec, H3G 1N2 —Telephone: (514) 283-7377

A/Regional Director .... G. Mezzetta
A/Chief Air Pollution Control Officer .... D. Pilon
Supplementary personnel — full time: 1 administrator, 2 engineers, 1 technician

ATMOSPHERIC ENVIRONMENT SERVICE

Department of the Environment, Atmospheric Environment Service, 4905 Dufferin Street, Downsview, Ontario, M3H 5T4, — Telephone: (416) 667-4760

Assistant Deputy Minister .... H. Ferguson
Director, Air Quality and Inter-Environmental Research Branch .... J. W. S. Young
Supplementary personnel — full time: 2 administrators, 1 chemist, 10 meteorologists, 6 office staff, 3 post doctorate fellows, 32 research scientists, 16 technicians, 2 computer analysts

Department of the Environment, Terrasses de la Chaudière, Ottawa, Ontario K1A 0H3

Minister of the Environment .... Suzanne Blais-Grenier

CANADIAN PROVINCIAL AND LOCAL AGENCIES

■ ALBERTA

PROVINCIAL AGENCY

Alberta Environment, Environmental Protection Service, Oxbridge Place, 9820-106 St., Edmonton, Alberta T5K 2J6 — Telephone: (403) 427-5893/427-5872

Assistant Deputy Minister .... Carl Primus
Head, Air Quality, Pollution Control .... J. C. Lack
Head, Air Quality, Standards and Approval .... S. L. Dobko
Supplementary personnel — full time: 1 administrator, 1 chemist, 1 data processing specialist, 12 engineers, 2 meteorologists, 3 office staff, 30 technicians; part time: 1 attorney, 1 public information specialist

■ BRITISH COLUMBIA

PROVINCIAL AGENCY

Waste Management Branch, 810 Blanshard Street, Victoria, BC V8W 1X5 — Telephone: (604) 387-4321

Director, Waste Management .... R. H. Ferguson, P.Eng.
Assistant Director, Air Program .... R. G. Wilson, Ph.D.
Supplementary personnel — full time: 2 administrators, 13 engineers, 15 office staff, 9 technicians, 2 data processing specialists, 8 meteorologists, 1 climatologist

Environmental Appeal Board, 557 Superior Street, Room 218, Victoria, BC V8V 1X5 — Telephone: (604) 387-3464

Chairperson .... Frank Hillyer

VANCOUVER

Great Vancouver Regional District, 4330 Kingsway, Burnaby, BC V5H 4G8 — Telephone: (604) 432-6200

Regional Manager .... D. L. MacKay
District Director, Pollution Control .... J. B. Mills
Supplementary personnel — full time: 3 administrators, 1 engineer, 8 inspectors, 1 office staff, 6 technicians; part time: 1 attorney, 2 chemists, 1 data processing specialist, 1 public information specialist

Regional Board of the Greater Vancouver Regional District (same address and phone as above)

Chairperson .... D. A. Ross

■ MANITOBA

PROVINCIAL AGENCY

Environmental Management Division, Building #2, 139 Tuxedo Avenue, Winnipeg, Manitoba, R3C 0V8 — Telephone: (204) 945-7100

Director, Environment Control .... C. Orcutt
Head, Air Standards and Studies Section .... D. Bezak
Head, Air Pollution Control Section .... A. Jackson
Supplementary personnel — full time: 1 data processing specialist, 12 environmental officers, 3 technicians; part time: 1 administrator, 1 attorney, 1 public information specialist

Manitoba Clean Environment Commission (same address and phone as above)

Chairperson .... Stanley Eagleton

■ NEW BRUNSWICK

PROVINCIAL AGENCY

Pollution Control Branch, Department of the Environment, P.O. Box 6000, Fredericton, NB E3B 5H1 — Telephone: (506) 453-2861

Director, Pollution Control Branch .... D. R. Silliphant
Chief, Air Quality Section .... J. F. L. Knight
Supplementary personnel — full time: 2 engineers, 1 inspector, 1 technician; part time: 1 administrator, 1 attorney, 1 chemist, 6 inspectors, 1 meteorologist, 2 office staff, 1 public information specialist

Environment New Brunswick (same address and phone as above)

Minister .... The Hon. C. W. Harmer

■ NEWFOUNDLAND

PROVINCIAL AGENCY

Department of Environment, Industrial Environmental Engineering Division, Elizabeth Towers, St. John's, Newfoundland A1C 5T7 — Telephone: (709) 737-2555

Minister .... The Hon. Hal Andrews
Branch Manager .... Les Hulett
Supplementary personnel — full time: 1 chemist, 2 engineers, 1 technician; part time: 3 administrators, 1 attorney, 1 data processing specialist, 10 inspectors, 2 technicians, 1 office staff, 1 public information specialist

■ NOVA SCOTIA

PROVINCIAL AGENCY

Nova Scotia Department of the Environment, P.O. Box 2107, Halifax, NS B3J 3B7 — Telephone: (902) 424-5300

Minister .... Hon. George Moody
Director .... John Turner
Regional Supervisor .... Clive Oldreive
Supplementary personnel — full time: 1 engineer, 3 technicians

■ ONTARIO

PROVINCIAL AGENCY

Air Resources Branch, Ministry of the Environment, 880 Bay St., Toronto, Ontario M5S 1Z8 — Telephone: (416) 965-6343

Director .... Dr. D. Balsillie
Manager of Air Quality and Meteorology Section ... L. Shenfield

Manager of Phytotoxicology Section .... **Dr. S. Linzon**
Manager of Emission Technology and Regulation Development Section .... **C. E. Duncan**
Manager of Administrative Services.............. **J. Prince**

Supplementary personnel — full time: 4 meteorologists, 9 scientists, 24 engineers, 5 data processing specialists, 4 mathematical modellers, 12 phytotoxicologists, 30 technicians, 6 data clerks, 1 statistician, 13 office staff, 2 project analysts, 10 chemists; part time: 5 attorneys, 3 public information specialists

Ontario Ministry of the Environment, 135 St. Clair Avenue, West, Toronto, Ontario M4V 1P5

Minister .... **Andy S. Brandt**
Deputy Minister .... **Dr. A. E. Dyer**

REGIONAL OFFICE: for chemical laboratory services and industrial abatement

Laboratory Services Branch, Resources Road, Rexdale, Ontario M9W 5L1 — Telephone: (416) 248-3001

Director .... **G. C. Ronan**

CENTRAL REGION, 150 Ferrand Drive, Don Mills, Ontario M3C 3C3 — Telephone: (416) 424-3000

Regional Director .... **G. Mierzynski**

NORTHEASTERN REGION, 199 Larch St., Sudbury, Ontario, P3E 5P9 — Telephone: (705) 675-4501

Regional Director .... **W. J. Gibson**

NORTHWESTERN REGION, 435 James St. S., Thunder Bay, Ontario P7C 5G1 — Telephone: (807) 475-1205

Regional Director .... **W. M. Vrooman**

SOUTHEASTERN REGION, 133 Dalton St., Kingston, Ontario, K7L 4X6 — Telephone: (613) 549-4000

Regional Director .... **D. N. Jeffs**

SOUTHWESTERN REGION, 985 Adelaide St. S., London, Ontario N6E 1V3 — Telephone: (519) 681-3600

Regional Director .... **D. A. McTavish**

WEST-CENTRAL REGION, 119 King St. W., Hamilton, Ontario L8N-3Z9 — Telephone: (416) 521-7640

Regional Director .... **G. H. Mills**

■ SASKATCHEWAN

PROVINCIAL AGENCY

Saskatchewan Department of the Environment — Air Pollution Control Branch, Walter Scott Building, 3085 Albert St., Regina, Saskatchewan S4S 0B1 — Telephone: (306) 565-6195

Deputy Minister .... **R. S. van Es**
Director, Air Pollution Control Branch .... **L. J. Lechner**

Supplementary personnel — full time: 2 engineers, 3 technicians, 1 office clerk; part time: 2 administrators, 1 attorney, 1 data processing specialist, 4 office staff, 1 public information specialist

■ QUEBEC

PROVINCIAL AGENCY

Ministere de L'Environment, 2360 Chemin Ste-Foy, Ste-Foy, Quebec G1V 4H2 — Telephone: (418) 643-7860

Sous-Ministre .... **Pierre B. Meunier**
Directeur Assainissement de l'air .... **Andre Harvey**

Supplementary personnel — full time: 1 administrator, 1 chemist, 1 data processing specialist, 10 engineers, 3 office staff, 5 technicians, 2 meteorologists; part time: 1 attorney, 1 public information specialist, 1 chemist, 2 technicians

l'A.R.M.A.T. (same address and phone as above)

Directeur General .... **Jean A. Roy**

MONTREAL

Service de l'assainissement de l'air et de l'inspection des aliments, 9150 Boul. de l'Acadie, Suite 101, Montreal, Quebec H4N 2T2 — Telephone: (514) 872-6848

Directeur .... **Jean Marier, ing., M.Sc.**

Supplementary personnel — full time: 7 administrators, 4 chemists, 7 engineers, 17 inspectors, 6 office staff, 17 technicians; part time: 2 attorneys, 4 office staff

Communaute Urbaine de Montreal, 2, Complexe Desjardins, Case postale, 129, Montreal, Quebec H5B 1E6

President .... **Pierre Des Marais II**

INTERNATIONAL JOINT COMMISSION

INTERNATIONAL JOINT COMMISSION (Canadian Section)

100 Metcalfe Street, Ottawa, Ontario K1P 5M1 — Telephone: (613) 995-2984

Chairperson .... **P.-Andre Bissonnette**
Secretary .... **David G. Chance**
For Air Pollution Affairs, contact .... **Geoffrey Thornburn**

(United States Section)

Washington, DC 20440 (Zip Code is unique to the Agency. The delivery address is 2nd Floor, 2001 S Street, N.W., Washington, DC) — Telephone: (202) 673-6222

Chairperson .... **Robert C. McEwen**
Secretary .... **David A. LaRoche**
For Air Pollution Affairs, contact .... **Dr. Joel L. Fisher**

The International Joint Commission was established in 1909, by a Treaty between the United States of America and Great Britain. This Treaty was confirmed and sanctioned by the International Boundary Waters Treaty Act (S.C., 1911, Chapter 28, as amended). (Canadian Ratification.) The Commission deals with the use, obstruction, and diversion of boundary waters and rivers crossing the boundary between Canada and the United States. Structures which would increase the levels of water at the Boundary require the prior approval of the Commission before work can proceed. In addition, the Commission carries out extensive investigations on questions arising between the Canadian and United States Governments along the "common frontier" and reports its findings with recommendations to the Governments. The Commission consists of six members, three from the Government of Canada appointed by the Governor General of Canada with approval of the Privy Council, and three from the Government of the United States appointed by the President of the United States with the consent of the Senate.

International Air Pollution Advisory Board

Chairperson, Canadian Section .... **Dr. James Young**
Chairperson, U.S. Section .... **Dr. Courtney Riordan**

The International Air Pollution Advisory Board has the function of alerting the Commission to problems of air pollution along the United States-Canadian boundary. In particular, the Board is concerned with the pollution problems which are of a transboundary nature, show the interaction of air quality and water quality in the boundary region, and with the general state of combined air pollution/water pollution monitoring in the transboundary region. The International Air Pollution Advisory Board functions under the general authority granted by the Boundary Waters Treaty of 1909.

Appendix E

State/Local Emissions Trading Programs*

Area	*Type of Rule*	*Pollutant*	*Status*
Alabama	Generic Bubble	VOC, TSP, SO_2, NO_X, & CO	Under development
Arizona (Pima County)	Banking	TSP	Adopted by county
California (Kern Co.)	Banking	VOC, TSP, SO_2, CO, & NO_X	Adopted by county
California (Kings Co.)	Banking	VOC, TSP, SO_2, CO, NO_X	Under development
California (Monterey Co.)	Banking	VOC, TSP, SO_2, CO, & NO_X	Adopted by county EPA proposed approval 11/15/83
California (San Diego Co.)	Banking	VOC, TSP, SO_2, CO, & NO_X	Under development

*As of December 1984.
Source: *Annual Report, Regulatory Reform Staff* (Washington, D.C., U.S. E.P.A., May 1, 1985).

Area	*Type of Rule*	*Pollutant*	*Status*
California (Ventura Co.)	Banking	VOC, TSP, SO_2, CO, & NO_X	Adopted by county (will undergo revision)
California San Francisco	Banking	VOC, SO_2, & TSP	Program in place (rule under revision)
California South Coast, LA	Banking	VOC, TSP, SO_2, CO, & NO_X	Adopted by county Being reviewed by EPA
Colorado	Emissions Trading: Bubble & Banking	VOC, SO_2, TSP, NO_X, & CO	Under consideration
Connecticut	Generic Bubble	VOC	EPA approved 6/7/82
Connecticut	Generic Bubble	VOC (Additional source categories)	EPA approved 10/19/84
Georgia	Generic Bubble	VOC, SO_2, TSP, NO_X, & CO	Disapproved by EPA 11/7/84
Idaho	Banking	VOC, TSP, SO_2, CO, & NO_X	EPA Proposed Approval 10/11/84
Idaho	Generic Bubble	VOC & NO_X	Under development
Illinois	Banking	VOC, TSP, SO_2, CO, & NO_X	Under consideration (open docket)
Illinois	Generic Bubble	VOC, SO_2, TSP, CO, & NO_X	Adopted by State Being reviewed by EPA
Indiana	Generic Bubble	VOC, SO_2, TSP, CO, & NO_X	Adopted by State Being reviewed by EPA

Area	*Type of Rule*	*Pollutant*	*Status*
Kentucky	Generic Bubble & Banking Rule	VOC, TSP, SO_2, NO_X, CO, & Pb	EPA proposed approval 6/28/82 Final being reviewed by EPA
Kentucky (Jefferson Co.)	Emissions Trading: Bubble & Banking	VOC, TSP, & SO_2	Banking program active Being reviewed by EPA
LRAPA (Eugene, OR)	Emissions Trading: Bubble & Banking	VOC, SO_2, TSP, CO, & NO_X	EPA approved 9/20/84
Louisiana	Generic Bubble	VOC, SO_2, TSP, CO, & NO_X	Being redrafted
Maryland	Generic Bubble	VOC	Under development
Maryland	Banking	VOC, SO_2, & TSP	Under development
Massachusetts	Generic Bubble for surface coating	VOC	EPA approved 3/29/82
Massachusetts	Generic Bubble	VOC	Proposed. Reproposal w/30-day averaging under EPA review.
Minnesota	Banking	TSP & SO_2	Original rule has been revised
Missouri	Banking	VOC, TSP, SO_2, CO, & NO_X	EPA approved 6/22/82
Missouri	Generic Bubble	VOC	On Hold
New Jersey	Generic Bubble	VOC	EPA approved 4/6/81
New Jersey	Generic Bubble	SO_2	EPA approved 7/8/83
New Jersey (Middlesex Co.)	Banking	VOC, TSP, SO_2, CO, & NO_X	Adopted by county
North Carolina	Generic Bubble	VOC	EPA approved 7/26/82
North Dakota	Emissions Trading	SO_2	Under consideration

Area	*Type of Rule*	*Pollutant*	*Status*
Oklahoma	Generic Bubble	VOC	Returned to State Under development
Oregon	Emission Trading: Bubble & Banking	VOC, SO_2, TSP, CO, NO_X	EPA approved 8/13/32
Pennsylvania	Generic Bubble for Two Source Types	VOC	EPA approved 1/19/83
Pennsylvania	Generic Bubble	VOC, SO_2, TSP, NO_X, & CO	Under development
Pennsylvania (Allegheny Co.)	Banking	VOC, SO_2, TSP, CO, & NO_X	EPA approved 10/21/81
Pennsylvania (Philadelphia)	Banking	VOC, TSP, SO_2, NO_X, & CO	Under consideration
Rhode Island	Emissions Banking	VOC, TSP, SO_2, CO, & NO_X	EPA approved 6/28/83
Rhode Island	Gener: surface coating/graphic arts/general	VOC	EPA approved 7/6/83
South Carolina	Generic Bubble	VOC, SO_2, TSP	EPA approved 9/3/82
Tennessee	Generic Bubble	VOC	Under development
Virginia	Generic Bubble	VOC, TSP, SO_2, NO_X, & CO	Under consideration
Washington	Banking	VOC, SO_2, TSP, CO, & NO_X	EPA proposed approval 10/9/84
Washington	Generic Bubble	VOC & NO_X	Under development

Area	*Type of Rule*	*Pollutant*	*Status*
Washington Seattle/Tacoma	Banking	VOC, CO, & TSP	Active program (deposits, no withdrawals yet)
Wisconsin	Generic Bubble	VOC	Under consideration (waiting for final ETPS)

Index